Phytochemicals in Plants—Recent Developments on the Occurrence, Composition, Stability, Health, Food and Pharmaceutical Applications

Phytochemicals in Plants—Recent Developments on the Occurrence, Composition, Stability, Health, Food and Pharmaceutical Applications

Guest Editor

Nicoleta-Gabriela Hădărugă

MDPI

Basel • Beijing • Wuhan • Barcelona • Belgrade • Novi Sad • Cluj • Manchester

Guest Editor
Nicoleta-Gabriela Hădărugă
Department of Food Science
University of Life Sciences
"King Mihai I" from Timişoara
Timişoara
Romania

Editorial Office
MDPI AG
Grosspeteranlage 5
4052 Basel, Switzerland

This is a reprint of the Special Issue, published open access by the journal *Plants* (ISSN 2223-7747), freely accessible at: https://www.mdpi.com/journal/plants/special_issues/DDBE9Z71HN.

For citation purposes, cite each article independently as indicated on the article page online and as indicated below:

Lastname, A.A.; Lastname, B.B. Article Title. *Journal Name* **Year**, *Volume Number*, Page Range.

ISBN 978-3-7258-7128-5 (Hbk)
ISBN 978-3-7258-7129-2 (PDF)
https://doi.org/10.3390/books978-3-7258-7129-2

Contents

About the Editor

Nicoleta-Gabriela Hădărugă

Nicoleta-Gabriela Hădărugă is a full professor, habilitated doctor and director of the Council for Doctoral University Studies at the University of Life Sciences "King Mihai I" from Timişoara. She coordinates the Safety, Food Security and Public Health master's program at the Faculty of Food Engineering. She is also the leader of the Food Quality and Stability Research Group within the university. She has expertise in the isolation, purification, and derivatization of antioxidants/polyphenols, as well as other natural compounds and systems with biological activity. The stability, protection and controlled release of such compounds are enhanced by micro- and nanoencapsulation, using especially natural and modified cyclodextrins. Studies on these topics have seen the acclaim of research grants, articles in highly esteemed journals (*Food Chemistry* and *Carbohydrate Polymers* from Elsevier, *Environmental Chemistry Letters* from Springer, or *Plants* from MDPI) and book chapters in international publishing houses (Elsevier, Springer Nature and Wiley).

plants

MDPI

Review

Exploring the Nutritional Potential and Functionality of Hemp and Rapeseed Proteins: A Review on Unveiling Anti-Nutritional Factors, Bioactive Compounds, and Functional Attributes

Marina Axentii and Georgiana Gabriela Codină *

Faculty of Food Engineering, Ştefan cel Mare University of Suceava, 720229 Suceava, Romania; axentii.marina@outlook.com

* Correspondence: codina@fia.usv.ro; Tel.: +40-7-4546-0727

Abstract: Plant-based proteins, like those derived from hemp and rapeseed can contribute significantly to a balanced diet and meet human daily nutritional requirements by providing essential nutrients such as protein, fiber, vitamins, minerals, and antioxidants. According to numerous recent research papers, the consumption of plant-based proteins has been associated with numerous health benefits, including a reduced risk of chronic diseases such as heart disease, diabetes, and certain cancers. Plant-based diets are often lower in saturated fat and cholesterol and higher in fiber and phytonutrients, which can support overall health and well-being. Present research investigates the nutritional attributes, functional properties, and potential food applications of hemp and rapeseed protein for a potential use in new food-product development, with a certain focus on identifying anti-nutritional factors and bioactive compounds. Through comprehensive analysis, anti-nutritional factors and bioactive compounds were elucidated, shedding light on their impact on protein quality and digestibility. The study also delves into the functional properties of hemp and rapeseed protein, unveiling their versatility in various food applications. Insights from this research contribute to a deeper understanding of the nutritional value and functional potential of hemp and rapeseed protein, paving the way for their further utilization in innovative food products with enhanced nutritional value and notable health benefits.

Keywords: hemp; rapeseed; protein; functionality; anti-nutritional factors; bioactive compounds; food applications; digestibility; innovative products

1. Introduction

The escalating demand for supplementary products, especially among the young and elderly population [1,2], a surge in lactose intolerance and gluten sensitivity [3], a rise in the adoption of vegan or vegetarian diets [4,5], and an increasing demand for sustainable and environmentally friendly protein sources [6,7] are drivers that move plant-based protein products to a new level and make them clearly an important source of nutrients for the human diet. In the realm of food product development, conditioned by the exploration of alternative protein sources, the use of rapeseed and hemp proteins has gained significant traction, fueled by the increasing demand for sustainable and nutritious options, but also for relatively cheap alternatives for the human diet.

Among food trends and modern consumer requirements nowadays, hemp and rapeseed proteins stand out as promising contenders, offering plenty of nutritional benefits and functional properties conducive to innovative food formulations [8]. Hemp and rapeseed proteins are renowned for their complete amino acid profiles [9], boasting a rich array of essential nutrients for the human diet, overall health, and well-being. These plant-based proteins are not only abundant in protein content but also provide valuable dietary fiber, vitamins, minerals, and antioxidants, rendering them invaluable assets in promoting balanced diets and meeting daily nutritional requirements [10,11].

Of note, both hemp and rapeseed are oilseed crops, meaning they are grown in many parts of the world in order to produce biofuels, animal livestock, and oil for food use, although the amount of protein these sources provide among other oilseed crops should be taken into consideration as they can be successfully implemented into high-protein product recipes, with an enhanced texture and flavor [12,13]. According to the United States Department of Agriculture (USDA) data (Figure 1), the most protein-rich oilseed crop is soybean (36.4 g of protein per 100 g), followed by hemp (31.6 g of protein per 100 g) and mustard (26 g of protein per 100 g). In contrast, 100 g of rapeseed contains only 18.6 g of protein [14].

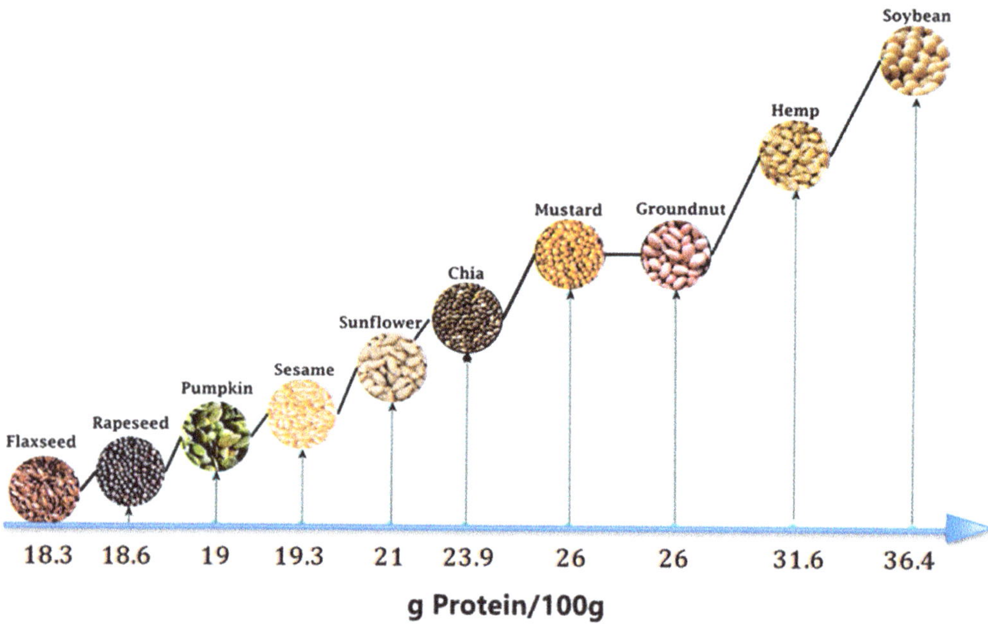

Figure 1. Food-use oilseed protein content per 100 g.

Beyond their nutritional prowess, hemp and rapeseed proteins harbor an array of bioactive compounds with potential health-promoting effects [15–18]. These compounds, ranging from polyphenols to phytosterols, contribute to the holistic health benefits associated with consuming these plant-based protein sources [16,19].

However, despite their nutritional virtues, hemp and rapeseed proteins also contain anti-nutritional factors that may impede optimal nutrient absorption and utilization. Understanding the presence and impact of these anti-nutritional factors is crucial for mitigating their effects and maximizing the nutritional benefits of hemp and rapeseed proteins [20,21].

Moreover, the digestibility and protein quality of hemp and rapeseed proteins play pivotal roles in determining their suitability for various food applications (Figure 2). Assessing factors such as amino acid composition, protein solubility, and protein digestibility is essential for elucidating the functional attributes and potential limitations of these plant-based protein sources in food product development.

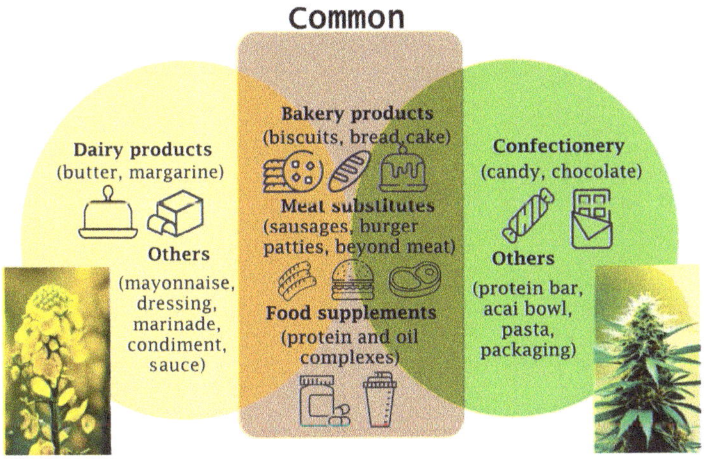

Figure 2. Hemp and rapeseed food applications.

In light of these considerations, this review paper aims to provide a comprehensive analysis of hemp and rapeseed proteins as viable options for food product development. By examining their nutritional value, bioactive compounds, anti-nutritional factors, digestibility, and protein quality, this study seeks to uncover the potential of hemp and rapeseed proteins as versatile ingredients in the creation of innovative and nutritious food products. Through rigorous exploration and analysis, this research endeavors to contribute valuable insights to the burgeoning field of plant-based food science and pave the way for the development of novel and health-enhancing food formulations.

2. Evaluating Hemp and Rapeseed Protein Quality

Protein quality is measured using various methods that assess the amino acid composition and digestibility of the protein. Some of the most effective and commonly used methods include calculating two scores: the Protein Digestibility Corrected Amino Acid Score (PDCAAS) and/or the Digestible Indispensable Amino Acid Score (DIAAS) [22].

The PDCAAS evaluates protein quality by comparing the amino acid profile of a protein to a reference protein pattern established by the Food and Agriculture Organization (FAO) and the World Health Organization (WHO) [23]. This method takes into account both the amino acid content and digestibility of the protein, whereas the DIAAS is a newer method developed by the FAO that considers only the digestibility of the protein, specifically the digestible indispensable amino acid content [24]. It provides a more accurate assessment of protein quality compared to the PDCAAS, especially for proteins that are less well-digested. Both scores may be useful for evaluating the nutritional quality of animal or plant-based proteins and guiding food formulation and dietary recommendations. Each method has its strengths and limitations, and researchers often use multiple approaches to assess protein quality comprehensively.

As hemp's most predominant amino acids are edestin and 2S albumin, which are two highly digestible globular proteins [25], the PDCAAS score for the industrial hemp and hemp by-products ranges between 48% for hemp seed meal and 66% for dehulled hemp seeds, respectively. Compared to casein's 100% PDCAAS or beef with a PDCAAS of 92%, hemp protein has a relatively low digestibility. However, it has been scientifically proved that a lower digestibility of plant-based proteins, such as grains or legumes, including hemp, is strictly correlated with the presence of anti-nutritional factors, such as phytic acid and trypsin inhibitors identified in hemp seed [26,27]. In comparison with other plant-based sources of protein, hemp has a similar PDCAAS to that of black and pinto beans, as well as lentils [28].

While both hemp and rapeseed contain all nine essential amino acids, meaning they are both complete sources of protein, the specific PDCAAS values vary depending on different research based on factors such as crop, protein concentration, extraction method, processing method, and digestibility [29–31].

To improve hemp's protein digestibility score, various modern treatment techniques were applied. Ultrasonication and Ph shifting were proven to have a positive effect on the hemp's protein structure by increasing the solubility, emulsifying activity, and foam capacity [32–39], as well as increasing the amphipathic property of the proteins by altering their globular structure [38,40,41]. There is also another method named supercritical CO_2 extraction that allows a gentle extraction of all important bioactive compounds and minimizes the use of solvents, making it a safer alternative to traditional extraction methods [42]. Moreover, in order to provide suitable amounts of all nine amino acids, hemp protein can be paired with pea, soy, oat, and/or microalgae [43–45]. These combinations enable formulators to develop new products with a high protein quality, chemically close to those of animal origin [46].

The nutritional quality of rapeseed protein, in terms of PDCAAS, is higher than hemp protein and it is comparable to that of soy protein (Table 1). However, the amounts of sulfur-containing amino acids, methionine, and cysteine in canola protein exceed the nutritional requirements for both children and adults, while soy protein and casein both fall short of the requirements for infant nutrition and only casein meets the requirements for higher-age categories [47]. As well as hemp protein, rapeseed protein contains anti-nutritional factors such as erucic and phytic acid, glucosinolates, and some protease inhibitors. Different extraction techniques, including heat treatment [35], ultrafiltration [36], nanofiltration [37], alkali extraction combined with membrane processing [38], and protein micellar mass processing [39], are employed to decrease the levels of these compounds, thereby enhancing the PDCAAS score.

The DIAAS places hemp in the same range as cooked kidney beans, lentils, wheat and whey. These protein sources are classified in the no-quality-claim category (DIAAS < 75). However, by associating hemp protein with rapeseed or soy, the nutritional requirements of different ages can be met. The increased DIAAS values obtained from mixtures show the potential to achieve protein nutritional efficiency with sustainable protein sources. Nutritional efficiency lies in meeting physiological requirements with a minimal intake of high-quality protein, as opposed to a higher protein intake of low-quality protein [32].

It is to be noted that, besides digestibility, the amino acid profile of cooked protein food can differ greatly from the raw form due to a potential loss of soluble protein fractions into the boiling liquid and through the formation of amino acid derivatives [48–50]. With an altered amino acid composition, the DIAAS value of the cooked protein, and possibly its limiting amino acid, may differ strongly. Furthermore, animal-based proteins are also subject to variations in protein quality resulting from processing.

The understanding of the DIAAS method offers an opportunity to enhance no-quality sources of plant-based proteins such as hemp, fava bean, oats, peas, and lentils with high-quality ones, such as rapeseed or soy, therefore creating mixtures with an enhanced nutritional value and improved functionality.

Table 1. PDCAAS and DIAAS evaluation of plant-based and animal proteins.

Food Source	PDCAAS [1] (%)	Ref.	Food Source	DIAAS [1] (%)	Ref.
Casein	100	[51,52]	Milk protein concentrate	118	[52,53]
Egg white	100	[51]	Egg	101	[32]
Soy protein concentrate	99	[54]	Soy protein concentrate	91.5	[55]
Dehulled hemp seed	66	[56]	Defatted hemp hearts	45	[56]
Rapeseed protein concentrate	93	[54]	Rapeseed protein isolate, heat-treated	100–110	[57]
Soy protein isolate	92	[54]	Soy protein isolate	90	[58]

Table 1. *Cont.*

Food Source	PDCAAS [1] (%)	Ref.	Food Source	DIAAS [1] (%)	Ref.
Beef	92	[54,56]	Beef	111.6	[55]
Rapeseed protein isolate	83	[54]	Rapeseed protein isolate	76–83	[42]
Hemp seed	51	[56]	Hemp seed	54	[59]
Hemp seed meal	48	[56]	Hemp seed meal	-	
Pea protein concentrate	72	[54]	Pea protein isolate	82	[58]
Kidney beans	64.8	[53]	Cooked kidney beans	58.8	[56]
Peas	50–64	[60]	Peas	64.7	[55]
Pinto beans	57–63	[56]	Cooked pinto beans	83	[61]
Rolled oats	67	[53]	Cooked rolled oats	54.2	[53]
Black beans	53	[62]	Black beans	49	[63]
Peanuts	52	[54]	Roasted peanuts	43.34	[53]
Split red lentils	53.8	[64]	Split red lentils	50	[64]
Chickpeas	52	[52]	Cooked chickpeas	67	[52]
Whole wheat	40	[56]	Wheat	40–48	[55,58]

[1] PDCAAS—Protein Digestibility Corrected Amino Acid Score, %; DIAAS—Digestible Indispensable Amino Acid Score, %; Ref.—reference.

3. Amino Acid Profile

Both rapeseed and hemp proteins are considered complete proteins, meaning they provide all essential amino acids required by the human body [9,65]. However, the composition and quantity of amino acids vary between the two (Table 2).

Table 2. Amino acid profile of hemp and rapeseed meals [66,67].

Amino Acid	Content (g/100g)	
	Hemp Meal	Rapeseed Meal
Arginine	5.56	2.02
Histidine	4.40	0.93
Isoleucine	1.79	1.41
Leucine	3.11	2.42
Lysine	5.92	1.94
Methionine	1.15	0.67
Phenylalanine	2.11	1.37
Threonine	1.68	1.45
Trypthofan	0.58	0.42
Valine	2.28	1.84
	Non-essential amino acid	
Alanine	2.07	1.51
Aspartic acid	6.64	2.51
Cysteine	0.80	0.76
Glutamic acid	8.44	5.29
Glycine	2.15	1.75
Proline	1.94	1.97
Serine	2.44	1.21
Tyrosine	1.54	0.97

As indicated by Sobhy Ahmed El-Sohaimy et al. [68], hemp contains a substantial quantity of edestin, which is the predominant protein type in hemp seeds, followed by a significant proportion of globulin (67–75%) and globular albumin, ranging between 25% and 37% [69].

Overall, several researchers discovered that hemp seeds contain about 181 different proteins, which makes hemp a valuable source of biologically active compounds, polyunsaturated fatty acids, enzymes, and micronutrients beneficial for the human body [68]. Moreover, some of the enzymes found in the hemp seeds can be used to produce hydrolyzed proteins that can be successfully used in nutraceuticals or as a co-ingredient in various functional foods, as potential hypotensive agents, and as antioxidants [61,62,68,70,71]. From the chemical point of view, edestin, which is a hexameric 11S protein, and easily digestible, contains significant amounts of all essential amino acids, especially sulfur amino acids and arginine [72]. Foods formulated with the addition of arginine contained in the hemp seeds have been proven to prevent or help treat cardiovascular diseases, by regulating blood pressure [73]. Despite high-quality storage proteins, hemp seeds also contain other health-promoting amino acids, such as sulfur-containing methionine and cysteine (3.5–5.9%). When it comes to valuable/important amounts of fatty acids, hempseed oil contains over 80% polyunsaturated fatty acids (PUFAs), making it a valuable source of essential fatty acids (EFAs), specifically linoleic acid (18:2 omega-6) and alpha-linolenic acid (18:3 omega-3). The omega-6 to omega-3 ratio (n6/n3) in hempseed oil typically falls between 2:1 and 3:1, which is considered optimal for human health. Additionally, hemp seed oil contains the biological metabolites gamma-linolenic acid (18:3 omega-6; 'GLA') and stearidonic acid (18:4 omega-3; 'SDA') [74].

Nevertheless, hemp seed oil comprises a substantial quantity of tocopherols and tocotrienols (ranging from 100 to 150 mg per 100 g of oil), along with phytosterols, phospholipids, carotenes, and minerals [75,76].

In summary, the well-balanced and diverse amino acid profile of hemp protein makes it an excellent and nutritionally rich choice to support human health and diet and meet the essential needs of the health-conscious modern consumers.

According to both amino acid profiles, hemp protein stands out for its well-balanced profile, is rich in essential fatty acids, and contains a higher concentration of arginine (Table 2), an amino acid with potential cardiovascular benefits. Rapeseed protein, on the other hand, has a more favorable lysine-to-arginine ratio [77].

Compared to hemp protein, the rapeseed protein profile consists of over 45 different proteins, which is much less than hemp: 20 weakly acidic, approximately 20 neutral, and 5 basic. It has a distinct profile among other plant-based sources of protein. Rapeseed proteins are known to have structural, catalytic, and storage functions. [78]. The major storage proteins contained in rapeseed seeds are the 12S globulin (cruciferin) and the 2S albumin (napin), making up more than 70% of total rapeseed proteins [79].

Napins, characterized by their small-molecular-weight albumins (15–17 kDa), exhibit solubility across a broad pH range and possess remarkable heat stability (with a denaturation temperature Td $\geq$ 100 °C) [76]. The hydrophobic regions of napins are concentrated on a single side of the protein, resembling a Janus particle [78].

On the other hand, cruciferins are globulins with a higher molecular weight (300 kDa) and adopt a hexamer structure composed of two trimers [79]. The hydrophobic domains of cruciferins are distributed extensively over the protein's surface and are also nestled within the trimeric units of cruciferins [80]. Compared to napins, the hexameric structure of cruciferins is more susceptible to undergoing structural alterations and unfolding in response to changes in temperature and pH [75,76].

Oleosin is a minor protein (1–4% by weight) present in canola seed, which functions as a stabilizer at the surface of the oil bodies so that the oil remains in the form of discrete droplets in the oil seed [80]. Nevertheless, rapeseed also consists of lipid transfer proteins and other minor proteins of a non-storage nature [79].

Overall, rapeseed contains high glutamine, glutamic acid, arginine, and leucine contents and low amounts of sulfur-containing amino acids, which are relatively altered during the industrial oil-extraction process. Indeed, the amino acid composition depends on the process used for protein extraction from the canola meal residue, which is usually up to 30% of the total [81].

Although there are well-established techniques to separate the soy protein or flax-seed protein, there are a lack of well-studied extraction methods for rapeseed protein, and therefore alternative technologies and conditions are needed, due to the differences in seed chemistry and protein profile. Details of rapeseed as a protein source in the human diet, based on research data, are also rarely found.

Although not studied enough, the viability of incorporating rapeseed meal proteins into food processing is proved by the well-balanced amino acid profile. Moreover, rapeseed/canola protein outperforms lentils, beans, and hemp protein and meets the amino acid requirements recommended by the FAO, WHO, and United Nations University (UNU) for both adults and children [82].

A widely acknowledged correlation already exists between the incorporation of plant proteins into functional foods, nutraceuticals, and various natural health products, due to their amino acid profile and nutritional value, contributing to health promotion and the positive effect it has on preventing various diseases. Plant proteins play a significant role in the food industry, and among them, rapeseed and hemp are recognized as promising co-ingredients suitable for human nutrition/diet.

4. Conventional and Alternative Processing Techniques

Protein concentrate production typically relies on two main extraction methods: dry fractionation or wet extraction [83].

Dry fractionation involves the separation of seeds based on their physical properties [84]. In the context of protein production, this method isolates protein-rich fractions from plant sources by utilizing mechanical techniques such as grinding, milling, sieving, or air classification [85–87]. These methods separate the different components based on particle size, density, or other physical characteristics.

A study conducted by Laguna et al. describes a dry-fractionation method applied to rapeseed and sunflower meals. In order to mechanically separate the protein-rich fractions, ultrafine milling was applied, followed by two separation technologies based on either particle charge (electrostatic sorting—ES) or density (turbo-separation—TS) to increase the mass yield. For rapeseed meal, electrostatic sorting was found to increase the protein and phenolic contents by 50–55% and 80–100%, respectively [88].

Dry fractionation is favored for its simplicity, cost-effectiveness, and avoidance of solvent use, making it an attractive option for manufacturers who are interested in producing protein concentrates from plant sources [89]. However, it may not yield protein purity as high as that achieved by wet-extraction methods [86].

Similarly, hemp and rapeseed protein concentrates are derived from their meals through a wet-extraction process, which involves using a liquid solvent, to dissolve and separate the protein from the insoluble portion. The resulting extract is then concentrated via centrifugation, ultrafiltration [90,91] ion-exchange, pH precipitation [92,93] or comparable methods to yield a highly pure protein with minimal carbohydrates and fat.

Can Karaca reported an efficient wet-extraction method using an isoelectric and salt precipitation on canola meal. Salt-extracted proteins, therefore collected via a centrifugation process, showed enhanced solubility and interfacial activity [94]. Another tandem wet method represented by alkaline extraction plus isoelectric precipitation offered higher efficiency (80% or more) for hemp and rapeseed protein extraction [95]. Alkaline extraction followed by acid or isoelectric precipitation and micellization extraction or reverse micellization [96] are commonly used to prepare hemp and rapeseed protein isolates with high protein content [9,97]. However, these methods were already reported to have a negative impact on hemp's sensory properties, expressed by a non-appealing greenish color [62,98,99].

Novel protein-extraction techniques offer the potential to produce protein extracts with enriched essential amino acids and improved physicochemical and functional properties [38–42]. However, despite these advancements, many companies continue to rely on conventional dry methods due to concerns regarding processing costs and sustainability factors.

5. The Anti-Nutritional Factors of Hemp and Rapeseed Proteins

Hemp, as well as rapeseed, contain a list of off-flavor, allergenic and anti-nutritional factors. These compounds directly affect the global acceptability of the oilseeds and their by-products used in human nutrition; however, the taste and chemical behavior can be improved by protein modification. Processing techniques, such as heat treatment, are often employed to mitigate these undesirable compounds [100].

Brassica oilseeds, grains and legumes such as hemp and rapeseed but also mustard, Brussels sprouts, cabbage, cauliflower, horseradish, kale and other crops contain a series of anti-nutritive elements, which can behave both beneficially or harmlessly (differently) for animal and human food use, depending on processing conditions [93]. Anti-nutritional factors mainly refer to the glucosinolates and their derived forms, isothiocyanates, naturally produced by Brassica plants as a self-defense mechanism through various metabolic processes or mechanisms in order to avoid being eaten.

However, for food product development, other anti-nutritional elements must be considered, such as low-molecular-weight phenolic compounds, polyphenolic tannins, enzyme inhibitors, phytates, and phytic acid, as they influence the nutritional and sensory characteristics of the new products [101].

The development of new products using hemp meal requires careful consideration due to the potential presence of anti-nutritional components. One primary reason for the limited large-scale manufacturing of hemp products is the inherent risk of contamination in the final product. Hemp naturally contains variable amounts of condensed tannins, trypsin inhibitors, phytic acid, and saponins, which significantly diminish protein availability either by precipitating it or by inhibiting digestive enzymes [69,102–105]. Processes such as chelation or complex formation (involving phytic acid, condensed tannins, and saponins) can further decrease the absorption of mineral elements and vitamins [25,102,103,106].

Condensed tannins found in both hemp and rapeseed are considered anti-nutritional factors due to their ability to form undesirable complexes with proteins and other macromolecules such as starch, thus reducing the nutritional value of the product [107]. However, condensed tannins are generically phenolic compounds and can have a beneficial effect on human health due to their antioxidant capacity associated with the phenolic rings presented in their structure [108].

Another important anti-nutritional factor found predominantly in nuts, legumes and seeds such as hemp is phytic acid, the stored form of phosphorus, which works in the human body as a mineral inhibitor promoting mineral deficiency by chelating bivalent minerals like calcium, iron, zinc, and copper, as well as starch, protein, and enzymes [109]. As well as condensed tannins, ingestion in low amounts may provide several beneficial effects, proven by several recent studies due to antitumor and antioxidant capacities. It is also known that phytates can interact with proteins due to the affinity of phosphate groups for cationic amino acids. This interaction can be harmful, disrupting protein digestion, but it can also protect the human body from harmful effects of specific proteins, such as oxidases and pathogenic proteins and several diseases like diabetes [110,111].

The third group of anti-nutritional factors called saponins, biologically represent surfactants with lipophilic aglycone and hydrophilic glycosyl groups. Saponins have been shown to affect the protein digestibility of hemp by binding the activity of certain metabolic catalysts such as trypsin and chymotrypsin, thus decreasing the physiological availability of nutrients and enzymes [111].

It is important to note that the presence and levels of these antinutritional factors can vary among different varieties of hemp and may be influenced by factors such as growing conditions and processing methods. Researchers and nutritionists assess these factors to understand their impact on the nutritional quality of hemp products and to develop strategies to mitigate their effects. Processing techniques such as heat treatment or fermentation can sometimes help reduce the levels of anti-nutritional factors in hemp products. Additionally, proper preparation methods, such as soaking and cooking, can contribute to minimizing the impact of these factors when incorporating hemp into diets.

The main anti-nutritional factors contained in rapeseed are glucosinolates, sinapine, sinapic acid, tannin, phytic acid, and mucilage [112].

The problem associated with Glucosinolates, which are glycosides of β-d-thioglucose, and myrosinase activity are the most serious factors influencing the quality of rapeseed oil and meal and limiting the increased utilization of these products.

Rapeseed meals that contain intact glucosinolates cannot be used as a protein source for human nutrition due to the possibility of hydrolysis during digestion or the presence of myrosinase in other foods. Recently, several procedures have been devised for the complete aqueous extraction of glucosinolates from rapeseed meal [113], ground rapeseed, and, by diffusion, from intact rapeseed [114–116].

Common rapeseed meal contains significant amounts of glucosinolates, with the main five types being 3-butenylglucosinolates, 4-pentenylglucosinolates, 2-hydroxy-3-butenylglucosinolates, 2-hydroxy-4-pentenylglucosinolates, and 2-allylglucosinolates. Brassica napus meal typically contains 3-butenylglucosinolates and 2-hydroxy-3-butenylglucosinolates, while Brassica rape meal includes 3-butenylglucosinolates. It is essential to limit the total glucosinolate content in rapeseed meal for animal feed to 2.5 μmol/g, as elevated levels of these antinutritional compounds can lead to physiological disorders in animals, including hemorrhagic liver. Prolonged consumption of glucosinolate-rich foods may result in liver, kidney, and thyroid-gland enlargement in humans [108].

Moreover, the bitterness associated with certain glucosinolates in rapeseed meal can impact the acceptability of the final product. Nevertheless, glucosinolates are recognized for their natural anti-microbial and anti-carcinogenic properties [117].

Sinapine, another anti-nutritional factor presented in 70–80% of rapeseed polyphenols, is an acetylcholinesterase inhibitor [118]. Sinapic acid (SA, 4-hydroxy-3,5-dimethoxy-cinnamic acid) is the predominant free phenolic acid found in rapeseed [119]. Sinapine, also known as sinapoylcholine, is a compound found in various plants, including mustard seeds, rapeseeds, and certain other cruciferous vegetables. It is a type of alkaloid and is commonly present in the seed coats of these plants. While sinapine is not typically considered highly toxic for humans, it is important to note that excessive consumption or exposure to high levels of any compound can potentially lead to adverse effects.

Erucic acid, a monounsaturated omega-9 fatty acid, mainly common in rapeseed and mustard seeds can constitute about 30–60% of the total fatty acids of natural rapeseed [120]. The toxic effect of erucic acid was studied on animals, as exposure to diets with oils containing excessive erucic acid may lead to adverse health effects for the heart as the principal target organ. The most common effect in experimental animals is myocardial lipidosis, an accumulation of lipids in heart muscle fibers that may reduce the contractile force of heart muscles. So far, nevertheless, no evidence that dietary exposure to erucic acid is correlated to myocardial lipidosis has been established yet in humans [121]. However, there are no data to prove any heart disease-associated effects associated with the human diet. Also, the crops that are used for food product development are cultivated specifically with low erucic-acid content. Although natural forms of rapeseed and mustard contain high levels of erucic acid (over 40% of total fatty acids), levels in rapeseed cultivated for food use are typically below 0.5% [122]. Erucic acid is regulated by the European Food Safety Authority (EFSA) and the US Food and Drug Administration (FDA). In 2016, the European Food Safety Authority (EFSA) proposed a lower maximum content of erucic acid in edible oils of 2% and also recommended a tolerable daily intake of 7 mg of erucic acid per kg body weight [123]. However, there exist special crops of rapeseed named high erucic acid rapeseed (HEAR), which is a specialty rapeseed, selected for its high erucic content. It has over 50% erucic acid, and is grown as a key ingredient for plastics, personal-care products, and pharmaceuticals [124].

It is important to mention that nowadays advancements in breeding and processing techniques have been employed to reduce the levels of these anti-nutritional factors in rapeseed and its products, making them safer for consumption. Canola oil, for example, is derived from low-erucic-acid rapeseed varieties and has a favorable nutritional profile [125].

6. Bioactive Compounds

The rising number of health-conscious consumers and the significant growth in modern consumers with enthusiasm for functional foods have led to an increased demand for plant-based products that provide natural remedies for various diseases [126,127]. Both hemp and rapeseed have important bioactive compounds with nutraceutical properties.

Hemp seeds are notably rich in polyphenols (mainly flavonoids, stilbenoids, and lignanamides), alkaloids, cannabinoids, and terpenoids [128]. Flavonoids represent secondary plant metabolites composed of polyphenolic compounds with antioxidant properties that directly contribute to the hemp's color, flavor, and potential health benefits. Hemp seeds contain unique types of flavonoids that cannot be found in any other plants, such as cannflavin A, cannflavin B, and cannflavin C [129,130]. Other kinds of flavonoids contained in hemp include quercetin, apigenin, and kaempferol. Quercetin is studied due to its anti-inflammatory properties preventing chronic inflammation linked to various health conditions, including heart disease, diabetes, and certain types of cancer [131]. Several studies also suggest that quercetin may have positive effects on cardiovascular health. It may help lower blood pressure, reduce cholesterol levels, and improve overall heart function [132]. Moreover, the flavonoids present in hemp act in a synergistic action with other compounds, such as the cannabinoids and terpenes, to produce antioxidant, antidepressant, anti-inflammatory, and disease-fighting properties [133].

Cannabinoids are a class of chemical compounds found in the cannabis plant, including hemp (*Cannabis sativa*). Hemp is known for its relatively high levels of certain cannabinoids, with a particular focus on Cannabidiol (CBD) and Tetrahydrocannabinol (THC) [134]. The effect of administrating Cannabidiol is currently investigated for its use in mood disorders such as anxiety, control for chronic pain, anti-inflammatory diseases, neurodegenerative diseases such as Alzheimer's and Parkinson's disease, and for its anti-tumorigenic properties. Of note, Cannabidiol is successfully used as a medicine against the seizure disorders Lennox–Gastaut syndrome and Dravet syndrome [135]. Industrial hemp contains no more than 0.3% concentration (on a dry-weight basis) of the psychoactive compound delta-9-tetrahydrocannabinol (THC), due to applicable European and federal law [136].

It is important to note that the therapeutic effects and potential health benefits of cannabinoids are an active area of research, and more studies are needed to fully understand their mechanisms and applications.

In addition to phytocannabinoids, hemp also contains terpenes, aromatic compounds found in many varieties of plants, including hemp. They contribute to the plant's flavor and aroma profile [137,138]. Terpenes in hemp may also have potential therapeutic effects and can work synergistically with cannabinoids in what is known as the "entourage effect". So far, almost 120 different terpenes have been found in hemp, but the most prevalent are myrcene, pinene, limonene, and linalool [139].

Another source of bioactive compounds identified in hemp is phytosterols. They currently present high interest for pharmaceutical companies due to a wide range of beneficial actions. One of the primary uses of phytosterols is in managing cholesterol levels. By reducing LDL cholesterol levels, phytosterols contribute to overall cardiovascular health [140]. Some research suggests that phytosterols may have anti-inflammatory effects and can also modulate the immune system [141]. Of note, hemp is a nutritious plant that contains a variety of vitamins and minerals, such as Vitamin E, particularly gamma-tocopherol, Magnesium, Phosphorus, Potassium, Zinc, Iron, Cooper, Manganese, and B Vitamins, making it a valuable addition to a balanced diet [142].

Over the years, various health-promoting compounds have been identified in rapeseed, including polyphenols, phytosterols, carotenoids, and others. However, it is worth noting that the composition and quantity of bioactive compounds present in rapeseed can vary depending on several key factors, such as the crop species, growing conditions, and processing methods [143].

Phenolic acids and their derivatives, along with both soluble and insoluble tannins, constitute the primary phenolic bioactive compounds present in rapeseed. Reports indicate that rapeseed meal can contain as much as 6% tannins. Hence, utilizing hulls, post-dehulling, as a source of natural antioxidants offers a potential avenue for their practical utilization [106].

Research indicates that rapeseed possesses a higher phenolic-component content in comparison to other seeds within the oilseed category. The phenolic compounds extracted from rapeseed have proven efficacy as potent antioxidants, finding successful applications in the realms of food, cosmetics, and pharmaceuticals [144].

From the chemical point of view, there is a total of approximately 400 mg per kg of concentration of all identified phenolic compounds in rapeseed meal. Sinapic acid stands out as the predominant phenolic compound, constituting over 85% of all quantified phenolic compounds, with an average concentration of 357 ± 13 mg/kg and a range of 339–379 mg/kg [145]. Among hydroxycinnamic acid derivatives, sinapine emerges as the most abundant bioactive compound in rapeseed. Its noteworthy properties, including antitumor, neuroprotective, antioxidant, and hepatoprotective attributes, underscore its significance for promoting health [144]. Remarkably, polyphenolic compounds, which have attracted considerable interest in recent years, are thought to confer numerous health benefits by modulating metabolic disorders, potentially through interactions with the gut microbiota [140,141]. Among the most active antioxidant components discovered in the polar fraction of rapeseed extracts is canolol (4-vinylsyringol or 2,6-dimethoxy-4-vinylphenol). Canolol is formed through the decarboxylation of sinapic acid during the roasting process of rapeseed [146].

Furthermore, nutraceutical companies are keen on substituting synthetic antioxidants like butylated hydroxyanisole (BHA), butylated hydroxytoluene (BHT), propyl gallate (PG), and tert-butylhydroquinone (TBHQ) with natural plant alternatives, as the synthetic counterparts are carcinogenic. Several previous studies have illustrated the fact that the unsaturated fatty acids and phytosterols present in rapeseed oil can effectively reduce both total cholesterol and "bad" cholesterol (such as low-density lipoprotein, LDL), while simultaneously maintaining the levels of "good" cholesterol (e.g., high-density lipoprotein, HDL). This dual action not only decreases the susceptibility to cardiovascular diseases but also acts as a preventive measure against clotting and the proliferation of vascular smooth muscle. The positive outcomes are often attributed to the optimal ratio of omega-6 to omega-3 polyunsaturated fatty acids (PUFAs), which is approximately 2:1, as supported by research studies [147,148].

Beyond their lipid solubility, these micronutrients showcase a broad spectrum of biological properties, encompassing antioxidant, anti-inflammatory, and anticancer activities [147,148].

Omega-6 fatty acids, particularly linoleic acid and its derivatives such as γ-linolenic acid, are well-known for their beneficial effects on health and are abundant in rapeseed oil. Research indicates that a diet rich in γ-linolenic acid can effectively reduce elevated levels of blood lipids, lower high blood pressure, and regulate skin perspiration [149,150]. Moreover, γ-linolenic acid demonstrates various physiological effects including anti-cancer properties, anti-thrombotic effects on the cardio-cerebrovascular system, and advantages in managing diabetes [151,152]. Conversely, α-linolenic acid is linked to physiological functions such as anti-atherosclerotic effects, facilitation of weight loss, reduction in blood lipid levels, and prevention of cardiovascular and cerebrovascular diseases [153]. The distinctive fatty acid composition found in rapeseed oil contributes to a spectrum of biological functions that support human health [143].

Carotenoids and Vitamin E present in rapeseed protein and meals also have health-promoting benefits. The main type of carotenoid found in rapeseed is (all-E)-lutein; however, there is also a minor amount of (all-E)-zeaxanthin found in cold-pressed rapeseed oil [154]. Lutein and zeaxanthin work by protecting the retina of the eye from the effects of aging. These carotenoids may prevent macular degeneration [155]. Clinical studies have also shown that lutein's anti-oxidative and anti-inflammatory properties provide benefits by protecting and alleviating other ocular diseases like cataracts, diabetic retinopathy, myopia, and retinopathy of prematurity [156]. Rapeseed oil contains relatively high levels of tocopherols (Vitamin E) and moderate levels of vitamin K compared to other plant oils [143,157].

Overall, these bioactive compounds collectively contribute to the nutritional value and health-promoting properties of rapeseed and its derived products. Research suggests that the bioactive components in rapeseed, such as phytosterols with cholesterol-lowering effects and omega fatty acids with cardiovascular benefits, may have positive implications for human health. Furthermore, the antioxidant, anti-inflammatory, and potentially anticancer properties associated with polyphenolic compounds in rapeseed oil add to its overall health-promoting profile. As our understanding of these bioactive compounds continues to grow, incorporating rapeseed and its oil into a balanced diet may offer a range of potential health benefits.

7. Food Applications

The use of industrial hemp in the human diet has been a subject of debate for decades, primarily due to its chemical composition and the presence of cannabinoids, known for their potential psychotropic effects in larger quantities. In the United States, the Agriculture Improvement Act of 2018, commonly known as the 2018 Farm Bill, legalized the cultivation of industrial hemp. According to the bill, industrial hemp is defined as cannabis sativa plants containing 0.3% THC or less on a dry-weight basis [158]. Similarly, in the European Union, regulations regarding industrial hemp and THC content are established by individual member states; however, the THC levels mustnot exceed 0.3% [159,160]. Compliance with this THC threshold is crucial for hemp cultivation and the production of hemp-derived products.

Hemp protein, meal, and seed-based products (Figure 3) are readily available in various combinations on the global market. Thanks to its versatility, hemp can be consumed raw as hemp hearts (seeds) or in a processed form as powder, meal, oil, and flour. However, the most significant impact lies in serving hemp as a co-ingredient in various food products, elevating their nutritional value and asserting a compelling functional potential [19,161,162]. For the functional food market hemp stands out as an exceptionally fitting food ingredient. This is attributed to its noteworthy nutritional content and the array of health benefits it offers. Nowadays, manufacturers use industrial hemp in yogurt, snack bars, cookies, bread, pasta, milk, butter, ice cream, Beyond Meat, tofu, and other innovative/functionally improved products [65].

Seed processing techniques such as germination have been reported to promote changes in the phytochemical profile of seeds and have drawn interest to the commercial development of sprouts enriched in specific phytochemicals [163].

A study reported an increased polyphenol, flavonoid content, antioxidant activity, and protein concentration of hemp sprouts produced under a blue (B) light-led emitting diode compared to raw seed [164]. Additionally, it has been proven that hemp sprouts possess no hallucinogenic effects, do not contain high delta-9-tetrahydrocannabinol, and thus can be safely consumed without any concerns of a negative health impact [165].

Hemp seed can be also processed and manufactured into high-moisture meat analogs and hemp milk.

A high-quality and nutritious hemp milk was developed from seeds. Such milk contains about 25–30% protein and 35% fatty acids, with an optimum essential omega-3 and omega-6 fatty acid content [166,167].

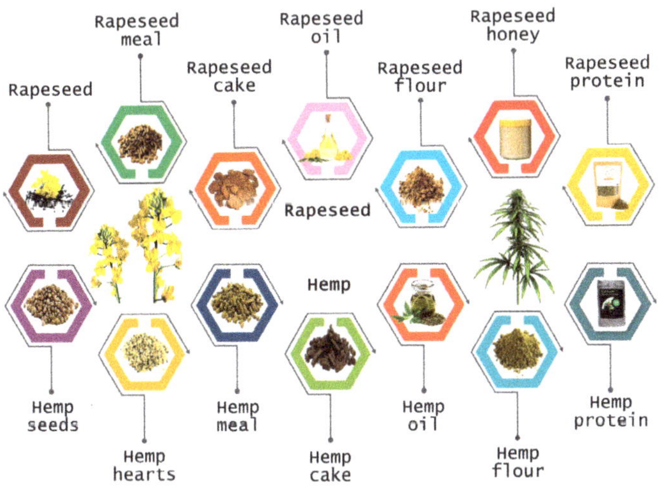

Figure 3. Hemp and rapeseed by-products used in the food industry.

Collectively, these findings demonstrate that hemp can be an excellent nutritional source of important bioactive compounds and can be successfully used for new food product development, based on accepted knowledge and new experiments.

As with hemp, rapeseed, available in various forms on the market, such as protein powder, meal, cake, flour or oil (Figure 3) may also be incorporated into some food products, particularly those designed to boost nutritional content [168].

Rapeseed has been consumed by humans as a condiment for about 3000 years. The original use of rapeseed-mustard was to mask the taste of degraded perishables. The spiciness of rapeseed-mustard is caused by a group of compounds called isothiocyanates [19].

The most used rapeseed product is rapeseed oil, also known as canola oil, a popular choice worldwide for cooking oils due to its neutral flavor, high smoke point, and healthy fat profile. It is commonly used for frying, sautéing, baking, and salad dressings.

Rapeseed oil is also used as a key ingredient in the production of margarine and spreads, providing a plant-based alternative to butter [169]. Another plant-based alternative based on rapeseed oil is mayonnaise, as rapeseed oil contributes to its creamy texture and rich flavor. When it comes to the preparation of salad dressing, marinades, and sauces, adding rapeseed oil gives a smooth texture and mild flavor.

Rapeseed cake, protein, and oil are gaining popularity as co-ingredients for baking goods, such as cakes, bread, biscuits, and gluten-free pasta. The use of rapeseed-based products is proven to enhance moisture retention and texture, extending the shelf life of the product but also improving the nutritional value of the final products [170].

Another study suggests that rapeseed flour, made from ground rapeseed meal, can be used as a gluten-free alternative in baking and cooking. Functional foods made with rapeseed flour are suitable for individuals with celiac disease or gluten intolerance. Several patents are proposed to patent the use of rapeseed protein for the production of diary-free alternatives such as plant-based milk, yogurts, and cheeses, to provide protein and improve mouthfeel [171].

With a biorefinery approach, rapeseed proteins may be extracted and recovered for high-end uses to fully exploit their nutritional and functional properties [172]. Several studies have been carried out in recent years on the applications of rapeseed/canola proteins in food products as the partial or total replacement of animal proteins. Depending on the food application tested, rapeseed proteins have been proposed as a thickener ingredient or as an emulsifier, binder, foaming, or gelling agent able to modify texture or simply fortify the protein content of a product. The range of possible food applications for rapeseed/canola proteins includes bakery and dairy products, meat, confectionery, and beverages, as well as dressings, sauces, snacks, or flavorings [172–175].

Meat alternatives incorporating rapeseed protein have been deemed acceptable based on their nutritional composition, and deemed sufficient to fulfill human requirements for essential amino acids [176]. However, the taste of rapeseed protein is reported to be unfavorable due to the presence of free and esterified phenolic acids [81]. Nonetheless, enhancements in taste can be achieved by steaming the protein concentrate and incorporating it into sausage formulations, resulting in improved taste, texture, and a distinctive aroma, thus rendering it comparable to soy-based alternatives [171,177,178].

Another study highlights the potential of using rapeseed protein isolate as a food supplement for elderly people with mastication and dysphagia problems. The obtained results show that the texture modification of food combined with rapeseed protein isolate supplementation may have a positive impact on the nutritional and sensory profile of the products [172].

Thus, hemp and rapeseed proteins offer a versatile and nutritious option for food product development, contributing to a wide range of culinary applications and meeting the needs of consumers seeking plant-based alternatives with enhanced health benefits.

8. Conclusions

Both rapeseed and hemp protein offer promising protein quality, with hemp showcasing a rich content of edestin and rapeseed demonstrating a high PCDAAS score. Further research is needed to comprehensively evaluate their protein digestibility and amino acid bioavailability in a specific functional food product, applying various processing methods and treatments that will meet, in the final instance, the market needs.

The presence of anti-nutritional factors in rapeseed and hemp, such as condensed tannins, trypsin inhibitors, and phytic acid, underscores the importance of careful processing techniques to mitigate their effects. Advances in extraction and processing methods offer opportunities to minimize these compounds while retaining nutritional integrity.

Both rapeseed and hemp contain bioactive compounds with potential health benefits, including polyphenols, phytosterols, and omega fatty acids. Harnessing these compounds in food formulations could contribute to functional foods targeting specific health outcomes, such as cardiovascular health and inflammation management.

Rapeseed and hemp proteins hold promise for a wide range of food applications, including plant-based meat substitutes, dairy alternatives, and baked goods. Their functional properties, such as emulsification and foaming capabilities, make them suitable for various formulations, catering to the growing demand for alternative protein sources.

The future perspective of utilizing rapeseed and hemp proteins in food product development hinges on continued research and innovation. Addressing challenges related to taste, texture, and consumer acceptance will be crucial for mainstream adoption. Additionally, exploring novel processing techniques and genetic modifications could unlock further potential for enhancing nutritional quality and functionality.

In conclusion, rapeseed and hemp proteins offer exciting opportunities for the development of innovative and sustainable food products. With careful consideration of protein quality, anti-nutritional factors, bioactive compounds, and food applications, these plant-based protein sources hold immense promise for meeting the evolving demands of the modern food industry and promoting human health and well-being.

Author Contributions: M.A. and G.G.C. contributed equally to the study design, collection of data, development of the sampling, the analyses, interpretation of results and preparation of the paper. All authors have read and agreed to the published version of the manuscript.

Funding: This work was funded by the Ministry of Research, Innovation and Digitalization within Program 1—Development of national research and development system, Subprogram 1.2—Institutional Performance—RDI excellence funding projects, under contract no. 10PFE/2021.

Conflicts of Interest: The authors declare no conflicts of interest.

References

1. Sicińska, E.; Pietruszka, B.; Januszko, O.; Kałuża, J. Different Socio-Demographic and Lifestyle Factors Can Determine the Dietary Supplement Use in Children and Adolescents in Central-Eastern Poland. *Nutrients* **2019**, *11*, 658. [CrossRef] [PubMed]
2. Djaoudene, O.; Romano, A.; Bradai, Y.D.; Zebiri, F.; Ouchene, A.; Yousfi, Y.; Amrane-Abider, M.; Sahraoui-Remini, Y.; Madani, K. A Global Overview of Dietary Supplements: Regulation, Market Trends, Usage during the COVID-19 Pandemic, and Health Effects. *Nutrients* **2023**, *15*, 3320. [CrossRef] [PubMed]
3. Savarese, M.; Wismer, W.; Graffigna, G. Conceptualizing "Free-from" Food Consumption Determinants: A Systematic Integrative Literature Review Focused on Gluten and Lactose. *Food Qual. Prefer.* **2021**, *90*, 104170. [CrossRef]
4. Bakaloudi, D.R.; Halloran, A.; Rippin, H.L.; Oikonomidou, A.C.; Dardavesis, T.I.; Williams, J.; Wickramasinghe, K.; Breda, J.; Chourdakis, M. Intake and Adequacy of the Vegan Diet. A Systematic Review of the Evidence. *Clin. Nutr.* **2021**, *40*, 3503–3521. [CrossRef] [PubMed]
5. Clem, J.; Barthel, B. A Look at Plant-Based Diets. *Missouri Med.* **2021**, *118*, 233–238. [PubMed]
6. Sobczak, P.; Grochowicz, J.; Łusiak, P.; Żukiewicz-Sobczak, W. Development of Alternative Protein Sources in Terms of a Sustainable System. *Sustainability* **2023**, *15*, 12111. [CrossRef]
7. Safdar, L.B.; Foulkes, M.J.; Kleiner, F.H.; Searle, I.R.; Bhosale, R.A.; Fisk, I.D.; Boden, S.A. Challenges Facing Sustainable Protein Production: Opportunities for Cereals. *Plant Commun.* **2023**, *4*, 100716. [CrossRef]
8. Dr Erdogan Ceylan Challenges and Opportunities in Plant-Based Protein Innovation. Available online: https://na.mxns.com/blog/challenges-and-opportunities-in-plant-based-protein-innovation (accessed on 12 April 2024).
9. Chen, H.; Xu, B.; Wang, Y.; Li, W.; He, D.; Zhang, Y.; Zhang, X.; Xing, X. Emerging Natural Hemp Seed Proteins and Their Functions for Nutraceutical Applications. *Food Sci. Hum. Wellness* **2023**, *12*, 929–941. [CrossRef]
10. Lu, R.-R.; Qian, P.; Sun, Z.; Zhou, X.-H.; Chen, T.-P.; He, J.-F.; Zhang, H.; Wu, J. Hempseed Protein Derived Antioxidative Peptides: Purification, Identification and Protection from Hydrogen Peroxide-Induced Apoptosis in PC12 Cells. *Food Chem.* **2010**, *123*, 1210–1218. [CrossRef]
11. Oomah, B.D.; Busson, M.; Godfrey, D.V.; Drover, J.C. Characteristics of Hemp (*Cannabis sativa* L.) Seed Oil. *Food Chem.* **2002**, *76*, 33–43. [CrossRef]
12. Abiodun, O.A. The Role of Oilseed Crops in Human Diet and Industrial Use. In *Oilseed Crops*; Wiley: Hoboken, NJ, USA, 2017; pp. 249–263.
13. Bernard, J.K. Feed Ingredients | Feed Concentrates: Oilseed and Oilseed Meals. In *Encyclopedia of Dairy Sciences*; Elsevier: Amsterdam, The Netherlands, 2011; pp. 349–355.
14. Ash, M.; Knisley, S. *Oil Crops Outlook; OCS-16-H*; United States Department of Agriculture (USDA): Washington, DC, USA, 2016.
15. Frankowski, J.; Przybylska-Balcerek, A.; Graczyk, M.; Niedziela, G.; Sieracka, D.; Stuper-Szablewska, K. The Effect of Mineral Fertilization on the Content of Bioactive Compounds in Hemp Seeds and Oil. *Molecules* **2023**, *28*, 4870. [CrossRef] [PubMed]
16. Tileuberdi, N.; Turgumbayeva, A.; Yeskaliyeva, B.; Sarsenova, L.; Issayeva, R. Extraction, Isolation of Bioactive Compounds and Therapeutic Potential of Rapeseed (*Brassica napus* L.). *Molecules* **2022**, *27*, 8824. [CrossRef] [PubMed]
17. Nandasiri, R.; Eskin, N.A.M. Canolol and Its Derivatives: A Novel Bioactive with Antioxidant and Anticancer Properties. In *Advances in Food and Nutrition Research*; Academic Press: Cambridge, MA, USA, 2022; pp. 109–129.
18. SzydłOwska-Czerniak, A. Rapeseed and Its Products—Sources of Bioactive Compounds: A Review of Their Characteristics and Analysis. *Crit. Rev. Food Sci. Nutr.* **2013**, *53*, 307–330. [CrossRef] [PubMed]
19. Aloo, S.O.; Mwiti, G.; Ngugi, L.W.; Oh, D.-H. Uncovering the Secrets of Industrial Hemp in Food and Nutrition: The Trends, Challenges, and New-Age Perspectives. *Crit. Rev. Food Sci. Nutr.* **2022**, *28*, 1–20. [CrossRef]
20. Rizzo, G.; Storz, M.A.; Calapai, G. The Role of Hemp (*Cannabis sativa* L.) as a Functional Food in Vegetarian Nutrition. *Foods* **2023**, *12*, 3505. [CrossRef] [PubMed]
21. Bell, J.M. Factors Affecting the Nutritional Value of Canola Meal: A Review. *Can. J. Anim. Sci.* **1993**, *73*, 689–697. [CrossRef]
22. Mathai, J.K.; Liu, Y.; Stein, H.H. Values for Digestible Indispensable Amino Acid Scores (DIAAS) for Some Dairy and Plant Proteins May Better Describe Protein Quality than Values Calculated Using the Concept for Protein Digestibility-Corrected Amino Acid Scores (PDCAAS). *Br. J. Nutr.* **2017**, *117*, 490–499. [CrossRef] [PubMed]
23. Hertzler, S.R.; Lieblein-Boff, J.C.; Weiler, M.; Allgeier, C. Plant Proteins: Assessing Their Nutritional Quality and Effects on Health and Physical Function. *Nutrients* **2020**, *12*, 3704. [CrossRef] [PubMed]
24. FAO. *Dietary Protein Quality Evaluation in Human Nutrition: Report of an FAO Expert Consultation*; Food and Agriculture Organization of the United Nations: Rome, Italy, 2013; p. 92.
25. Galasso, I.; Russo, R.; Mapelli, S.; Ponzoni, E.; Brambilla, I.M.; Battelli, G.; Reggiani, R. Variability in Seed Traits in a Collection of *Cannabis sativa* L. Genotypes. *Front. Plant Sci.* **2016**, *7*, 688. [CrossRef]
26. Sarwar, G. The Protein Digestibility–Corrected Amino Acid Score Method Overestimates Quality of Proteins Containing Antinutritional Factors and of Poorly Digestible Proteins Supplemented with Limiting Amino Acids in Rats. *J. Nutr.* **1997**, *127*, 758–764. [CrossRef]
27. Sosulski, F.W.; Minja, L.A.; Christensen, D.A. Trypsin Inhibitors and Nutritive Value in Cereals. *Plant Foods Hum. Nutr.* **1988**, *38*, 23–34. [CrossRef] [PubMed]
28. Kahraman, O.; Petersen, G.E.; Fields, C. Physicochemical and Functional Modifications of Hemp Protein Concentrate by the Application of Ultrasonication and PH Shifting Treatments. *Foods* **2022**, *11*, 587. [CrossRef] [PubMed]

29. Ashkar, F.; Wu, J. Effects of Food Factors and Processing on Protein Digestibility and Gut Microbiota. *J. Agric. Food Chem.* **2023**, *71*, 8685–8698. [CrossRef] [PubMed]
30. Marinangeli, C.P.F.; House, J.D. Potential Impact of the Digestible Indispensable Amino Acid Score as a Measure of Protein Quality on Dietary Regulations and Health. *Nutr. Rev.* **2017**, *75*, 658–667. [CrossRef] [PubMed]
31. Singh, R.; Langyan, S.; Sangwan, S.; Rohtagi, B.; Khandelwal, A.; Shrivastava, M. Protein for Human Consumption From Oilseed Cakes: A Review. *Front. Sustain. Food Syst.* **2022**, *6*, 856401. [CrossRef]
32. Herreman, L.; Nommensen, P.; Pennings, B.; Laus, M.C. Comprehensive Overview of the Quality of Plant-and Animal-sourced Proteins Based on the Digestible Indispensable Amino Acid Score. *Food Sci. Nutr.* **2020**, *8*, 5379–5391. [CrossRef]
33. Jambrak, A.R.; Mason, T.J.; Lelas, V.; Herceg, Z.; Herceg, I.L. Effect of Ultrasound Treatment on Solubility and Foaming Properties of Whey Protein Suspensions. *J. Food Eng.* **2008**, *86*, 281–287. [CrossRef]
34. Malik, M.A.; Sharma, H.K.; Saini, C.S. High Intensity Ultrasound Treatment of Protein Isolate Extracted from Dephenolized Sunflower Meal: Effect on Physicochemical and Functional Properties. *Ultrason. Sonochem.* **2017**, *39*, 511–519. [CrossRef] [PubMed]
35. Sun, Y.; Chen, J.; Zhang, S.; Li, H.; Lu, J.; Liu, L.; Uluko, H.; Su, Y.; Cui, W.; Ge, W.; et al. Effect of Power Ultrasound Pre-Treatment on the Physical and Functional Properties of Reconstituted Milk Protein Concentrate. *J. Food Eng.* **2014**, *124*, 11–18. [CrossRef]
36. Shen, X.; Fang, T.; Gao, F.; Guo, M. Effects of Ultrasound Treatment on Physicochemical and Emulsifying Properties of Whey Proteins Pre- and Post-Thermal Aggregation. *Food Hydrocoll.* **2017**, *63*, 668–676. [CrossRef]
37. Amiri, A.; Sharifian, P.; Soltanizadeh, N. Application of Ultrasound Treatment for Improving the Physicochemical, Functional and Rheological Properties of Myofibrillar Proteins. *Int. J. Biol. Macromol.* **2018**, *111*, 139–147. [CrossRef] [PubMed]
38. Wang, Y.; Wang, Y.; Li, K.; Bai, Y.; Li, B.; Xu, W. Effect of High Intensity Ultrasound on Physicochemical, Interfacial and Gel Properties of Chickpea Protein Isolate. *LWT* **2020**, *129*, 109563. [CrossRef]
39. Ranjha, M.M.A.N.; Irfan, S.; Lorenzo, J.M.; Shafique, B.; Kanwal, R.; Pateiro, M.; Arshad, R.N.; Wang, L.; Nayik, G.A.; Roobab, U.; et al. Sonication, a Potential Technique for Extraction of Phytoconstituents: A Systematic Review. *Processes* **2021**, *9*, 1406. [CrossRef]
40. Jiang, J.; Wang, Q.; Xiong, Y.L. A PH Shift Approach to the Improvement of Interfacial Properties of Plant Seed Proteins. *Curr. Opin. Food Sci.* **2018**, *19*, 50–56. [CrossRef]
41. Mao, C.; Wu, J.; Zhang, X.; Ma, F.; Cheng, Y. Improving the Solubility and Digestibility of Potato Protein with an Online Ultrasound-Assisted PH Shifting Treatment at Medium Temperature. *Foods* **2020**, *9*, 1908. [CrossRef] [PubMed]
42. Bartončíková, M.; Lapčíková, B.; Lapčík, L.; Valenta, T. Hemp-Derived CBD Used in Food and Food Supplements. *Molecules* **2023**, *28*, 8047. [CrossRef]
43. Eilam, Y.; Khattib, H.; Pintel, N.; Avni, D. Microalgae—Sustainable Source for Alternative Proteins and Functional Ingredients Promoting Gut and Liver Health. *Glob. Challenges* **2023**, *7*, 2200177. [CrossRef]
44. Shanthakumar, P.; Klepacka, J.; Bains, A.; Chawla, P.; Dhull, S.B.; Najda, A. The Current Situation of Pea Protein and Its Application in the Food Industry. *Molecules* **2022**, *27*, 5354. [CrossRef]
45. Rafique, H.; Dong, R.; Wang, X.; Alim, A.; Aadil, R.M.; Li, L.; Zou, L.; Hu, X. Dietary-Nutraceutical Properties of Oat Protein and Peptides. *Front. Nutr.* **2022**, *9*, 950400. [CrossRef]
46. Hoffman, J.R.; Falvo, M.J. Protein—Which Is Best? *J. Sports Sci. Med.* **2004**, *3*, 118–130.
47. Kleba, S.; Ismail, B.P. *Report on Minnesota Plant Based Protein for Food*; Agriculture Utilization Research Institute: Crookston, MN, USA, 2019.
48. Carbonaro, M.; Maselli, P.; Nucara, A. Relationship between Digestibility and Secondary Structure of Raw and Thermally Treated Legume Proteins: A Fourier Transform Infrared (FT-IR) Spectroscopic Study. *Amino Acids* **2012**, *43*, 911–921. [CrossRef] [PubMed]
49. Nierle, W. Views on the Amino Acid Composition of Grain and the Influence of Processing. In *Amino Acid Composition and Biological Value of Cereal Proteins*; Springer: Dordrecht, The Netherlands, 1985; pp. 371–382.
50. Struthers, B.J. Lysinoalanine: Production, Significance and Control in Preparation and Use of Soya and Other Food Proteins. *J. Am. Oil Chem. Soc.* **1981**, *58*, 501–503. [CrossRef]
51. van Vliet, S.; Burd, N.A.; van Loon, L.J. The Skeletal Muscle Anabolic Response to Plant- versus Animal-Based Protein Consumption. *J. Nutr.* **2015**, *145*, 1981–1991. [CrossRef] [PubMed]
52. Phillips, S.M. Current Concepts and Unresolved Questions in Dietary Protein Requirements and Supplements in Adults. *Front. Nutr.* **2017**, *4*, 13. [CrossRef] [PubMed]
53. Rutherfurd, S.M.; Fanning, A.C.; Miller, B.J.; Moughan, P.J. Protein Digestibility-Corrected Amino Acid Scores and Digestible Indispensable Amino Acid Scores Differentially Describe Protein Quality in Growing Male Rats. *J. Nutr.* **2015**, *145*, 372–379. [CrossRef] [PubMed]
54. Petrusán, J.I.; Rawel, H.; Huschek, G. Protein-Rich Vegetal Sources and Trends in Human Nutrition: A Review. *Curr. Top. Pept. Protein Res.* **2016**, *17*, 1–19.
55. Ertl, P.; Knaus, W.; Zollitsch, W. An Approach to Including Protein Quality When Assessing the Net Contribution of Livestock to Human Food Supply. *Animal* **2016**, *10*, 1883–1889. [CrossRef] [PubMed]
56. House, J.D.; Neufeld, J.; Leson, G. Evaluating the Quality of Protein from Hemp Seed (*Cannabis sativa* L.) Products Through the Use of the Protein Digestibility-Corrected Amino Acid Score Method. *J. Agric. Food Chem.* **2010**, *58*, 11801–11807. [CrossRef] [PubMed]

57. Bailey, S. The Amino Acid Digestibility and Digestible Indispensable Amino Acid Score for Rapeseed Protein Isolate Increases after Moderate Heating Resulting in a Protein Quality Similar to Whey Protein Isolate. Available online: https://nutrition.ansci.illinois.edu/node/1787 (accessed on 21 January 2024).
58. Sim, S.Y.J.; SRV, A.; Chiang, J.H.; Henry, C.J. Plant Proteins for Future Foods: A Roadmap. *Foods* **2021**, *10*, 1967. [CrossRef]
59. Schmidt, J.A.; Rinaldi, S.; Scalbert, A.; Ferrari, P.; Achaintre, D.; Gunter, M.J.; Appleby, P.N.; Key, T.J.; Travis, R.C. Plasma Concentrations and Intakes of Amino Acids in Male Meat-Eaters, Fish-Eaters, Vegetarians and Vegans: A Cross-Sectional Analysis in the EPIC-Oxford Cohort. *Eur. J. Clin. Nutr.* **2016**, *70*, 306–312. [CrossRef]
60. Gan, Y.; Liang, C.; Wang, X.; McConkey, B. Lowering Carbon Footprint of Durum Wheat by Diversifying Cropping Systems. *Field Crop. Res.* **2011**, *122*, 199–206. [CrossRef]
61. Girgih, A.T.; Udenigwe, C.C.; Aluko, R.E. In Vitro Antioxidant Properties of Hemp Seed (*Cannabis sativa* L.) Protein Hydrolysate Fractions. *J. Am. Oil Chem. Soc.* **2011**, *88*, 381–389. [CrossRef]
62. Girgih, A.T.; He, R.; Malomo, S.; Offengenden, M.; Wu, J.; Aluko, R.E. Structural and Functional Characterization of Hemp Seed (*Cannabis sativa* L.) Protein-Derived Antioxidant and Antihypertensive Peptides. *J. Funct. Foods* **2014**, *6*, 384–394. [CrossRef]
63. Nosworthy, M.G.; House, J.D. Factors Influencing the Quality of Dietary Proteins: Implications for Pulses. *Cereal Chem.* **2017**, *94*, 49–57. [CrossRef]
64. Nosworthy, M.G.; Neufeld, J.; Frohlich, P.; Young, G.; Malcolmson, L.; House, J.D. Determination of the Protein Quality of Cooked Canadian Pulses. *Food Sci. Nutr.* **2017**, *5*, 896–903. [CrossRef] [PubMed]
65. Axentii, M.; Stroe, S.-G.; Codină, G.G. Development and Quality Evaluation of Rigatoni Pasta Enriched with Hemp Seed Meal. *Foods* **2023**, *12*, 1774. [CrossRef] [PubMed]
66. Maison, T.; Stein, H.H. Amino Acid Digestibility in Canola Meal, 00-Rapeseed Meal, and 00-Rapeseed Expellers Fed to Growing Pigs. *J. Anim. Sci.* **2014**, *92*, 3502–3514. [CrossRef]
67. Wang, Q.; Xiong, Y.L. Processing, Nutrition, and Functionality of Hempseed Protein: A Review. *Compr. Rev. Food Sci. Food Saf.* **2019**, *18*, 936–952. [CrossRef]
68. El-Sohaimy, S.A.; Androsova, N.V.; Toshev, A.D.; El Enshasy, H.A. Nutritional Quality, Chemical, and Functional Characteristics of Hemp (*Cannabis sativa* ssp. *Sativa*) Protein Isolate. *Plants* **2022**, *11*, 2825. [CrossRef]
69. Guillamón, E.; Pedrosa, M.M.; Burbano, C.; Cuadrado, C.; Sánchez, M.d.C.; Muzquiz, M. The Trypsin Inhibitors Present in Seed of Different Grain Legume Species and Cultivar. *Food Chem.* **2008**, *107*, 68–74. [CrossRef]
70. Girgih, A.T.; Alashi, A.; He, R.; Malomo, S.; Aluko, R.E. Preventive and Treatment Effects of a Hemp Seed (*Cannabis sativa* L.) Meal Protein Hydrolysate against High Blood Pressure in Spontaneously Hypertensive Rats. *Eur. J. Nutr.* **2014**, *53*, 1237–1246. [CrossRef] [PubMed]
71. Girgih, A.T.; Udenigwe, C.C.; Aluko, R.E. Reverse-Phase HPLC Separation of Hemp Seed (*Cannabis sativa* L.) Protein Hydrolysate Produced Peptide Fractions with Enhanced Antioxidant Capacity. *Plant Foods Hum. Nutr.* **2013**, *68*, 39–46. [CrossRef] [PubMed]
72. Aiello, G.; Fasoli, E.; Boschin, G.; Lammi, C.; Zanoni, C.; Citterio, A.; Arnoldi, A. Proteomic Characterization of Hempseed (*Cannabis sativa* L.). *J. Proteomics* **2016**, *147*, 187–196. [CrossRef] [PubMed]
73. Wu, G.; Meininger, C.J. Regulation of Nitric Oxide Synthesis by Dietary Factors. *Annu. Rev. Nutr.* **2002**, *22*, 61–86. [CrossRef] [PubMed]
74. Callaway, J.C. Hempseed as a Nutritional Resource: An Overview. *Euphytica* **2004**, *140*, 65–72. [CrossRef]
75. Mikulcová, V.; Kašpárková, V.; Humpolíček, P.; Buňková, L. Formulation, Characterization and Properties of Hemp Seed Oil and Its Emulsions. *Molecules* **2017**, *22*, 700. [CrossRef] [PubMed]
76. Śmiarowska, M.; Białecka, M.; Machoy-Mokrzyńska, A. Cannabis and Cannabinoids: Pharmacology and Therapeutic Potential. *Neurol. Neurochir. Pol.* **2022**, *56*, 4–13. [CrossRef] [PubMed]
77. Leslie, A.J.; Summers, J.D.; Grandhi, R.; Leeson, S. Arginine-Lysine Relationship in Rapeseed Meal. *Poult. Sci.* **1976**, *55*, 631–637. [CrossRef] [PubMed]
78. Sarv, V. A Comparative Study of Camelina, Canola and Hemp Seed Processing and Products. Master's Thesis, University of Toronto, Toronto, ON, Canada, 2017.
79. Wanasundara, J.P.D.; McIntosh, T.C.; Perera, S.P.; Withana-Gamage, T.S.; Mitra, P. Canola/Rapeseed Protein-Functionality and Nutrition. *OCL* **2016**, *23*, D407. [CrossRef]
80. Tzen, J.; Cao, Y.; Laurent, P.; Ratnayake, C.; Huang, A. Lipids, Proteins, and Structure of Seed Oil Bodies from Diverse Species. *Plant Physiol.* **1993**, *101*, 267–276. [CrossRef]
81. Aider, M.; Barbana, C. Canola Proteins: Composition, Extraction, Functional Properties, Bioactivity, Applications as a Food Ingredient and Allergenicity—A Practical and Critical Review. *Trends Food Sci. Technol.* **2011**, *22*, 21–39. [CrossRef]
82. Tan, S.H.; Mailer, R.J.; Blanchard, C.L.; Agboola, S.O. Canola Proteins for Human Consumption: Extraction, Profile, and Functional Properties. *J. Food Sci.* **2011**, *76*, R16–R28. [CrossRef] [PubMed]
83. Eze, C.R.; Kwofie, E.M.; Adewale, P.; Lam, E.; Ngadi, M. Advances in Legume Protein Extraction Technologies: A Review. *Innov. Food Sci. Emerg. Technol.* **2022**, *82*, 103199. [CrossRef]
84. Pulivarthi, M.K.; Buenavista, R.M.; Bangar, S.P.; Li, Y.; Pordesimo, L.O.; Bean, S.R.; Siliveru, K. Dry Fractionation Process Operations in the Production of Protein Concentrates: A Review. *Compr. Rev. Food Sci. Food Saf.* **2023**, *22*, 4670–4697. [CrossRef] [PubMed]

85. Wockenfuss, L.; Lammers, V.; Heinz, V.; Sozer, N.; Silventoinen-Veijalainen, P. Two Steps of Dry Fractionation: Comparison and Combination of Air Classification and Electrostatic Separation for Protein Enrichment from Defatted Rapeseed Press Cake. *J. Food Eng.* **2023**, *357*, 111623. [CrossRef]
86. Assatory, A.; Vitelli, M.; Rajabzadeh, A.R.; Legge, R.L. Dry Fractionation Methods for Plant Protein, Starch and Fiber Enrichment: A Review. *Trends Food Sci. Technol.* **2019**, *86*, 340–351. [CrossRef]
87. Rempel, C.; Li, X.; Geng, X.; Liu, Q.; Zhang, Y. Manufacture of Defatted Canola Meal with Enhanced Nutritive Composition by Air Classification on an Industrial Scale. *J. Sci. Food Agric.* **2020**, *100*, 764–774. [CrossRef] [PubMed]
88. Laguna, O.; Barakat, A.; Alhamada, H.; Durand, E.; Baréa, B.; Fine, F.; Villeneuve, P.; Citeau, M.; Dauguet, S.; Lecomte, J. Production of Proteins and Phenolic Compounds Enriched Fractions from Rapeseed and Sunflower Meals by Dry Fractionation Processes. *Ind. Crops Prod.* **2018**, *118*, 160–172. [CrossRef]
89. Hopf, A.; Dehghani, F.; Buckow, R. Dry Fractionation of Plant-Based Proteins for Better Meat Analogue Applications. *Curr. Food Sci. Technol. Rep.* **2023**, *1*, 91–98. [CrossRef]
90. Ghodsvali, A.; Khodaparast, M.H.H.; Vosoughi, M.; Diosady, L.L. Preparation of Canola Protein Materials Using Membrane Technology and Evaluation of Meals Functional Properties. *Food Res. Int.* **2005**, *38*, 223–231. [CrossRef]
91. Malomo, S.A.; Aluko, R.E. Conversion of a Low Protein Hemp Seed Meal into a Functional Protein Concentrate through Enzymatic Digestion of Fibre Coupled with Membrane Ultrafiltration. *Innov. Food Sci. Emerg. Technol.* **2015**, *31*, 151–159. [CrossRef]
92. Ahlström, C.; Thuvander, J.; Rayner, M.; Matos, M.; Gutiérrez, G.; Östbring, K. The Effect of Precipitation PH on Protein Recovery Yield and Emulsifying Properties in the Extraction of Protein from Cold-Pressed Rapeseed Press Cake. *Molecules* **2022**, *27*, 2957. [CrossRef] [PubMed]
93. Deng, Q.; Zinoviadou, K.G.; Galanakis, C.M.; Orlien, V.; Grimi, N.; Vorobiev, E.; Lebovka, N.; Barba, F.J. The Effects of Conventional and Non-Conventional Processing on Glucosinolates and Its Derived Forms, Isothiocyanates: Extraction, Degradation, and Applications. *Food Eng. Rev.* **2015**, *7*, 357–381. [CrossRef]
94. Karaca, A.C.; Low, N.; Nickerson, M. Emulsifying Properties of Canola and Flaxseed Protein Isolates Produced by Isoelectric Precipitation and Salt Extraction. *Food Res. Int.* **2011**, *44*, 2991–2998. [CrossRef]
95. Cruz-Solis, I.; Ibarra-Herrera, C.C.; Rocha-Pizaña, M.d.R.; Luna-Vital, D. Alkaline Extraction–Isoelectric Precipitation of Plant Proteins. In *Green Protein Processing Technologies from Plants*; Springer International Publishing: Cham, The Netherlands, 2023; pp. 1–29.
96. Fang, B.; Gu, Z.; Ohm, J.-B.; Chen, B.; Rao, J. Reverse Micelles Extraction of Hemp Protein Isolate: Impact of Defatting Process on Protein Structure, Functionality, and Aromatic Profile. *Food Hydrocoll.* **2023**, *135*, 108158. [CrossRef]
97. Tzeng, Y.; Diosady, L.L.; Rubin, L.J. Production of Canola Protein Materials by Alkaline Extraction, Precipitation, and Membrane Processing. *J. Food Sci.* **1990**, *55*, 1147–1151. [CrossRef]
98. Hadnađev, M.; Dapčević-Hadnađev, T.; Lazaridou, A.; Moschakis, T.; Michaelidou, A.-M.; Popović, S.; Biliaderis, C.G. Hempseed Meal Protein Isolates Prepared by Different Isolation Techniques. Part I. Physicochemical Properties. *Food Hydrocoll.* **2018**, *79*, 526–533. [CrossRef]
99. Teh, S.-S.; Bekhit, A.E.-D.; Carne, A.; Birch, J. Effect of the Defatting Process, Acid and Alkali Extraction on the Physicochemical and Functional Properties of Hemp, Flax and Canola Seed Cake Protein Isolates. *J. Food Meas. Charact.* **2014**, *8*, 92–104. [CrossRef]
100. Zhang, M.; Wang, O.; Cai, S.; Zhao, L.; Zhao, L. Composition, Functional Properties, Health Benefits and Applications of Oilseed Proteins: A Systematic Review. *Food Res. Int.* **2023**, *171*, 113061. [CrossRef]
101. Singh, P.; Pandey, V.K.; Sultan, Z.; Singh, R.; Dar, A.H. Classification, Benefits, and Applications of Various Anti-Nutritional Factors Present in Edible Crops. *J. Agric. Food Res.* **2023**, *14*, 100902. [CrossRef]
102. Urbano, G.; López-Jurado, M.; Aranda, P.; Vidal-Valverde, C.; Tenorio, E.; Porres, J. The Role of Phytic Acid in Legumes: Antinutrient or Beneficial Function? *J. Physiol. Biochem.* **2000**, *56*, 283–294. [CrossRef] [PubMed]
103. Schlemmer, U.; Frølich, W.; Prieto, R.M.; Grases, F. Phytate in Foods and Significance for Humans: Food Sources, Intake, Processing, Bioavailability, Protective Role and Analysis. *Mol. Nutr. Food Res.* **2009**, *53*, S330–S375. [CrossRef] [PubMed]
104. Russo, R.; Reggiani, R. Antinutritive Compounds in Twelve *Camelina sativa* Genotypes. *Am. J. Plant Sci.* **2012**, *03*, 1408–1412. [CrossRef]
105. Cheeke, P.R. Nutritional And Physiological Implications Of Saponins: A Review. *Can. J. Anim. Sci.* **1971**, *51*, 621–632. [CrossRef]
106. Amarowicz, R.; Naczk, M.; Shahidi, F. Antioxidant Activity of Crude Tannins of Canola and Rapeseed Hulls. *J. Am. Oil Chem. Soc.* **2000**, *77*, 957–961. [CrossRef]
107. Mattila, P.H.; Pihlava, J.-M.; Hellström, J.; Nurmi, M.; Eurola, M.; Mäkinen, S.; Jalava, T.; Pihlanto, A. Contents of Phytochemicals and Antinutritional Factors in Commercial Protein-Rich Plant Products. *Food Qual. Saf.* **2018**, *2*, 213–219. [CrossRef]
108. de Hoyos-Martínez, P.L.; Merle, J.; Labidi, J.; Charrier-El Bouhtoury, F. Tannins Extraction: A Key Point for Their Valorization and Cleaner Production. *J. Clean. Prod.* **2019**, *206*, 1138–1155. [CrossRef]
109. Lee, H.; Yildiz, G.; dos Santos, L.C.; Jiang, S.; Andrade, J.E.; Engeseth, N.J.; Feng, H. Soy Protein Nano-Aggregates with Improved Functional Properties Prepared by Sequential PH Treatment and Ultrasonication. *Food Hydrocoll.* **2016**, *55*, 200–209. [CrossRef]
110. Pires, S.M.G.; Reis, R.S.; Cardoso, S.M.; Pezzani, R.; Paredes-Osses, E.; Seilkhan, A.; Ydyrys, A.; Martorell, M.; Sönmez Gürer, E.; Setzer, W.N.; et al. Phytates as a Natural Source for Health Promotion: A Critical Evaluation of Clinical Trials. *Front. Chem.* **2023**, *11*, 1174109. [CrossRef]

111. Lee, S.-H.; Park, H.-J.; Chun, H.-K.; Cho, S.-Y.; Cho, S.-M.; Lillehoj, H.S. Dietary Phytic Acid Lowers the Blood Glucose Level in Diabetic KK Mice. *Nutr. Res.* **2006**, *26*, 474–479. [CrossRef]
112. Feng, D.; Zuo, J. Nutritional and Anti-Nutritional Composition of Rapeseed Meal and Its Utilization as a Feed Ingredient for Animal. *Agric. Food Sci.* **2015**, 265–270. Available online: https://www.gcirc.org/fileadmin/documents/Proceedings/IRCWuhan2007%20vol5/Pages_de_vol-5-37.pdf (accessed on 10 February 2024).
113. Ballester, D.; Rodrigo, R.; Nakouzi, J.; Chichester, C.; Yáñez, E.; Mönckeberg, F. Rapeseed meal III—A simple method for detoxification. *J. Sci. Food Agric.* **1970**, *21*, 143–144. [CrossRef]
114. Sosulski, F.W.; Soliman, F.S.; Bhatty, R.S. Diffusion Extraction of Glucosinolates from Rapeseed. *Can. Inst. Food Sci. Technol. J.* **1972**, *5*, 101–104. [CrossRef]
115. Sosulski, F.W.; Bakal, A. Isolated Proteins from Rapeseed, Flax and Sunflower Meals. *Can. Inst. Food Technol. J.* **1969**, *2*, 28–32. [CrossRef]
116. Kozlowska, H.; Sosulski, F.W.; Youngs, C.G. Extraction of Glucosinolates from Rapeseed. *Can. Inst. Food Sci. Technol. J.* **1972**, *5*, 149–154. [CrossRef]
117. Wieczorek, D.; Żyszka-Haberecht, B.; Kafka, A.; Lipok, J. Phosphonates as Unique Components of Plant Seeds—A Promising Approach to Use Phosphorus Profiles in Plant Chemotaxonomy. *Int. J. Mol. Sci.* **2021**, *22*, 11501. [CrossRef] [PubMed]
118. Li, Y.; Li, J.; Cao, P.; Liu, Y. Sinapine-Enriched Rapeseed Oils Reduced Fatty Liver Formation in High-Fat Diet-Fed C57BL/6J Mice. *RSC Adv.* **2020**, *10*, 21248–21258. [CrossRef]
119. Qiao, H.Y.; Dahiya, J.P.; Classen, H.L. Nutritional and Physiological Effects of Dietary Sinapic Acid (4-Hydroxy-3,5-Dimethoxy-Cinnamic Acid) in Broiler Chickens and Its Metabolism in the Digestive Tract. *Poult. Sci.* **2008**, *87*, 719–726. [CrossRef] [PubMed]
120. Suen, L. *Erucic Acid in Edible Fats and Oils. Food Safety Focus*; Food Safety Platform, Center for Food Safety: Washington, DC, USA, 2020; Volume 162.
121. Galanty, A.; Grudzińska, M.; Paździora, W.; Paśko, P. Erucic Acid—Both Sides of the Story: A Concise Review on Its Beneficial and Toxic Properties. *Molecules* **2023**, *28*, 1924. [CrossRef]
122. Knutsen, H.K.; Alexander, J.; Barregård, L.; Bignami, M.; Brüschweiler, B.; Ceccatelli, S.; Dinovi, M.; Edler, L.; Grasl-Kraupp, B.; Hogstrand, C.; et al. Erucic Acid in Feed and Food. *EFSA J.* **2016**, *14*, e04593. [CrossRef]
123. Vetter, W.; Darwisch, V.; Lehnert, K. Erucic Acid in Brassicaceae and Salmon—An Evaluation of the New Proposed Limits of Erucic Acid in Food. *NFS J.* **2020**, *19*, 9–15. [CrossRef]
124. Russo, M.; Yan, F.; Stier, A.; Klasen, L.; Honermeier, B. Erucic Acid Concentration of Rapeseed (*Brassica Napus* L.) Oils on the German Food Retail Market. *Food Sci. Nutr.* **2021**, *9*, 3664–3672. [CrossRef] [PubMed]
125. Beszterda, M.; Nogala-Kałucka, M. Current Research Developments on the Processing and Improvement of the Nutritional Quality of Rapeseed (*Brassica Napus* L.). *Eur. J. Lipid Sci. Technol.* **2019**, *121*, 1800045. [CrossRef]
126. Baker, M.T.; Lu, P.; Parrella, J.A.; Leggette, H.R. Consumer Acceptance toward Functional Foods: A Scoping Review. *Int. J. Environ. Res. Public Health* **2022**, *19*, 1217. [CrossRef] [PubMed]
127. Shah, P.; Dhir, A.; Joshi, R.; Tripathy, N. Opportunities and Challenges in Food Entrepreneurship: In-Depth Qualitative Investigation of Millet Entrepreneurs. *J. Bus. Res.* **2023**, *155*, 113372. [CrossRef]
128. Farinon, B.; Molinari, R.; Costantini, L.; Merendino, N. The Seed of Industrial Hemp (*Cannabis sativa* L.): Nutritional Quality and Potential Functionality for Human Health and Nutrition. *Nutrients* **2020**, *12*, 1935. [CrossRef] [PubMed]
129. Tomko, A.M.; Whynot, E.G.; Ellis, L.D.; Dupré, D.J. Anti-Cancer Potential of Cannabinoids, Terpenes, and Flavonoids Present in Cannabis. *Cancers* **2020**, *12*, 1985. [CrossRef] [PubMed]
130. Rea, K.A.; Casaretto, J.A.; Al-Abdul-Wahid, M.S.; Sukumaran, A.; Geddes-McAlister, J.; Rothstein, S.J.; Akhtar, T.A. Biosynthesis of Cannflavins A and B from *Cannabis sativa* L. *Phytochemistry* **2019**, *164*, 162–171. [CrossRef] [PubMed]
131. Ullah, A.; Munir, S.; Badshah, S.L.; Khan, N.; Ghani, L.; Poulson, B.G.; Emwas, A.-H.; Jaremko, M. Important Flavonoids and Their Role as a Therapeutic Agent. *Molecules* **2020**, *25*, 5243. [CrossRef] [PubMed]
132. Zahedi, M.; Ghiasvand, R.; Feizi, A.; Asgari, G.; Darvish, L. Does Quercetin Improve Cardiovascular Risk Factors and Inflammatory Biomarkers in Women with Type 2 Diabetes: A Double-Blind Randomized Controlled Clinical Trial. *Int. J. Prev. Med.* **2013**, *4*, 777–785.
133. Atkins, P.L. Sample processing and preparation considerations for solid cannabis products. *J. AOAC Int.* **2019**, *102*, 427–433. [CrossRef]
134. Glivar, T.; Eržen, J.; Kreft, S.; Zagožen, M.; Čerenak, A.; Čeh, B.; Tavčar Benković, E. Cannabinoid Content in Industrial Hemp (*Cannabis sativa* L.) Varieties Grown in Slovenia. *Ind. Crops Prod.* **2020**, *145*, 112082. [CrossRef]
135. Meissner, H.; Cascella, M. *Cannabidiol (CBD)*; StatPearls: Treasure Island, FL, USA, 2020.
136. Visković, J.; Zheljazkov, V.D.; Sikora, V.; Noller, J.; Latković, D.; Ocamb, C.M.; Koren, A. Industrial Hemp (*Cannabis sativa* L.) Agronomy and Utilization: A Review. *Agronomy* **2023**, *13*, 931. [CrossRef]
137. Thurman, E.M. Analysis of Terpenes in Hemp (*Cannabis sativa*) by Gas Chromatography/Mass Spectrometry: Isomer Identification Analysis. In *Comprehensive Analytical Chemistry*; Elsevier: Amsterdam, The Netherlands, 2020; pp. 197–233.
138. Chen, C.; Pan, Z. Cannabidiol and Terpenes from Hemp—Ingredients for Future Foods and Processing Technologies. *J. Futur. Foods* **2021**, *1*, 113–127. [CrossRef]

139. Jokić, S.; Jerković, I.; Pavić, V.; Aladić, K.; Molnar, M.; Kovač, M.J.; Vladimir-Knežević, S. Terpenes and Cannabinoids in Supercritical CO2 Extracts of Industrial Hemp Inflorescences: Optimization of Extraction, Antiradical and Antibacterial Activity. *Pharmaceuticals* **2022**, *15*, 1117. [CrossRef]
140. Blasi, F.; Tringaniello, C.; Verducci, G.; Cossignani, L. Bioactive Minor Components of Italian and Extra-European Hemp Seed Oils. *LWT* **2022**, *158*, 113167. [CrossRef]
141. Kwon, Y.; Kim, K.; Heo, H.; Lee, J.; Sung, J. Vitamin E, Phytosterol, and Carotenoid Contents of Hemp (*Cannabis sativa* L.) Seed. *J. Korean Soc. Food Sci. Nutr.* **2023**, *52*, 975–981. [CrossRef]
142. Alonso-Esteban, J.I.; Torija-Isasa, M.E.; Sánchez-Mata, M.d.C. Mineral Elements and Related Antinutrients, in Whole and Hulled Hemp (*Cannabis sativa* L.) Seeds. *J. Food Compos. Anal.* **2022**, *109*, 104516. [CrossRef]
143. Shen, J.; Liu, Y.; Wang, X.; Bai, J.; Lin, L.; Luo, F.; Zhong, H. A Comprehensive Review of Health-Benefiting Components in Rapeseed Oil. *Nutrients* **2023**, *15*, 999. [CrossRef] [PubMed]
144. Miłek, M.; Ciszkowicz, E.; Sidor, E.; Hęclik, J.; Lecka-Szlachta, K.; Dżugan, M. The Antioxidant, Antibacterial and Anti-Biofilm Properties of Rapeseed Creamed Honey Enriched with Selected Plant Superfoods. *Antibiotics* **2023**, *12*, 235. [CrossRef]
145. Di Lena, G.; Sanchez del Pulgar, J.; Lucarini, M.; Durazzo, A.; Ondrejíčková, P.; Oancea, F.; Frincu, R.-M.; Aguzzi, A.; Ferrari Nicoli, S.; Casini, I.; et al. Valorization Potentials of Rapeseed Meal in a Biorefinery Perspective: Focus on Nutritional and Bioactive Components. *Molecules* **2021**, *26*, 6787. [CrossRef]
146. Zago, E.; Lecomte, J.; Barouh, N.; Aouf, C.; Carré, P.; Fine, F.; Villeneuve, P. Influence of rapeseed meal treatments on its total phenolic content and composition in sinapine, sinapic acid and canolol. *Ind. Crops Prod.* **2015**, *76*, 1061–1070. [CrossRef]
147. Chew, S.C. Cold Pressed Rapeseed (*Brassica napus*) Oil. In *Cold Pressed Oils*; Elsevier: Amsterdam, The Netherlands, 2020; pp. 65–80.
148. Chen, L.; Cao, H.; Xiao, J. Polyphenols. In *Polyphenols: Properties, Recovery, and Applications*; Elsevier: Amsterdam, The Netherlands, 2018; pp. 45–67.
149. Tso, P.; Caldwell, J.; Lee, D.; Boivin, G.P.; DeMichele, S.J. Comparison of Growth, Serum Biochemistries and N−6 Fatty Acid Metabolism in Rats Fed Diets Supplemented with High-Gamma-Linolenic Acid Safflower Oil or Borage Oil for 90days. *Food Chem. Toxicol.* **2012**, *50*, 1911–1919. [CrossRef]
150. Tasset-Cuevas, I.; Fernández-Bedmar, Z.; Lozano-Baena, M.D.; Campos-Sánchez, J.; de Haro-Bailón, A.; Muñoz-Serrano, A.; Alonso-Moraga, Á. Protective Effect of Borage Seed Oil and Gamma Linolenic Acid on DNA: In Vivo and In Vitro Studies. *PLoS ONE* **2013**, *8*, e56986. [CrossRef] [PubMed]
151. Blondeau, N.; Lipsky, R.H.; Bourourou, M.; Duncan, M.W.; Gorelick, P.B.; Marini, A.M. Alpha-Linolenic Acid: An Omega-3 Fatty Acid with Neuroprotective Properties—Ready for Use in the Stroke Clinic? *Biomed Res. Int.* **2015**, *2015*, 519830. [CrossRef]
152. Zhao, X.; Xiang, X.; Huang, J.; Ma, Y.; Sun, J.; Zhu, D. Studying the Evaluation Model of the Nutritional Quality of Edible Vegetable Oil Based on Dietary Nutrient Reference Intake. *ACS Omega* **2021**, *6*, 6691–6698. [CrossRef]
153. Bertoni, C.; Abodi, M.; D'Oria, V.; Milani, G.P.; Agostoni, C.; Mazzocchi, A. Alpha-Linolenic Acid and Cardiovascular Events: A Narrative Review. *Int. J. Mol. Sci.* **2023**, *24*, 14319. [CrossRef]
154. Franke, S.; Fröhlich, K.; Werner, S.; Böhm, V.; Schöne, F. Carotenoids and Vitamin E in Seed, Press Cake and Oil of Rapeseed, Sunflower, Flax and Safflower—Comparison of HPLC and Photometric Determination of Carotenoids. In Proceedings of the 13th International Rapeseed Congress, Prague, Czech Republic, 5–9 June 2011.
155. Mrowicka, M.; Mrowicki, J.; Kucharska, E.; Majsterek, I. Lutein and Zeaxanthin and Their Roles in Age-Related Macular Degeneration—Neurodegenerative Disease. *Nutrients* **2022**, *14*, 827. [CrossRef] [PubMed]
156. Tanumihardjo, S.A. Carotenoids: Health Effects. In *Encyclopedia of Human Nutrition*; Elsevier: Amsterdam, The Netherlands, 2013; pp. 292–297.
157. Matthaus, B.; Özcan, M.M.; Juhaimi, F. Al Some rape/canola seed oils: Fatty acid composition and tocopherols. *Zeitschrift fur Naturforsch. Sect. C J. Biosci.* **2016**, *71*, 73–77. [CrossRef]
158. Adesso, M.; Laser, P.; Mills, A. An overview of industrial hemp law in the United States. *UDC/DCSL Law Rev.* **2019**, *22*, 85.
159. Regulation (EU) No 1308/2013 of the European Parliament and of the Council of 17 December 2013 Establishing a Common Organisation of the Markets in Agricultural Products and Repealing Council Regulations (EEC) No 922/72, (EEC) No 234/79, (EC) No 1037/2001 and (EC) No 1234/2007. Available online: https://www.legislation.gov.uk/eur/2013/1308/contents (accessed on 10 February 2024).
160. Foti, V.T.; Scuderi, A.; Bellia, C. Actuality and future prospects of *Cannabis sativa* L. crops. features and problems. *Qual. Access Success* **2019**, *20*, 255–262.
161. Nissen, L.; Casciano, F.; Babini, E.; Gianotti, A. Industrial Hemp Foods and Beverages and Product Properties. In *Industrial Hemp*; Elsevier: Amsterdam, The Netherlands, 2022; pp. 219–246.
162. Trovato, E.; Arena, K.; La Tella, R.; Rigano, F.; Laganà Vinci, R.; Dugo, P.; Mondello, L.; Guarnaccia, P. Hemp Seed-Based Food Products as Functional Foods: A Comprehensive Characterization of Secondary Metabolites Using Liquid and Gas Chromatography Methods. *J. Food Compos. Anal.* **2023**, *117*, 105151. [CrossRef]
163. Werz, O.; Seegers, J.; Schaible, A.M.; Weinigel, C.; Barz, D.; Koeberle, A.; Allegrone, G.; Pollastro, F.; Zampieri, L.; Grassi, G.; et al. Cannflavins from Hemp Sprouts, a Novel Cannabinoid-Free Hemp Food Product, Target Microsomal Prostaglandin E2 Synthase-1 and 5-Lipoxygenase. *PharmaNutrition* **2014**, *2*, 53–60. [CrossRef]

164. Livadariu, O.; Raiciu, D.; Maximilain, C.; Căpitanu, E. Studies Regarding Treatments of LED-s Emitted Light on Sprouting Hemp (*Cannabis sativa* L.). *Rom. Biotechnol. Lett.* **2019**, *24*, 485–490. [CrossRef]
165. Moon, Y.; Cha, Y.; Lee, J.; Kim, K.; Kwon, D.; Kang, Y. Investigation of Suitable Seed Sizes, Segregation of Ripe Seeds, and Improved Germination Rate for the Commercial Production of Hemp Sprouts (*Cannabis sativa* L.). *J. Sci. Food Agric.* **2020**, *100*, 2819–2827. [CrossRef]
166. Zahari, I.; Ferawati, F.; Helstad, A.; Ahlström, C.; Östbring, K.; Rayner, M.; Purhagen, J.K. Development of High-Moisture Meat Analogues with Hemp and Soy Protein Using Extrusion Cooking. *Foods* **2020**, *9*, 772. [CrossRef] [PubMed]
167. Kwaśnica, A.; Pachura, N.; Masztalerz, K.; Figiel, A.; Zimmer, A.; Kupczyński, R.; Wujcikowska, K.; Carbonell-Barrachina, A.A.; Szumny, A.; Różański, H. Volatile Composition and Sensory Properties as Quality Attributes of Fresh and Dried Hemp Flowers (*Cannabis sativa* L.). *Foods* **2020**, *9*, 1118. [CrossRef] [PubMed]
168. Renzyaeva, T.V.; Renzyaev, A.O.; Reznichenko, I.U.; Popov, A.M.; Kravchenko, S.N.; Miller, E.S. Rapeseed Processing Products as a Component of Flour-Based Food for Gerontological Purpose. *IOP Conf. Ser. Earth Environ. Sci.* **2021**, *640*, 042004. [CrossRef]
169. Lin, L.; Allemekinders, H.; Dansby, A.; Campbell, L.; Durance-Tod, S.; Berger, A.; Jones, P.J. Evidence of Health Benefits of Canola Oil. *Nutr. Rev.* **2013**, *71*, 370–385. [CrossRef] [PubMed]
170. Raboanatahiry, N.; Li, H.; Yu, L.; Li, M. Rapeseed (*Brassica napus*): Processing, Utilization, and Genetic Improvement. *Agronomy* **2021**, *11*, 1776. [CrossRef]
171. Korus, J.; Chmielewska, A.; Witczak, M.; Ziobro, R.; Juszczak, L. Rapeseed Protein as a Novel Ingredient of Gluten-Free Bread. *Eur. Food Res. Technol.* **2021**, *247*, 2015–2025. [CrossRef]
172. Di Lena, G.; Schwarze, A.-K.; Lucarini, M.; Gabrielli, P.; Aguzzi, A.; Caproni, R.; Casini, I.; Ferrari Nicoli, S.; Genuttis, D.; Ondrejíčková, P.; et al. Application of Rapeseed Meal Protein Isolate as a Supplement to Texture-Modified Food for the Elderly. *Foods* **2023**, *12*, 1326. [CrossRef] [PubMed]
173. Hashmi, S.I.; Satwadhar, P.N.; Khotpal, R.R.; Deshpande, H.W.; Syed, K.A.; Vibhute, B.P. Rapeseed Meal Nutraceuticals. *J. Oilseed Brassica* **2010**, *1*, 43–54.
174. Salah, K.; Olkhovatov, E.A.; Aïder, M. Effect of Canola Proteins on Rice Flour Bread and Mathematical Modelling of the Baking Process. *J. Food Sci. Technol.* **2019**, *56*, 3744–3753. [CrossRef] [PubMed]
175. Vantreese, V.L. Hemp Support. *J. Ind. Hemp* **2002**, *7*, 17–31. [CrossRef]
176. Fleddermann, M.; Fechner, A.; Rößler, A.; Bähr, M.; Pastor, A.; Liebert, F.; Jahreis, G. Nutritional Evaluation of Rapeseed Protein Compared to Soy Protein for Quality, Plasma Amino Acids, and Nitrogen Balance—A Randomized Cross-over Intervention Study in Humans. *Clin. Nutr.* **2013**, *32*, 519–526. [CrossRef] [PubMed]
177. Mittermeier-Kleßinger, V.K.; Hofmann, T.; Dawid, C. Mitigating Off-Flavors of Plant-Based Proteins. *J. Agric. Food Chem.* **2021**, *69*, 9202–9207. [CrossRef]
178. Yoshie-Stark, Y.; Wada, Y.; Schott, M.; Wäsche, A. Functional and Bioactive Properties of Rapeseed Protein Concentrates and Sensory Analysis of Food Application with Rapeseed Protein Concentrates. *LWT Food Sci. Technol.* **2006**, *39*, 503–512. [CrossRef]

plants

Article

Phytochemical Profiling, Biological Activities, and In Silico Molecular Docking Studies of *Causonis trifolia* (L.) Mabb. & J.Wen Shoot

Samik Hazra [1,†], Anindya Sundar Ray [1,2,†], Swetarka Das [3,4], Arunava Das Gupta [3] and Chowdhury Habibur Rahaman [1,*]

1 Ethnopharmacology Laboratory, Department of Botany, Visva-Bharati University, Santiniketan 731235, West Bengal, India
2 Department of Animal Science, Kazi Nazrul University, Asansol 713340, West Bengal, India
3 Division of Microbiology, CSIR-Central Drug Research Institute, Lucknow 226031, Uttar Pradesh, India
4 Division of Microbiology and Immunology, College of Veterinary Medicine, Cornell University, Ithaca, NY 14853, USA
* Correspondence: habibur_cr@visva-bharati.ac.in
† These authors contributed equally to this work.

Abstract: *Causonis trifolia* (L.) Mabb. & J.Wen, commonly known as "fox grape", is an ethnomedicinally important twining herb of the Vitaceae family, and it is used by ethnic communities for its wide range of therapeutic properties. Our research aims to investigate the chemical composition; antioxidant, anti-inflammatory, and antidiabetic activities; and mechanisms of interaction between the identified selective chemical compounds and the target proteins associated with antioxidant, anti-inflammatory, and antidiabetic effects of the optimised phenolic extract of *Causonis trifolia* (L.) Mabb. & J.Wen, shoot (PECTS) to endorse the plant as a potential drug candidate for a future bioprospecting programme. Here, we employed the response surface methodology (RSM) with a Box–Behnken design to enrich the methanolic extract of *C. trifolia* shoot with phenolic ingredients by optimising three key parameters: solvent concentration (% *v/v*, methanol:water), extraction temperature (°C), and extraction duration (hours). From the quantitative phytochemical estimation, it was evident that the PECTS contained good amounts of phenolics, flavonoids, tannins, and alkaloids. During the HPLC analysis, we identified a total of eight phenolic and flavonoid compounds (gallic acid, catechin hydrate, chlorogenic acid, caffeic acid, p-coumaric acid, sinapic acid, coumarin, and kaempferol) and quantified their respective contents from the PECTS. The GC-MS analysis of the PECTS highlighted the presence of 19 phytochemicals. In addition, the bioactivity study of the PECTS showed remarkable potentiality as antioxidant, anti-inflammatory, and antidiabetic agents. In silico molecular docking and computational molecular modelling were employed to investigate the anti-inflammatory, antioxidant, and antidiabetic properties of the putative bioactive compounds derived from the PECTS using the GC-MS technique to understand the drug–receptor interactions, including their binding pattern. Out of the 19 phytocompounds identified by the GC-MS analysis, one compound, ergosta-5,22-dien-3-ol, acetate, (3β,22E), exhibited the best binding conformations with the target proteins involved in anti-inflammatory (e.g., Tnf-α and Cox-2), antioxidant (SOD), and antidiabetic (e.g., α-amylase and aldo reductase) activities. The nontoxic nature of this optimised extract was also evident during the in vitro cell toxicity assay against the Vero cell line and the in vivo acute toxicity study on BALB/c mice. We believe the results of the present study will pave the way for the invention of novel drugs efficacious for several ailments using the *C. trifolia* plant.

Keywords: *Causonis trifolia*; response surface methodology; phytochemical profiling; acute toxicity; molecular docking

1. Introduction

For thousands of years, nature has been a source of medicinal agents, and a remarkable number of modern medicaments have been obtained from natural sources, mainly from plants, and many of them are based on their usage in folk medicine systems. In the present scenario, the demand for herbal products is growing exponentially throughout the world, and scientists are paying attention to the screening of different pharmacological activities of plant extracts and isolating biologically active substances for their potential medicinal value.

Causonis trifolia (L.) Mabb. & J.Wen, commonly known as "fox grape" (synonym: *Cayratia trifolia* (L.) Domin), is an ethnomedicinally important twining herb belonging to the Vitaceae family. This perennial climber is native to Australia and several countries in Asia, including India. In India, this herb grows in the wild in Jammu, Rajasthan, Assam, Tripura, and West Bengal, extending into peninsular India by up to 600 m [1]. The broad-spectrum medicinal value of this herb is well documented in several important studies and practiced in various traditional communities for curing a wide range of health hazards. The whole plant is used as a diuretic and in tumours, neuralgia, and splenopathy [2,3]. Leaf decoctions, or the juice of the fresh leaves, are used to cure scurvy [4]. The roots are used as an astringent [5] and as an antidote to snake bites [6]. To control blood sugar levels, diabetic patients are administered with an aqueous extract of the tuber along with the infusion of seeds, whereas the powdered root tuber is administered orally with milk for the early healing of fractured bones [6,7]. Very few scientific explorations have been conducted previously to investigate the phytochemical and pharmacological potential of the studied plant parts. In 2018, Yusuf and coworkers investigated the in vitro antidiabetic potential of an ethanolic stem extract against a mice model. The study showed that the methanolic stem extract had a strong hypoglycaemic effect at a dose of 650 mg/kg of body weight [8]. The in vitro antioxidant and antimicrobial potentials of an n-hexane whole plant extract of this herb were highlighted in recent experiments by Meganathan et al. (2021) [9]. Moreover, several groups of scientists have performed a preliminary phytochemical screening and TLC and HPTLC studies of various crude extracts of the different parts of this plant [4,8,10,11]. However, there are some gaps in our knowledge of how the chemical compounds present in this plant are closely connected to the therapeutic properties ascribed to it and how they affect each other. In our research, we used an integrated in vitro–in silico strategy to predict the lead molecules that are mainly responsible for the *Causonis trifolia* plant's potential as an antioxidant, anti-inflammatory, and antidiabetic agent. Meanwhile, there are no previous reports on the optimisation of the extraction process for obtaining the most efficient condition for the extraction of the maximum yield of the bioactive phenols of this plant. Here, we made an attempt to optimise the extraction process using a suitable statistical tool.

An accurate extraction process is regarded as the most critical step behind the isolation of higher amounts of bioactive phytocompounds from herbal drugs. Like other phytochemical groups, the extraction of phenols from the powdered biomass of plants is highly influenced by the extraction environment, such as the temperature and solvent composition. Traditional phenol extraction methods employing one factor at a time have several drawbacks, such as a prolonged extraction time, unsatisfactory extraction efficiency, and breakdown of the thermolabile compounds. Furthermore, the interactions between the different factors in a single experimental setup are usually overlooked, resulting in a very low probability of attaining a true optimum. To address these issues, the response surface methodology (RSM) was developed to optimise the levels or values of the multiple extraction factors to achieve the best performing system for the maximum extraction of bioactive phenols from this plant's parts [12]. In the present study, we report an optimised extraction method for the extraction of total phenols from the shoots of *Causonis trifolia*.

The broad-spectrum therapeutic properties of this ethnomedicinal herb can be attributed to the large reservoir of bioactive secondary metabolites that occurs in this plant species. The HPLC and GC-MS techniques have firmly established themselves as key techniques for the phytochemical profiling of herbal drugs [13]. Phytochemical screening not only reveals the contents and types of constituents of plant extracts, but it also aids in the search for bioactive compounds that may be utilised in the development of effective drugs. Therefore, our present experiment focused on the quantification of major phytocompounds, such as phenol, flavonoid, tannin, and alkaloid, as well as the identification of bioactive compounds from an optimised phenolic extract of *Causonis trifolia* shoot (PECTS), employing HPLC and GC-MS analyses. Moreover, the in vitro biological activity studies provide basic and fundamental information on the pharmacological properties of drug elements, and these preliminary reports play a very significant role in the validation of the ethnomedicinal claims while also curing various diseases using this studied herb. Keeping this in mind, we also evaluated the therapeutic potentiality of PECTS, employing in vitro antioxidant, anti-inflammatory, and antidiabetic investigations.

Natural products with multiple pharmacological actions have been demonstrated as potential therapeutic candidates for treating a wide range of health complaints. Unfortunately, there are significant gaps in our understanding of which chemicals interact with which targets, and evaluating all probable interactions experimentally is impractical and also needs large amounts of time, expensive infrastructural facilities, and mammoth funding. Recent advances and developments in computational (in silico) approaches provide powerful tools for drug discovery, which can be applied to explore the poly-pharmacological profiles of phytochemicals found in medicinal plants [14]. Protein–ligand interactions are analogous to the lock-and-key paradigm in which the lock represents the protein and the key is correlated with the ligand (i.e., the bioactive compound). This entails docking a library of phytocompounds/ligands into a biological target and calculating the probability that a ligand will bind to the protein target using a scoring algorithm, which aids in the discovery of the most promising lead compounds for a biological target. In this regard, in silico molecular docking and computational molecular modelling were employed to investigate the anti-inflammatory, antioxidant, and antidiabetic properties of putative bioactive compounds derived from PECTS using GC-MS analysis.

Herbal products are often assumed to be safe or low in toxicity due to the fact of their long history of usage by humans [15]. Nevertheless, the latest studies indicate that many of these medicinal plants used for curing a wide range of diseases are potentially toxic, carcinogenic, or mutagenic when they are administered at high doses for prolonged periods [16–18]. As safety is still a major concern with the use of medicinal plants, screening for possible toxicity and the determination of permissible doses of herbal products are worthwhile endeavours. Even though *Causonis trifolia* is used in alternative medicine systems and enjoys large popularity globally due to the fact of its wide therapeutic applications, unfortunately, adequate information on its toxicity profile is still not available. In our present study, as part of a safety evaluation of PECTS, an in vitro cell toxicity assay against the Vero cell line and an in vivo acute toxicity study on BALB/c mice were performed.

In summary, our research objective was to statistically optimise the extraction procedure to obtain the maximum extraction of bioactive phenols, investigate the chemical composition, and explore the in vitro antioxidant, anti-inflammatory, and antidiabetic potential of the optimised plant extract and the mechanisms of the interactions between the identified chemical compounds via GC-MS analysis and the target proteins associated with the antioxidant, anti-inflammatory, and antidiabetic effects. Our investigation also explored both the in vitro and in vivo toxicity profiles of the optimised phenolic shoot extract to endorse the plant for a rational bioprospecting programme as a potential drug candidate.

2. Results

2.1. Statistical Optimisation of the Extraction Process for the Enhanced Extraction of Phenols

In our present study, linear, interactive (2 FI), quadratic, and cubic models were fitted to the experimental data to create regression models, and the results are provided in Supplementary Table S1. The quadratic model was the most adequate model for representing our experimental data set. An ANOVA was used to examine the RSM's statistical optimisation of the extraction parameters (analysis of variance). An ANOVA was also utilised for the regression analysis, results verification, and case statistics prediction (see Supplementary

Table S2). The Model F-value of 116.87 implies that the applied model is quite significant. A small F-value for the model is not desired, since it indicates that the variance was caused by random unexplained disturbances, referred to as noise. The probability of an F-value of 116.87 occurring due to the fact of noise is only 0.01%. *p*-Values less than 0.0500 indicate that the model terms are significant. In this case A, BC, A^2, B^2, and C^2 are significant model terms. If there are many insignificant terms (not counting those required to support the hierarchy), the model reduction can improve the model's quality. The lack-of-fit test was used as a supporting test for the adequacy of the fitted model. A lack-of fit F-value of 1.95 implies that the lack of fit is not significant. A nonsignificant lack of fit is good, as we want the model to fit. The model showed a standard deviation, R^2, and mean of 2.76, 0.9934, and 119.54, respectively. The predicted R^2 value of 0.9330 is in reasonable agreement with the adjusted R^2 of 0.9849, i.e., the difference was less than 0.2, which again indicates the validity of the model. "Adequate precision" measures the signal-to-noise ratio, and if it is greater than 4, then the model has a signal strong enough to be used for optimisation. The ratio of 25.237 indicated an adequate signal. Here, a lesser value for the coefficient of variation (CV%; 2.31) designates a better reliability of the model, whereas a correlation coefficient (R^2) of 0.9934 indicates a high degree of correlation between the experimental parameters and the response [19]. The three-dimensional response surface plots depict the interaction among the three parameters studied and demonstrate the combined influence of the parameters on the extraction of total phenolics from shoots of *C. trifolia*. The effect of their variables and interaction on the responses can be seen in Figure 1A–C. Therefore, an accurate optimisation of the level of each parameter is essential to obtain the maximum amount of phenolics. A normal probability plot is shown in Figure 1E, displaying scatters present closely along the line, indicating that the residuals followed a normal distribution, satisfying the assumptions of the ANOVA, which designated the accuracy of the RSM in optimising the three variables [20]. Similarly, as shown in Figure 1F, it is suggested that the actual and predicted values of the TPC were in close agreement, which again validates the model. Finally, the predictive extraction model suggests that a methanol concentration of 67.6754%, extraction temperature of 49.7753 °C, and extraction time of 12.5837 h are the most suitable and effective combinatory parameters for obtaining the highest yield of bioactive polyphenols (152.774 mg GAE/g dry tissue) with a 94.5% desirability, which is well accepted (Figure 2).

2.2. *Quantitative Estimation of Certain Groups of Phytochemicals of the PECTS*

The optimised values of the independent variables were used for the extraction, and the total phenolic content (i.e., response) was compared with the value generated by the RSM. The total phenolic content of the optimised phenolic extract of *Causonis trifolia* shoot (PECTS) was estimated to be 150.76 ± 0.02 mg GAE/g dry tissue, which is very close to the predicted value of the RSM model (152.774 mg GAE/g dry tissue). The contents of total flavonoid, total tannin, and total alkaloid of the PECTS were 20.7 ± 0.02 mg CE/g dry tissue, 23.21 ± 0.47 mg of TAE/g, and 13.37 ± 0.12 mg of PE/g, respectively.

2.3. *High-Performance Liquid Chromatography (HPLC) Analysis of the PECTS*

For the HPLC analysis, a total of 10 standards of phenol and flavonoid compounds (gallic acid, kaempferol, catechin hydrate, chlorogenic acid, caffeic acid, syringic acid, p-coumaric acid, sinapic acid, quercetin, and coumarin) were used. Qualitative and quantitative evaluations of these therapeutically important phytochemicals from the optimised shoot extract of *Causonis trifolia* were undertaken in the present experiment. this study identified a total of eight compounds of phenols and flavonoids (gallic acid, catechin hydrate, chlorogenic acid, caffeic acid, p-coumaric acid, sinapic acid, coumarin, and kaempferol), and the compounds were quantified through an HPLC analysis of the PECTS (Figure 3). This study also highlights that considerably higher amounts of sinapic acid (24.172 mg/g dry tissue), coumarin (13.211 mg/g dry tissue), and gallic acid (7.766 mg/g dry tissue) were estimated in the PECTS (Figure 4).

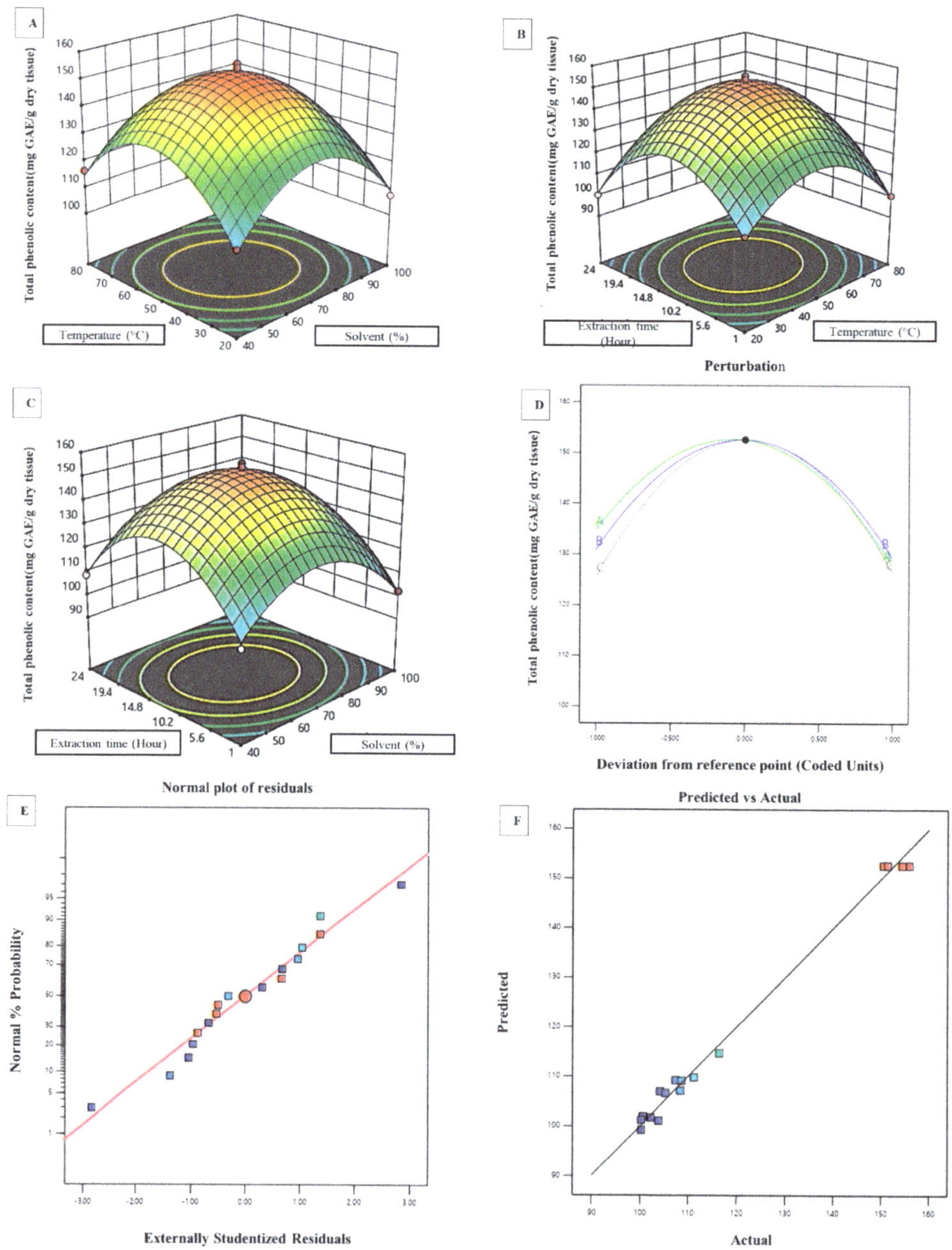

Figure 1. Statistical optimisation of the extraction parameters by the response surface methodology. The response surface plots demonstrate the combined effect of the (**A**) temperature (°C) and solvent (%); (**B**) extraction time (hour) and temperature (°C); (**C**) extraction time (hour) and solvent (%) on the yield of TPC. (**D**) Perturbation plot showing the relative influence of each parameter on the extraction of TPC. (**E**) Normal probability plot showing scatters along the standard curve supporting the prediction of the ANOVA analysis. (**F**) Comparison of the experimental and predicted results, showing a greater degree of agreement between them.

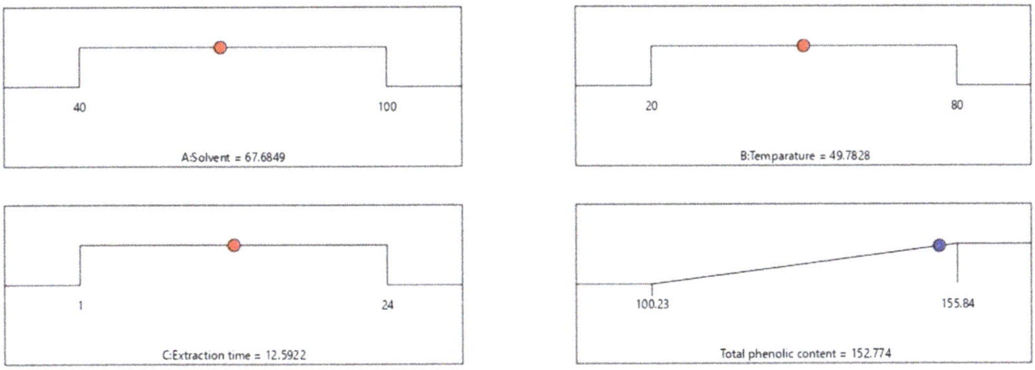

Figure 2. Predictive extraction model designed for the effective extraction of the highest amount of bioactive phenols. The model highlights the most suitable values of the three parameters, namely, methanol concentration (67.6754%), extraction temperature (49.7753 °C), and extraction time, (12.5837 h) predicted for the highest yield of bioactive phenols (152.774 mg GAE/g dry tissue).

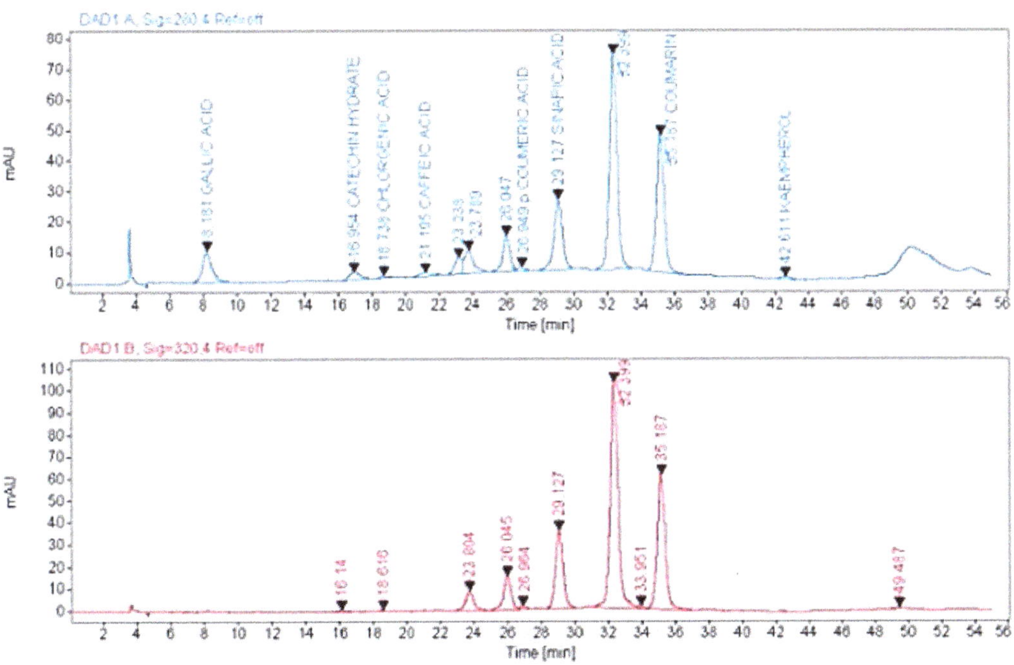

Figure 3. HPLC chromatogram obtained from the optimised phenolic extract of *Causonis trifolia* shoot (PECTS).

2.4. Gas Chromatography-Mass Spectrometry (GC–MS) Analysis of PECTS

The results pertaining to the GC-MS analysis led to the identification of 19 compounds from the GC fractions of the optimised phenolic extract of *Causonis trifolia* shoot (PECTS) (Figure 5). A detailed list of the identified compounds along with their chemical characters and peak areas are presented in Table 1. The 19 phytocompounds identified in the PECTS are 5,7-dodecadiyne-1,12-diol (1.123); ergosta-5,22-dien-3-ol, acetate, (3β,22E) (1.167); 1,2,4-metheno-1H-indene, octahydro-1,7a-dimethyl-5-(1-methylethyl)-

, (1S-(1α,2α,3aβ.,4 α.,5α,7aβ.,8S*)) (13.403); alpha-copaene (13.784); 1,5-cyclodecadiene, 1,5-dimethyl-8-(1-methylethenyl)-, (S-(Z,E)) (14.305); 3H-3a,7-methanozulene, 2,4,5,6,7,8-hexahydro-1,4,9,9-tetramethyl-, (3aR-(3aα,4β,7α)) (14.385); caryophyllene (14.981); 1H-cyclopropa[a]naphthalene; 1a,2,3,5,6,7,7a,7b-octahydro-1,1,7,7a-tetramethyl-, (1aR-(1aα,7α, 7aα,7bα)) (15.720); humulene (15.955); 1,4-methano-1H-indene, octahydro-4-methyl-8-methylene-7-(1-methylethyl)-, (1S-(1α,3aβ.,4α,7α,7aβ)) (16.397); 1H-3a,7-methanoazulene, octahydro-1,9,9-trimethyl-4-methylene-, (1.α.,3aα.,7 α.,8aβ) (16.534); germacrene D (16.781); naphthalene, 1,2,4a,5,6,8a-hexahydro-4,7-dimethyl-1-(1-methylethyl) (17.433); 1-isopropyl-4,7-dimethyl-1,2,3,5,6,8a-hexahydronaphthalene (18.089); 11,11-dimethyl-spiro [2,9] dodeca-3,7-dien (19.549); 2H-pyran, tetrahydro-4-methyl-2-(2-methyl-1-propenyl) (22.058); 3-methyl-mannoside (23.140); 3-O-methyl-d-glucose (23.573); and γ-elemene (38.188) (Figure 6). The highest and lowest peak areas were found in the case of germacrene D (16.755%) and 3-O-methyl-d-glucose (0.286%), respectively. The reported biological activity of all of the detected compounds are summarised in Table 2.

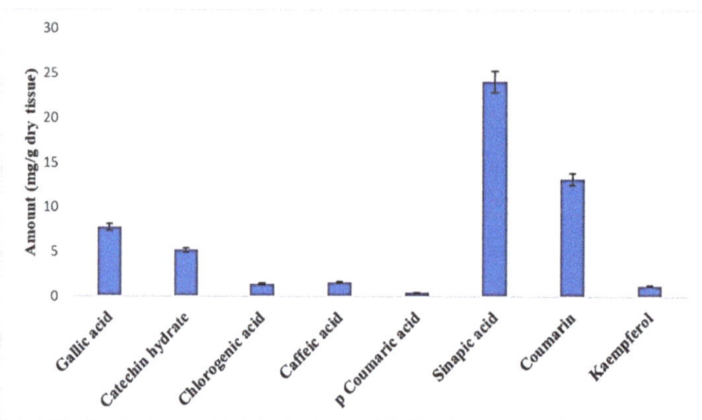

Figure 4. Phenolic and flavonoid profiles of the optimised phenolic extract of *Causonis trifolia* shoot (PECTS).

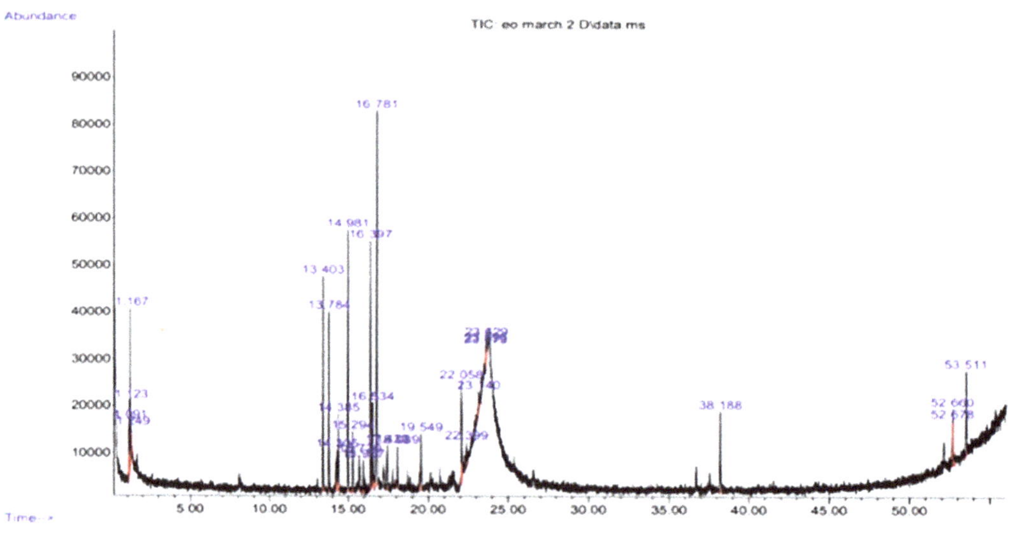

Figure 5. GC-MS chromatogram obtained from optimised phenolic extract of *Causonis trifolia* shoot.

Table 1. Phytochemical constituents (n = 19) identified in the optimised phenolic extract of *Causonis trifolia* shoot using GC-MS.

Sl. No.	Retention Time (Minute)	PubChem CID	Compound	Molecular Formula	Molecular Weight	Peak Area (%)
1	1.123	560878	5,7-Dodecadiyne-1,12-diol	$C_{12}H_{18}O_2$	194.13	0.979
2	1.167	5352877	Ergosta 5,22 dien 3 ol, acetate, (3β,22E)-	$C_{30}H_{48}O_2$	440.36	5.132
3	13.403	519960	1,2,4-Metheno-1H-indene, octahydro-1,7a-dimethyl-5-(1-methylethyl)-, [1S-(1α,2α,3aβ.,4 α.,5α,7aβ.,8S*)]-	$C_{15}H_{24}$	204.18	8.493
4	13.784	19725	α-Copaene	$C_{15}H_{24}$	204.18	7.422
5	14.305	20839485	1,5-Cyclodecadiene, 1,5-dimethyl 8 (1 methylethenyl)-, [S-(Z,E)]-	$C_{15}H_{24}$	204.18	1.508
6	14.385	99856	3H-3a,7-Methanozulene, 2,4,5,6,7,8-hexahydro-1,4,9,9-tetramethyl-, [3aR-(3aα,4β,7α)]	$C_{15}H_{24}$	204.18	2.771
7	14.981	5281515	Caryophyllene	$C_{15}H_{24}$	204.18	10.962
8	15.720	28481	1H-Cyclopropa[a]naphthalene. 1a,2,3,5,6,7,7a,7b-octahydro-1,1,7,7a-tetramethyl-, [1aR-(1aα,7α,7aα,7bα)]	$C_{15}H_{24}$	204.18	1.478
9	15.955	5281520	Humulene	$C_{15}H_{24}$	204.18	1.268
10	16.397	530427	1,4-Methano-1H-indene, octahydro-4-methyl-8-methylene-7-(1-methylethyl)-, [1S-(1α,3aβ.,4α,7α,7aβ)]-	$C_{15}H_{24}$	204.18	10.288
11	16.534	521302	1H-3a,7-Methanoazulene, octahydro-1,9,9-trimethyl-4-methylene-, (1.α.,3aα.,7 α.,8aβ)	$C_{15}H_{24}$	204.18	3.045
12	16.781	5317570	Germacrene D	$C_{15}H_{24}$	204.18	16.755
13	17.433	101708	Naphthalene, 1,2,4a,5,6,8a-hexahydro-4,7-dimethyl-1-(1-methylethyl)-	$C_{15}H_{24}$	204.18	2.023
14	18.089	10223	1-Isopropyl-4,7-dimethyl-1,2,3,5,6,8a-hexahydronaphthalene	$C_{15}H_{24}$	204.18	1.505
15	19.549	5367333	11,11-Dimethyl-spiro [2,9] dodeca-3,7-dien	$C_{14}H_{22}$	190.17	2.921
16	22.058	6432154	2H-Pyran, tetrahydro-4-methyl-2-(2-methyl-1-propenyl)-	$C_{10}H_{18}O$	154.13	5.438
17	23.140	247323	3-Methylmannoside	$C_7H_{14}O_6$	194.07	0.870
18	23.573	8973	3-O-Methyl-d-glucose	$C_7H_{14}O_6$	194.07	0.286
19	38.188	6432312	γ-Elemene	$C_{15}H_{24}$	204.18	3.857

Table 2. Chemical group, structure, and reported biological activity of the phytocompounds identified from the PECTS by GC-MS analysis.

Sl. No.	Compound	Chemical Group	Chemical Structure	Biological Activity	Reference
1	1,2,4-Metheno-1H-indene, octahydro-1,7a-dimethyl-5-(1-methylethyl)-, [1S-(1α,2α,3aβ.,4 α.,5α,7aβ.,8S*)]-	Sesquiterpene	CC(C)C1CCC2(C3C1C4C2 (C4C3)C)C	Antiproliferative, genotoxic, and oxidant activities	[21]
2	1,4-Methano-1H-indene, octahydro-4-methyl-8-methylene-7-(1-methylethyl)-, [1S-(1α,3aβ.,4α,7α,7aβ)]	Sesquiterpene	CC©C1CCC2(C3C1C (C2=C)CC3)C	Not reported	
3	11,11-Dimethyl-spiro [2,9] dodeca-3,7-dien	Spiro compound	CC1(CC12CCC=CC CC=CC2)C	Not reported	
4	1H-3a,7-Methanoazulene, octahydro-1,9,9-trimethyl-4-methylene-, (1.α.,3aα.,7 α.,8aβ)	Sesquiterpene	CC1CCC23C1CC© (C)C)CCC3=C	Not reported	
5	1H-Cyclopropa[a]naphthalene. 1a,2,3,5,6,7,7a,7b-octahydro-1,1,7,7a-tetramethyl-, [1aR-(1aα,7α,7aα,7bα)]-	Sesquiterpene	CC1CCC=C2C1(© (C3(C)C)CC2)C	Antibacterial	[21]
6	1-Isopropyl-4,7-dimethyl-1,2,3,5,6,8a-hexahydronaphthalene	Sesquiterpene	CC1=CC2C(CCC (=©C1) C)C(C)C	Not reported	
7	2H-Pyran, tetrahydro-4-methyl-2-(2-methyl-1-propenyl)	Monoterpene	CC©OC (C1)C=C(C)C	Anti-inflammatory	[22]
8	3H-3a,7-Methanozulene, 2,4,5,6,7,8-hexahydro-1,4,9,9-tetramethyl-, [3aR-(3aα,4β,7α)]	Sesquiterpene	CC1CC©C3=C(CCC 13C2(C)C)C	Not reported	
9	3-Methylmannoside	Carbohydrate	COC1C(C(OC (C1O)O)CO)O	Not reported	
10	3-O-Methyl-d-glucose	Carbohydrate	COC(C(C=O)O)C (C(CO)O)O	Anticancer and anti-inflammatory	[23]
11	5,7-Dodecadiyne-1,12-diol	Fatty acid	C(CCO)CC#CC #CCCCCO	Not reported	
12	1,5-Cyclodecadiene, 1,5-dimethyl-8-(1-methylethenyl)-, [S-(Z,E)]-	Terpenoid	CC1=CCCC(=CCC(CC1) C(=C)C)C	Not reported	
13	Caryophyllene	Sesquiterpene	CC1=C©(=C)C2CC (C2CC1)(C)C	Anticancer, antioxidant, and antimicrobial	[24]
14	Ergosta-5,22-dien-3-ol, acetate, (3β,22E)-	Sterol	CC(C)C(C)C=CC(C)C1CC C2C1(CCC3C2CC=C 4C3(CCC(C4)O)C)C	Not reported	
15	Germacrene D	Sesquiterpene	CC1=CCCC(=C)C=C C(CC1)C(C)C	Mosquitocidal and repellent activity against aphids and ticks	[25–27]
16	Humulene	Sesquiterpene	CC1=CCC(C=CCC (=CCC1)C)(C)C	Anti-inflammatory, analgesic, and antineoplastic	[28]
17	Naphthalene, 1,2,4a,5,6,8a-hexahydro-4,7-dimethyl-1-(1-methylethyl)-	Sesquiterpene	CC1=CC2C(CC1)C(=C CC2C(C)C)C	Not reported	
18	α-Copaene	Sesquiterpene	CC1=CCC2C3C1C2 (CCC3C(C)C)C	Antiproliferative and antioxidant	[21]
19	γ-Elemene	Sesquiterpene	CC(=C1CCC(C(C1) C(=C)C)(C)C=C)C	Insecticidal	[29]

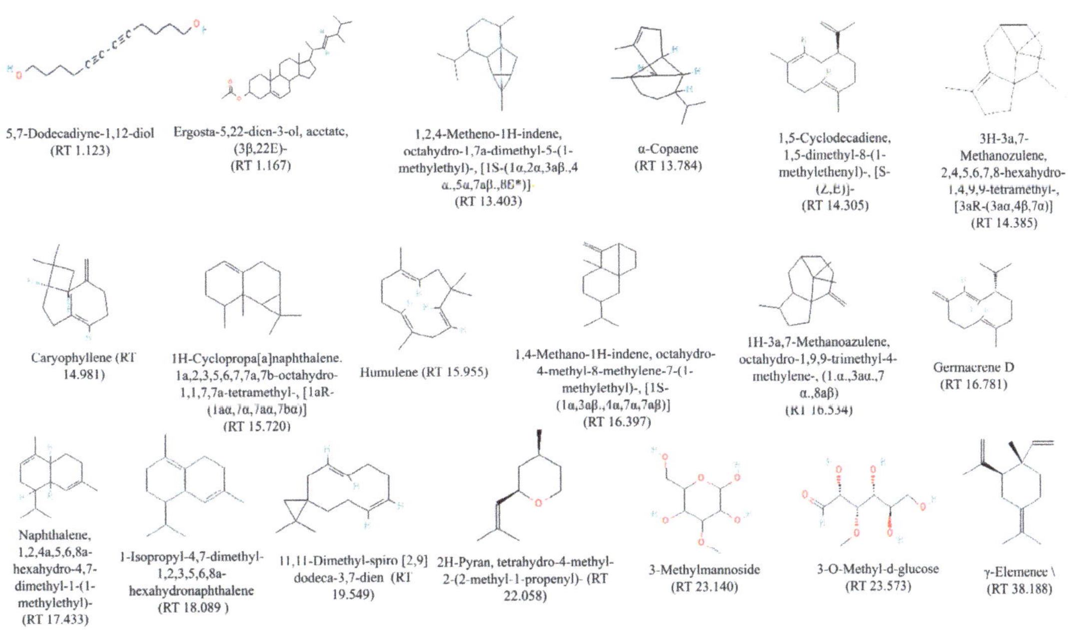

Figure 6. Chemical structure and retention time of the compounds identified from PECTS.

2.5. In Vitro Antioxidant Study of the PECTS

2.5.1. DPPH Radical Scavenging Activity

In this study, different concentrations of the optimised PECTS were subjected to a 2,2-diphenyl-1-picryl-hydrazyl-hydrate (DPPH) free radical scavenging assay. Figure 7A shows the DPPH free radical scavenging activity of the PECTS along with the standard of ascorbic acid. At a concentration of 100 μg/mL, the scavenging activity of the PECTS was 88.27 ± 0.57%, whereas, at the same concentration, 89.68 ± 3.14% DPPH radical scavenging activity was shown by the ascorbic acid. The IC_{50} values of the PECTS and ascorbic acid were calculated as 59.96 μg/mL and 28.99 μg/mL, respectively.

2.5.2. Hydrogen Peroxide Scavenging Assay

The hydroxyl radical scavenging activity of the PECTS sample was dose dependent (Figure 7B). At a concentration of 100 μg/mL, the hydroxyl radical scavenging activity of the PECTS and the standard ascorbic acid was 69.87 ± 2.29%, and 88.93 ± 2.47%, respectively. The IC_{50} value of the PECTS was estimated to be 68.13 μg/mL. An amount of 35.68 ± 1.16 μg/mL ascorbic acid was needed to obtain a 50% inhibition of the hydroxyl scavenging activity.

2.5.3. ABTS Radical Scavenging Activity

The relative antioxidant ability of the PECTS to scavenge the radical $ABTS^+$ was compared with the standard of ascorbic acid. The ABTS radical cation was produced in the stable form using potassium persulphate. After obtaining a stable absorbance, an extract was added to the reaction medium, and its antioxidant power was measured by studying the decolourisation of the medium. The IC_{50} value of the PECTS was estimated to be 47.94 μg/mL. At a concentration of 100 μg/mL, the PECTS exhibited 59.31 ± 0.21% $ABTS^+$ radical scavenging activity. At 100 μg/mL, the ascorbic acid showed the highest scavenging activity of 85.22 ± 1.91%, with an IC_{50} value of 33.51 μg/mL (Figure 7C).

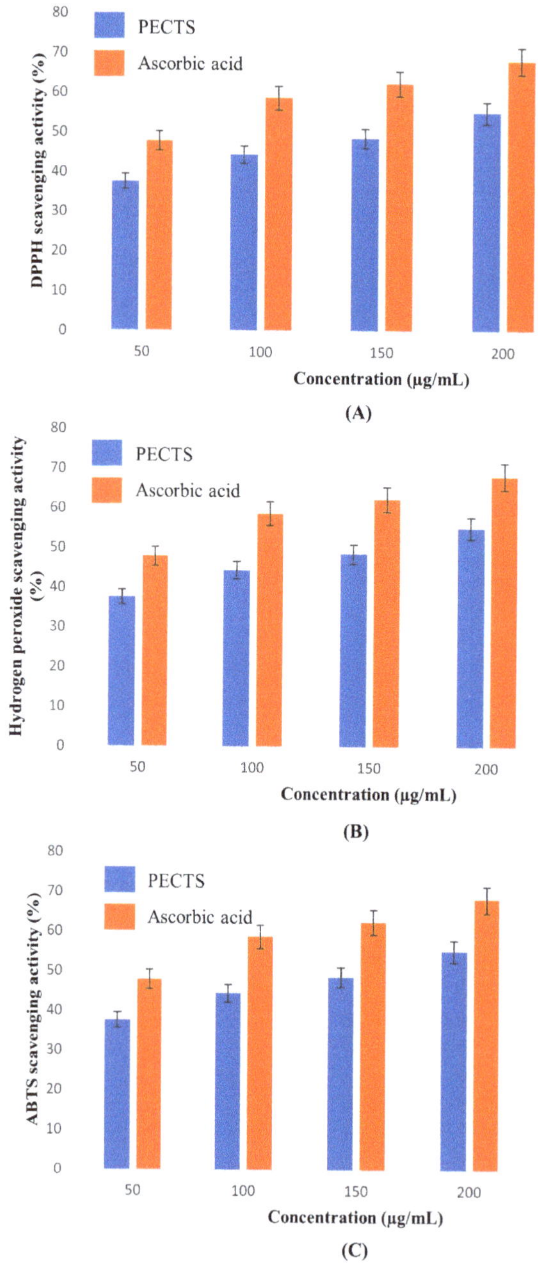

Figure 7. In vitro antioxidant activities exhibited by the PECTS in different radical scavenging assays and a comparison with ascorbic acid: (**A**) DPPH radical scavenging activity of PECTS (Y = 0.5907x + 14.557; IC_{50} = 59.96 μg/mL) and ascorbic acid (Y = 0.4368x + 37.336; IC_{50} = 28.99 μg/mL); (**B**) hydrogen peroxide scavenging activity of the PECTS (Y = 0.5288x + 13.969; IC_{50} = 68.13 μg/mL) and ascorbic acid (Y = 0.4526x + 33.847; IC_{50} = 35.68 μg/mL); (**C**) ABTS radical scavenging activity of the PECTS (Y = 0.58x − 2.0551; IC_{50} = 47.94 μg/mL) and ascorbic acid (Y = 0.4406x + 35.233; IC_{50} = 33.51 μg/mL).

2.5.4. Phosphomolybdenum Assay

The total antioxidant capacity (TAC) was based on the reduction of the valency of molybdenum from 6 to 5 by the extract and subsequent formation of the green phosphatemolybdenum complex. It was employed to evaluate the total antioxidant capacity of antioxidants soluble in water and fat. The total antioxidant capacity of the optimised phenolic extract of *Causonis trifolia* shoot was found to be 40.20 ± 1.57 mg AAE/g.

2.6. In Vitro Anti-Inflammatory Study of the PECTS

2.6.1. Inhibition of the Albumin Denaturation Assay

The denaturation of proteins is a well-known cause of cell inflammatory responses [30]. BSA was used as a reagent for the assay. BSA accounts for approximately 60% of the total protein in animal serum. BSA undergoes denaturation upon heating and starts expressing antigens associated with Type III hypersensitive reaction, which can lead to illness, such as glomerulonephritis and rheumatoid arthritis [31]. Thus, the inhibition of the BSA denaturation assay was used to evaluate the anti inflammatory potential of the optimised phenolic extract of *Causonis trifolia* shoot (PECTS).

The extract represented a dose-dependent inhibition of the BSA denaturation (Figure 8A). The IC_{50} values of the PECTS and standard drug (diclofenac sodium) were determined to be 306.78 μg/mL and 289.74 μg/mL, respectively.

2.6.2. Antiprotease Activity

The role of proteases in the pathophysiology of arthritis has been well documented. Neutrophils are reported to be a rich source of serine proteases, which are localised in lysosomal granules. Upon inflammatory stimulation, these serine proteases are released into extracellular spaces by exocytosis, and their uncontrolled release can lead to various chronic inflammatory disease conditions and cause damage to tissues. Protease inhibitors can provide substantial protection against this effect [32]. In our assay, trypsin was utilised as a serine protease enzyme and casein was used as a substrate.

The optimised extract of *Causonis trifolia* exhibited a concentration-dependent antiprotease action comparable to the antiprotease activity of the conventional drug at higher doses (Figure 8B). The IC_{50} value of the PECTS was 290.97 μg/mL, and it was nearer to the value of 270.19 μg/mL for diclofenac sodium, the standard drug.

2.6.3. Membrane Stabilisation Assay

As the membrane of human red blood cells (HRBCs) is similar to that of a lysosomal membrane in terms of integrity, it was employed to assess the membrane stabilising capacity of a drug or extract. If the extract can stabilise the HRBC membrane, it may be able to inhibit the release of lysosomal enzymes from the lysosome by stabilising its membrane system. Because heat induces membrane lysis and haemoglobin oxidation, the anti-inflammatory activity of the plant extract was studied using the membrane stability of HRBC against heat [33].

Heat-Induced Haemolysis Assay

The optimised phenolic extract of the PECTS exhibited a dose-dependent inhibition. The PECTS exhibited an IC_{50} value of 327.041 μg/mL, while the standard drug (diclofenac sodium) depicted more potency with an IC_{50} value of 253.48 μg/mL (Figure 8C).

2.7. In Vitro Antidiabetic Study of the PECTS

2.7.1. α-Amylase Inhibitory Assay

α-Amylase is an intestinal enzyme that degrades starch molecules into smaller units of two or three glucose monomers by acting upon the α-1,4-glycosidic bonds present in the starch polysaccharide. The suppression of digestive enzymes such as α-amylase causes a delay in the digestion of starch and oligosaccharides, which reduces glucose absorption and, as a result, lowers the blood glucose level [34]. The optimised phenolic extracts of *Causonis trifolia* (PECTS) were subjected to an α-amylase inhibitory assay along with acarbose as a standard (Figure 9A). The PECTS exhibited a percentage inhibition of 68.42 ± 0.13% at a 200 μg/mL concentration with an IC_{50} value of 117.38 μg /mL. The acarbose, a standard antidiabetic drug, showed an α-amylase inhibitory activity of 77.2 ± 0.27% at 200 μg/mL with an IC_{50} value of 84.48 μg/mL.

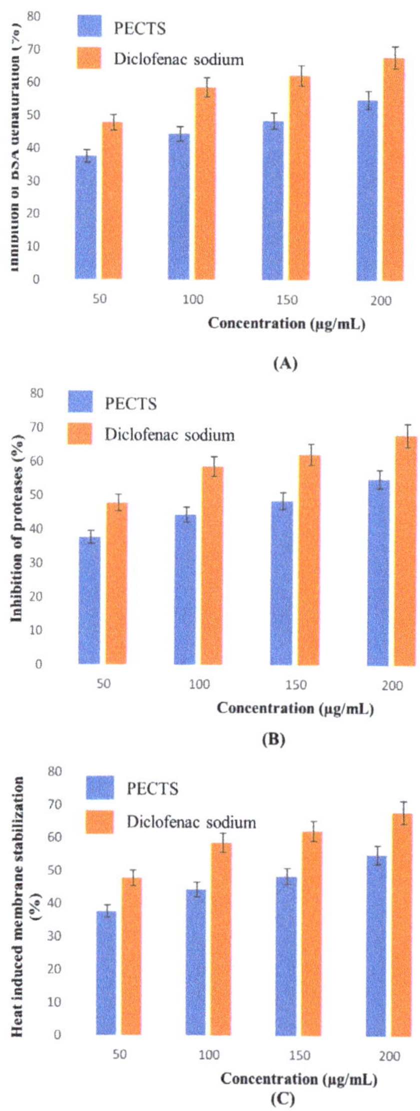

Figure 8. In vitro anti-inflammatory activities of the optimised phenolic extract of *Causonis trifolia* shoot and a comparison with the anti-inflammatory activity of the standard drug diclofenac sodium: (**A**) inhibition of the BSA denaturation activity of th ePECTS (Y = 0.1833x − 6.133; IC_{50} = 306.78 μg/mL) and diclofenac sodium (Y = 0.1852x − 3.66; IC_{50} = 288.74 μg/mL); (**B**) antiprotease activity of the PECTS (Y = 0.1541x + 5.161; IC_{50} = 290.97 μg/mL) and diclofenac sodium (Y = 0.1764x + 2.337; IC_{50} = 270.19 μg/mL); (**C**) inhibition of the heat-induced haemolysis activity of the PECTS (Y = 0.1547x − 0.594; IC_{50} = 327.04 μg/mL) and diclofenac sodium (Y = 0.177x + 5.131; IC_{50} = 253.48 μg/mL).

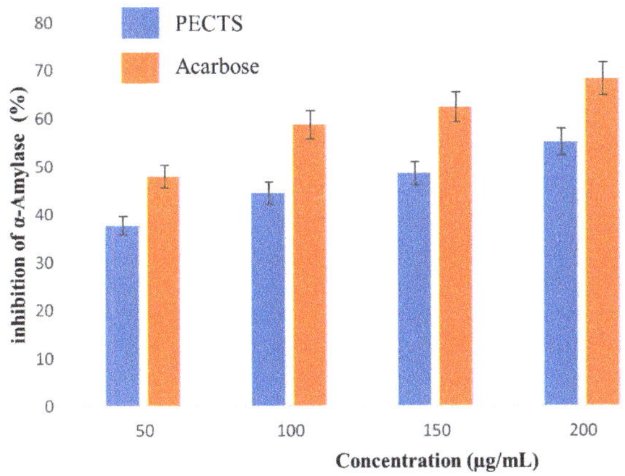

(**A**)

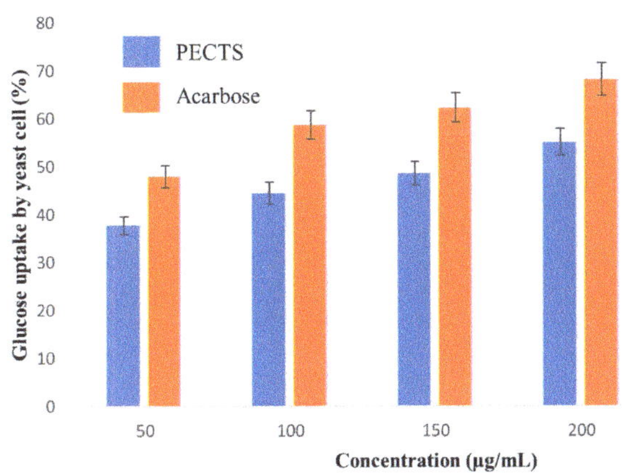

(**B**)

Figure 9. In vitro antidiabetic activity of the optimised phenolic extract of *Causonis trifolia* (PECTS) and a comparison with acarbose's activity: (**A**) inhibition of the α-amylase activity of the PECTS ($Y = 0.3021x - 14.55$; $IC_{50} = 117.38$ μg/mL) and acarbose ($Y = 0.2541x - 28.76$; $IC_{50} = 84.48$ μg/mL); (**B**) yeast cell glucose uptake activity of the PECTS ($Y = 0.1126x + 32.385$; $IC_{50} = 156.42$ μg/mL) and acarbose ($Y = 0.1284x + 43.2$; $IC_{50} = 52.95$ μg/mL).

2.7.2. Yeast Cell Glucose Uptake Assay

The glucose transport mechanism across the yeast cell membrane has received attention as a tool for the in vitro screening of different medicinal plant extracts for their hypoglycaemic effect. The amount of glucose that remains in the medium after a specific amount of time serves as a marker of the yeast cell glucose uptake. It has been reported that in yeast cells (*Saccharomyces cerevisiae*), glucose is transported by a facilitated diffusion process. Facilitated carriers are specific carriers that transport solutes from the higher concentration region to the lower concentration region. This means that effective transport will only be attained if there is the removal of intracellular glucose. In the present study, the optimised phenolic extract of *Causonis trifolia* (PECTS) were subjected to an in vitro yeast cell glucose uptake assay. The percentage of the increase in the glucose uptake in the yeast cells by the action of the PECTS was compared with the standard drug, acarbose. The increased concentration of the PECTS corresponded with the increased percentage of glucose uptake in the yeast cells. This result indicates that high concentrations of PECTS exhibit high glucose uptake (Figure 9B). The

PECTS exhibited the highest percentage of glucose uptake: 55.07 ± 0.25%, at a 200 μg/mL concentration. The IC_{50} value of the PECTS was estimated to be 156.42 μg/mL. With an IC_{50} value of 52.95 μg/mL, the acarbose showed a higher potency.

2.8. In Silico Molecular Docking Study

A total of 19 bioactive chemical compounds were identified through a GC–MS analysis of the optimised phenolic extract of the investigated plant. These 19 chemical compounds were then analysed for their activities against the target proteins associated with inflammatory, oxidative, and diabetic conditions. To determine the binding affinities of the 19 phytocompounds to five target proteins, docking studies were performed using the iGEMDOCK 2.1 software. Among the 19 phytocompounds, 1 compound, Ergosta-5,22-dien-3-ol, acetate, (3β,22E), exhibited the best binding conformations with the lowest binding energy values with all five target proteins, such as inflammatory (e.g., Tnf-α (−93.0843 kcal/mol), Cox-2 (−92.0437 kcal/mol)), oxidative (e.g., SOD −87.7343 kcal/mol), and diabetic (e.g., α-amylase (−108.844 kcal/mol) and aldo reductase (−115.66 kcal/mol)) proteins (see Supplementary Table S3). Our findings are in agreement with the earlier research findings, and they reveal that a lower binding energy score results in a greater protein–ligand binding stability [35,36].

The Autodock4.2 programme was used to determine the relative strengths of the binding interactions of the protein–ligand complex and to analyse the conformation and orientation (referred to together as the "pose") of the ligands into the binding site of a protein target. The top scoring phytocompound, Ergosta-5,22-dien-3-ol, acetate, (3β,22E), against each target protein was identified and subjected to further docking analysis using the Autodock 4.2 programme. The docking study revealed the potential binding affinity of the most potent phytocompound into the binding sites of the target proteins with minimum binding energies (ranging from −7.16 to −11.73 kcal/mol), ligand efficiency (ranging from −0.22 to −0.37 kcal/mol), and inhibition constant (5.62 to 964.03 nM) (Figure 10; Table 3). This compound also formed hydrogen bond interactions and the best possible binding pose with the residues of the targeted anti-inflammatory, antioxidant, and antidiabetic proteins, as shown by their corresponding 3D interaction models in Figures 11–15, respectively.

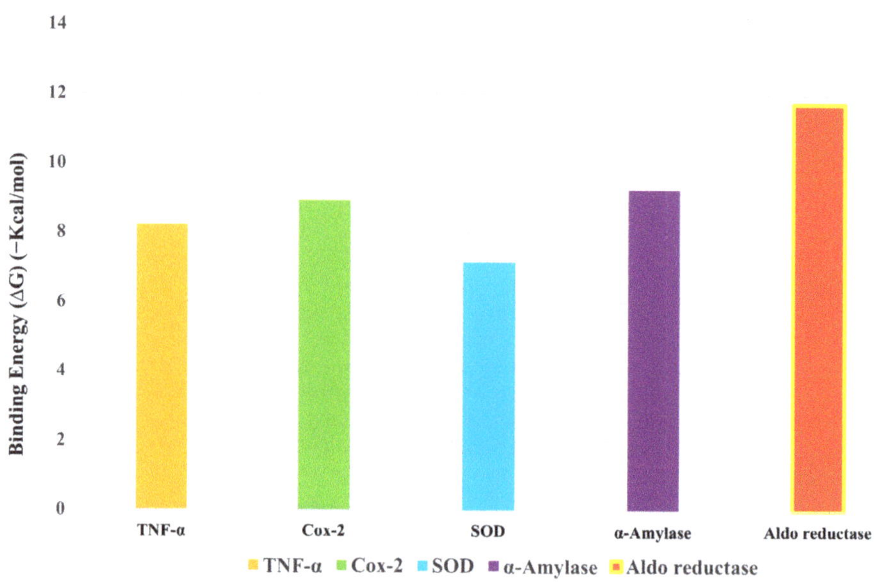

Figure 10. Binding energy (−Kcal/mol) of Ergosta-5,22-dien-3-ol, acetate, (3β,22E) with five target proteins.

Table 3. Results of the binding interactions of the compound Ergosta-5,22-dien-3-ol, acetate, (3β,22E), with the five target proteins.

Phytocompound	Target Proteins	Binding Energy (ΔG) (Kcal/mol)	Ligand Efficiency	Inhibition Constant ((Ki) (nM))
Ergosta-5,22-dien-3-ol, acetate, (3β,22E)	TNF-α	−8.21	−0.26	964.03
	Cox-2	−8.93	−0.28	282.86
	Superoxide dismutase	−7.16	−0.22	5.62
	α-Amylase	−9.27	−0.29	161.19
	Aldo reductase	−11.73	−0.37	2.53

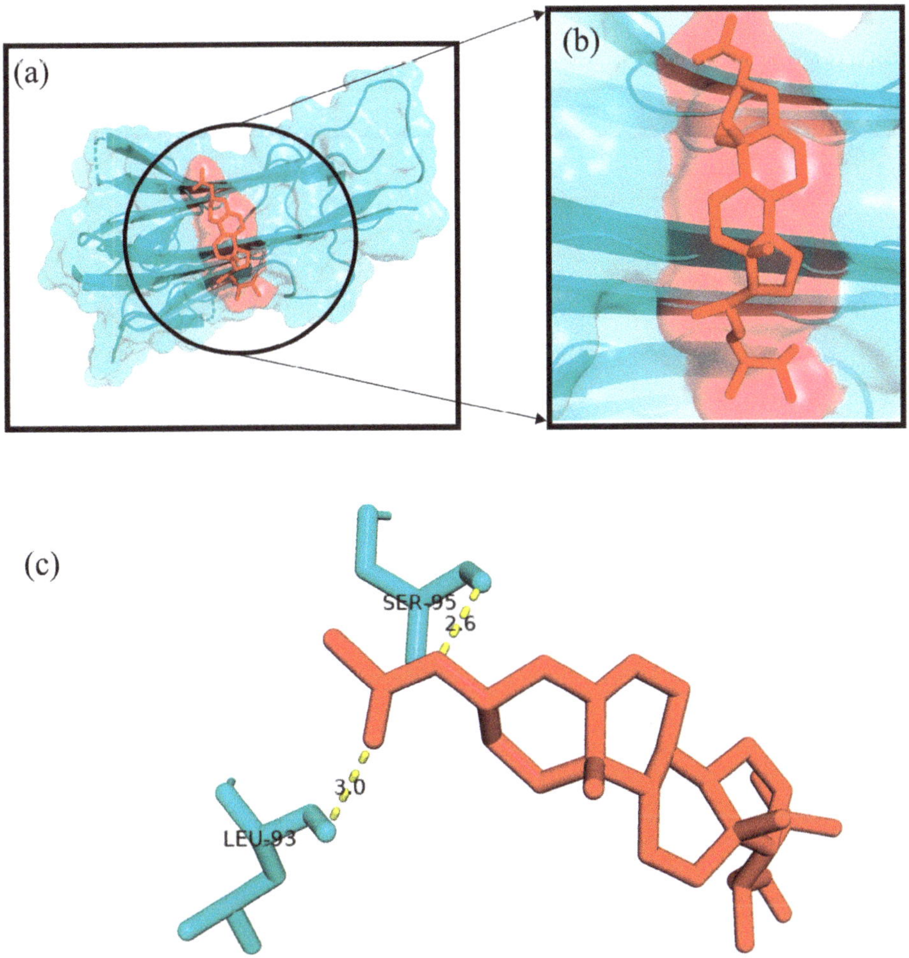

Figure 11. (**a**) Surface view of the docked complex of ergosta-5,22-dien-3-ol, acetate, (3β,22E), and an anti-inflammatory target protein (TNF-α); (**b**) binding pose of ergosta-5,22-dien-3-ol, acetate, (3β,22E); (**c**) ligand–protein docked complex showing the binding interaction of ergosta-5,22-dien-3-ol, acetate, (3β,22E), with TNF-α. Two hydrogen bonds (yellow, dotted line) formed between the ligand ergosta-5,22-dien-3-ol, acetate, (3β,22E), and two amino acid residues (Leu-93 (3.0 Å) and Ser-95 (2.6 Å)) of the target protein TNF-α.

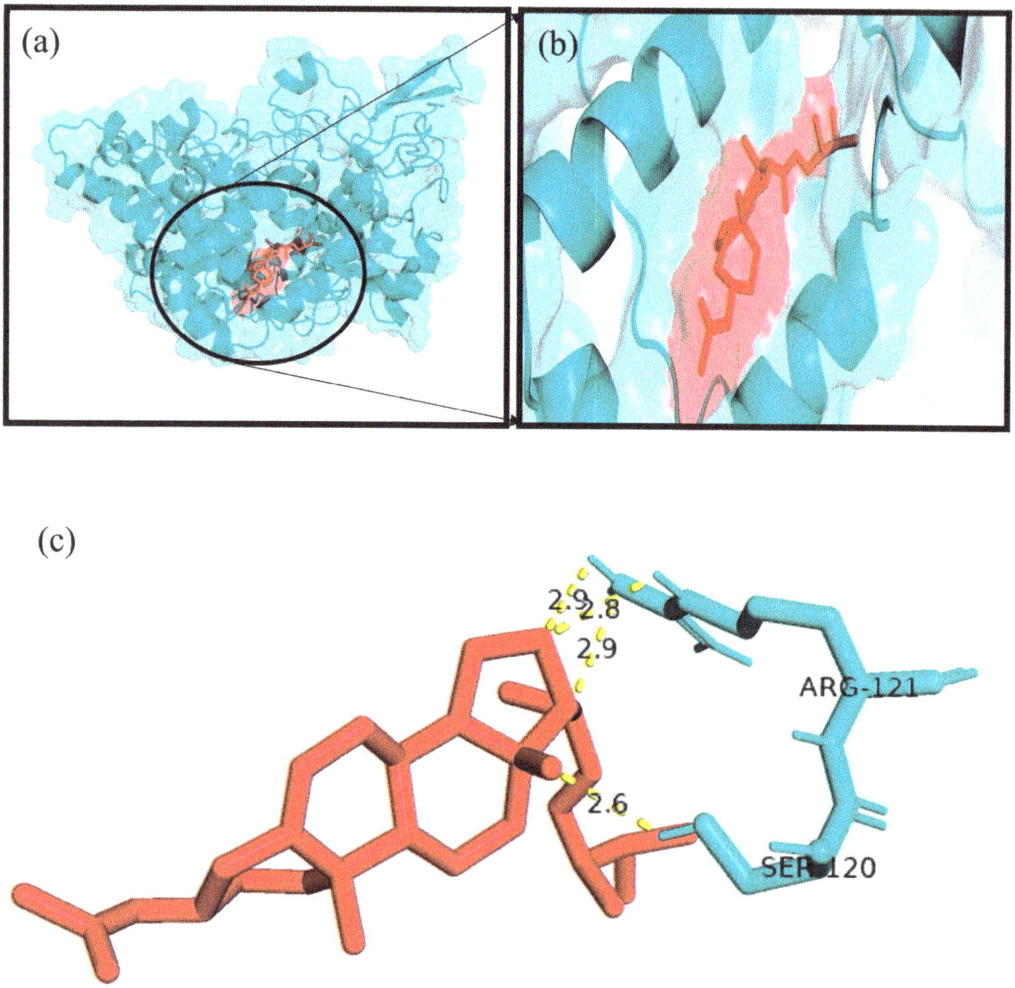

Figure 12. (**a**) Surface view of the docked complex of ergosta-5,22-dien-3-ol, acetate, (3β,22E), and the anti-inflammatory target protein (Cox-2); (**b**) binding pose of ergosta-5,22-dien-3-ol, acetate, (3β,22E); (**c**) ligand–protein docked complex showing the binding interaction of ergosta-5,22-dien-3-ol, acetate, (3β,22E), with Cox-2. Four hydrogen bonds (yellow, dotted line) formed between the ligand ergosta-5,22-dien-3-ol, acetate, (3β,22E), and two amino acid residues (Ser-120 and Arg-121) of the target protein Cox-2.

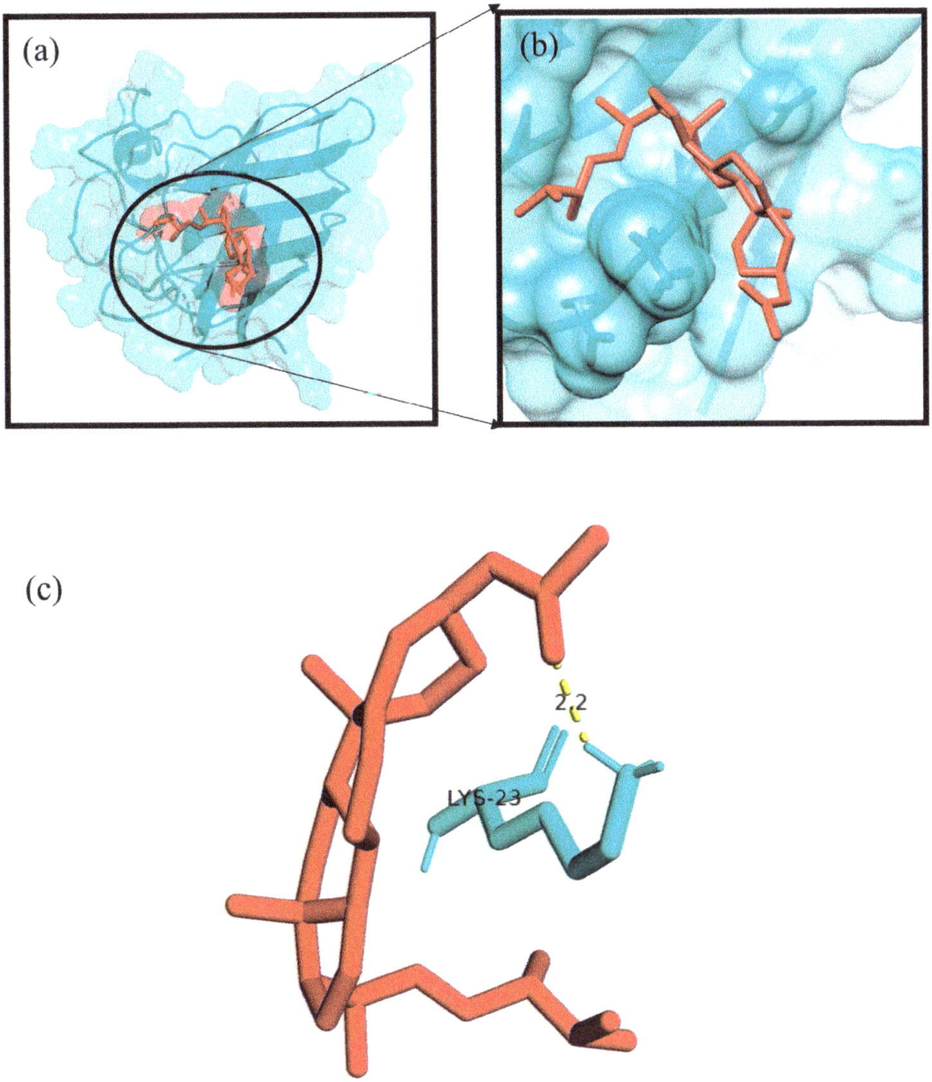

Figure 13. (**a**) Surface view of the docked complex of ergosta-5,22-dien-3-ol, acetate, (3β,22E), and the antioxidant target protein (SOD); (**b**) binding pose of ergosta-5,22-dien-3-ol, acetate, (3β,22E); (**c**) ligand–protein docked complex showing the binding interaction of ergosta-5,22-dien-3-ol, acetate, (3β,22E), with SOD. One hydrogen bond (yellow, dotted line) formed between the ligand ergosta-5,22-dien-3-ol, acetate, (3β,22E), and one amino acid residue (Lys-23 (2.2 Å)) of the target protein SOD.

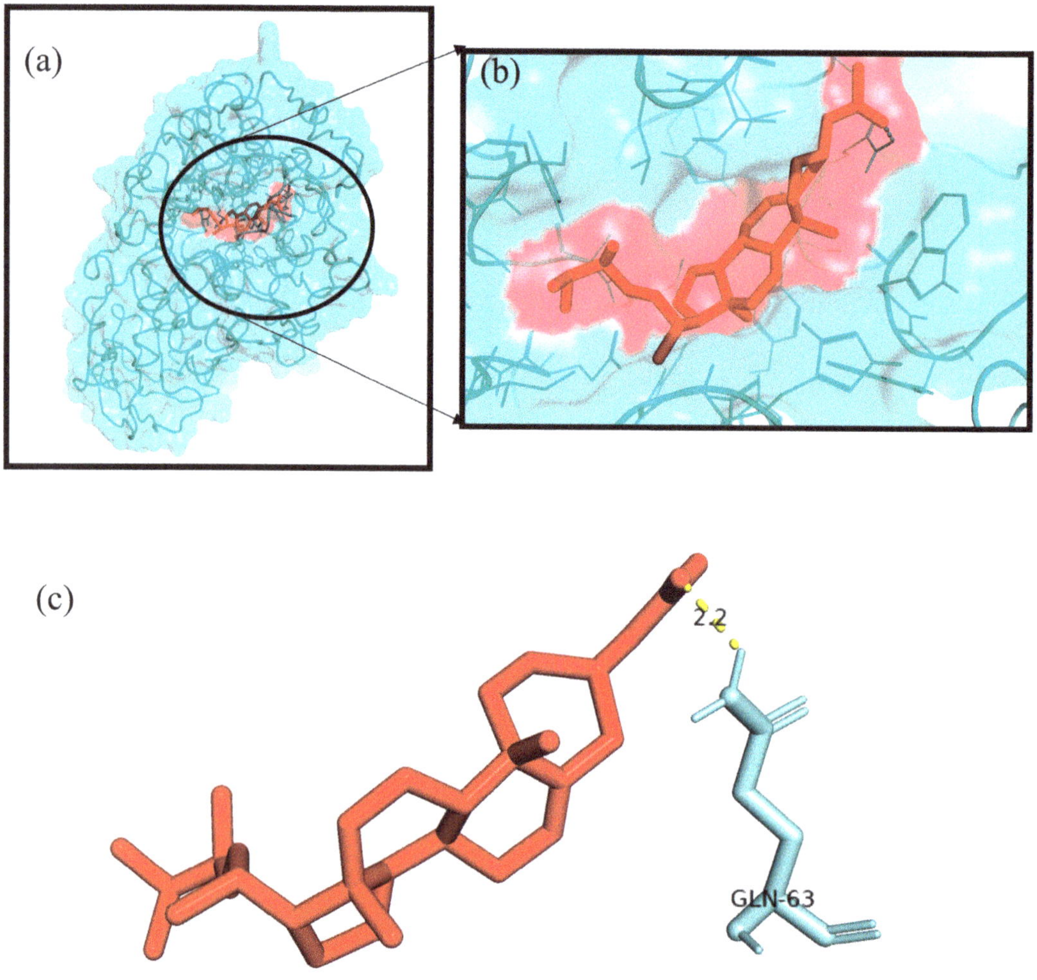

Figure 14. (**a**) Surface view of the docked complex of ergosta-5,22-dien-3-ol, acetate, (3β,22E), and the antidiabetic target protein (α-amylase); (**b**) binding pose of ergosta-5,22-dien-3-ol, acetate, (3β,22E); (**c**) ligand–protein docked complex showing the binding interaction of ergosta-5,22-dien-3-ol, acetate, (3β,22E), with α-amylase. One hydrogen bond (yellow, dotted line) formed between the ligand ergosta-5,22-dien-3-ol, acetate, (3β,22E), and one amino acid residue (Gln-63 (2.2 Å)) of the target protein α-amylase.

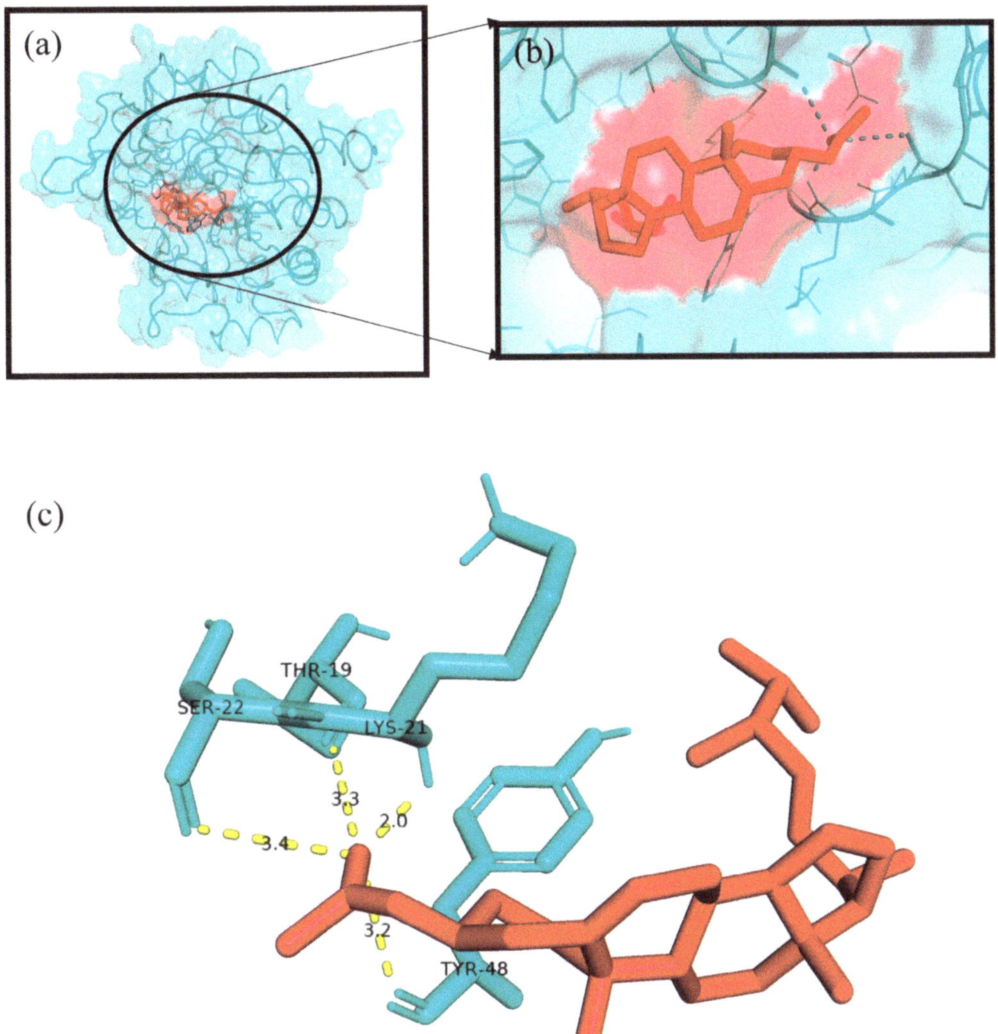

Figure 15. (**a**) Surface view of the docked complex of ergosta-5,22-dien-3-ol, acetate, (3β,22E), and the antidiabetic target protein (aldo reductase); (**b**) binding pose of ergosta-5,22-dien-3-ol, acetate, (3β,22E); (**c**) ligand–protein docked complex showing the binding interaction of ergosta-5,22-dien-3-ol, acetate, (3β,22E), with aldo reductase. Four hydrogen bonds (yellow, dotted line) formed between the ligand ergosta-5,22-dien-3-ol, acetate, (3β,22E), and four amino acid residues (Thr-19 (2.2 Å), Lys-21 (2.0 Å), Ser-22 (3.4 Å), and Tyr-48 (3.2 Å)) of the target protein aldo reductase.

2.9. Toxicity Analysis of the PECTS

2.9.1. In Vitro Cytotoxicity Assay

The in vitro cytotoxicity study of the optimised phenolic extract of *Causonis trifolia* (PECTS) in the Vero cell line (ATCC-CCL-81) did not show any significant cell death, even at the highest dose of 250 mg/L, and the cell morphology was also normal (Figure 16).

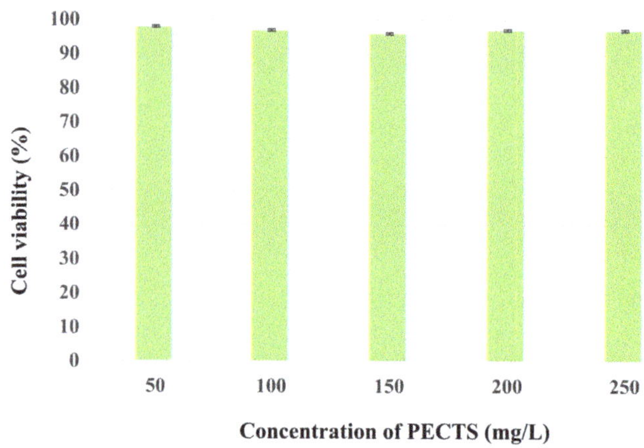

Figure 16. Cell viability (%) at different concentrations (50–250 mg/L) of the PECTS.

2.9.2. In Vivo Acute Toxicity Study

In the in vivo acute toxicity assay, the optimised phenolic extract of *Causonis trifolia* (PECTS) did not show any mortality or sign of toxicity in both males and females at a dose of 5000 mg/kg body weight over the observation period. Moreover, the treated mice did not show any characteristic change in body weight during the experimental period. As no deaths were observed in this experiment at the highest dose of administration, the LD_{50} of the PECTS could not be estimated and was considered to be greater than 5000 mg/kg body weight, an experimental upper limit, in the acute toxicity study (Figure 17A,B). Thus, the high LD_{50} value of this plant is a strong indicator of the fact that the extract could be considered safe.

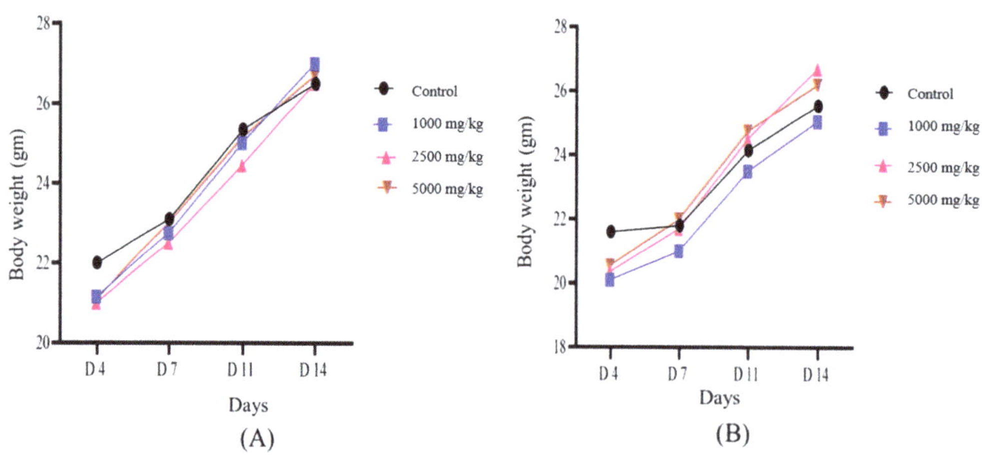

Figure 17. Changes in body weight of (**A**) female BALB/c and (**B**) male BALB/c mice over 14 days of acute exposure (oral administration) of the PECTS.

3. Discussion

The use of medicinal plants for curing diseases is as old as humankind itself. Contemporary science has recognised the active role of plants in the treatment of various diseases, and it has incorporated a variety of plant-derived drugs into modern pharmacotherapy that have been used throughout the millennia. The vast diversity of flora on the Indian subcontinent allows for many such plants to be discovered, and *Causonis trifolia* (L.) Mabb.

& J.Wen (Family: Vitaceae) is one such underexplored ethnomedicinal plants that has been used by various ethnic groups across the globe including in India.

The extraction procedure can be improved by optimising the different experimental factors that affect the extraction to achieve the highest possible output with bioactive phytoconstituents. Many works have been carried out to standardise an optimum method for the extraction of the highest amounts of bioactive phytoconstituents from a particular herbal drug. Some important examples can be highlighted here: optimisation of the extraction process of bioactive phytocompounds from the leaf of *Berberis vulgaris* [37], withanolides extraction process optimisation from the root of *Withania somnifera*, and optimisation for essential oils yield from the leaf of *Citrus latifolia* [38]. In our present study, major efforts were carried out to optimise the extraction process of *C. trifolia* shoots. As there is no previous study regarding the optimisation of the extraction conditions for the most efficient extraction of phenolic compounds from the shoot part of *C. trifolia*, this study can serve as a reference for obtaining the maximum yield of phenolic compounds from the studied part of the plant. We adopted an RSM-based optimisation using a Box–Behnken design to achieve this goal. The experimental outcomes suggest that the optimum phenolic content can be derived using the optimised extraction module with a 94.5% desirability. Our study has shown that the extraction of greater amount of active phenols is highly dependent on three parameters, namely, the solvent concentration (%), extraction temperature (°C), and duration of contact (hrs). Three-dimensional response surface plots depict the interaction among the three parameters studied and identify the combined influence of these three parameters on the extraction of total phenols from the dried shoot of the investigated plant (Figures 1 and 2). Moreover, the most effective extraction conditions for the highest yield of phenolics were derived from the RSM analysis: 67.6% methanol concentration, 49.7 °C extraction temperature, and 12.5 h extraction time.

Phenolic compounds contained in the medicinal plants and various food stuffs represent the most widely distributed (more than 8000 chemical structures known in the phenolic group) plant secondary metabolites exhibiting a plethora of beneficial effects for human health [39,40]. Epidemiological studies have shown that many phenolic derivatives possess strong antioxidant and radical scavenging properties and are found to be instrumental in reducing the risk of acute and chronic diseases, cardiovascular disorders, and certain types of cancer [40]. In our present experiment, significant amount of total phenols (150.76 $\pm$ 0.02 mg GAE/g dry tissue) was found in the optimised phenolic extract of *Causonis trifolia* shoot (PECTS). The phenolic content of our studied plant was higher than the contents of phenolics of many well recognised medicinal plants, such as myrobalan (134.47 mg GAE/g dry tissue), ginger rhizome (60.34 mg GAE/g dry tissue), and emblica fruit (81.5–126.0 mg GAE/g dry tissue) [41–43]. Among other phytochemical groups, flavonoids, alkaloids, and tannins have been recognised as therapeutically important groups with various key biological functions, such as free radical scavenging, anticarcinogenic, and anti-inflammatory effects [44,45]. From our study, it is evident that the analysed plant part contains quite a good amount of flavonoids (20.7 $\pm$ 0.02 mg CE/ g dry tissue), tannins (23.21 $\pm$ 0.47 mg GAE/g dry tissue), and alkaloids (13.37 $\pm$ 0.12 mg of PE/100 g dry tissue) as well. The presence of these therapeutically active phytochemicals in a very impressive amount also highlights the prospect of this plant as a good source of potent natural products with multiple therapeutic claims.

Moreover, the HPLC analysis of the optimised phenolic extract of the studied plant (PECTS) identified and quantified a total of eight phenolic and flavonoid compounds, namely, gallic acid, catechin hydrate, chlorogenic acid, caffeic acid, p-coumaric acid, sinapic acid, coumarin, and kaempferol. Hydroxycinnamic acid, synthesised by most of plant species, is a very significant member of phenolic acids with a bioactive carboxylic acid group, and it is present in food items, such as coffee, wine, tea, and cabbage, and in many popular herbal medicines, such as basil and turmeric [46,47]. This class of plant phenolics mainly includes caffeic acid, ferulic acid, sinapic acid, and coumarin. The perusal of the literature suggests that these compounds are capable of donating their hydrogen atom for the neutralisation of reactive free radicals. These phenolic acids and their derivatives are claimed to have several health benefits, including anti-inflammatory, cardioprotective, antiviral, and anticarcinogenic properties. A wide range of in vitro and in vivo studies have demonstrated the therapeutic potencies of many compounds in the phenolic acid group [47]. From the HPLC study, it was noticed that the PECTS is a very potent source of two important phenolic acids, namely, sinapic acid (24.172 mg/g dry tissue) and coumarin (13.211 mg/g dry tissue). Sinapic acid is known as a potent antioxidant, and the antibacterial, anti-inflammatory, anticancer, and anti-anxiety properties of this compound are well documented [48]. On the other hand, coumarin and its derivatives have also shown very strong pharmacological effects, including antitumor, anti-inflammation, antiviral, and antibacterial activities [49]. Our HPLC analysis of the PECTS clearly validates certain ethnomedicinal claims of this studied plant by elucidating the considerable amounts of the eight therapeutically potent phenolic molecules, and it also brings this underexplored medicinal plant into the limelight of scientific communities for its thorough chemopharmacological exploration.

Furthermore, the GC–MS analysis revealed the existence of 19 phytochemicals in the optimised phenolic extract of *Causonis trifolia* shoot (PECTS), which might contribute to the medicinal characteristics of this plant species. Out of the 19 phytochemicals identified, 12 compounds fall under the chemical group of sesquiterpenes. Germacrene D; 1,2,4-Metheno-1H-indene; octahydro-1,7a-dimethyl-5-(1-methylethyl); [1S-(1α,2α,3aβ.,4α.,5α,7aβ.,8S*)]; α-copaene; caryophyllene; and humulene are some of the sesquiterpenes that were detected in the PECTS by the GC–MS study. Germacrene-D, a sesquiterpenoid compound extracted from the leaves of *Chloroxylon swietenia* DC., has been reported for its mosquitocidal, as well as repellent, activity against aphids and ticks. Germacrene-D also possesses antibacterial properties [25–27]. 1,2,4-Metheno-1H-indene, octahydro-1,7a-dimethyl-5-(1-methylethyl), popularly known as Cyclosativene, reported from an ethyl acetate extract of *Psidium guajava* leaves, exhibits strong antiproliferative and antioxidant potencies [21]. The presence of the phytocompound α-copaene are considered a chemotaxonomic marker of the genus *Annona*, since their occurrence is very common in species of this genus. α-Copaene, identified in essential oils of *Annona salzmannii* and *A. pickelii*, has been reported to have antiproliferative and antioxidant properties [21,50]. Caryophyllene is a bicyclic sesquiterpene, previously reported in the essential oil of *Aquilaria crassna*, and displayed anticancer, antioxidant, and antimicrobial properties [24]. Humulene is the primary terpene found in hops (*Humulus lupulus*), but it is also present in cannabis, sage, and ginseng. Humulene, commonly known as α-caryophyllene, has shown potent anti-inflammatory activity in the rat models. Humulene is an effective analgesic agent and is known to have antineoplastic effect by inducing apoptosis [28]. 2H-Pyran, tetrahydro-4-methyl-2-(2-methyl-1-propenyl), also known as rose oxide, is the only monoterpenoid compound identified in the PECTS through our GC-MS study. Nonato et al., in 2012, experimentally proved the anti-inflammatory activity of this monoterpenoid compound [22]. Many studies have apprised the remarkable pharmacological effects of plant sterols, acting as chemopreventive, anti-inflammatory, antioxidant, antidiabetic, and antiatherosclerotic agents [51]. Ergosta-5,22-dien-3-ol acetate (3,22E), a sterol compound, has previously been reported in soft coral, (*Subergorgia reticulata*), but its biological activity has yet to be documented [52].

Based on the findings presented above, it can be concluded that the *Causonis trifolia* shoot extract contains a good number of therapeutically important phytochemicals including phenolic aids that have exhibited significant levels of various pharmacological actions, such as antioxidant, antimicrobial, anti-inflammatory, antiproliferative, and anticancer activities. Identifying various phytochemical compounds provides insight into the medicinal values of the plant *Causonis trifolia*, which can be further evaluated through bio-prospecting studies to optimise how this plant can be utilised to explore its therapeutic potential. In addition, we have also provided experimental evidence obtained in the present study in support of the antioxidant, anti-inflammatory, and antidiabetic activities of the optimised phenolic extract of *Causonis trifolia* shoot (PECTS).

The quest for the identification of active principles is a lengthy, tedious process, not to mention its overburdening financial demands. According to a study conducted in 2014, the cost of developing a new drug was estimated to be USD 2.5 billion [53]. The high failure rate of drug candidates is a primary causes of this dramatic increase in cost [54]. Understanding the drug–receptor interaction is crucial for managing the difficulties scientists face regarding the cost and establishment of a new lead molecule. Fortunately, computational tools have unquestionably played a pivotal role in optimising, assuring, and accelerating the drug development process. Molecular docking is now considered an efficient and inexpensive technique for designing and testing drug candidates [35]. The strong antioxidant, anti-inflammatory, and antidiabetic activities of the phenol-enriched extract of *Causonis trifolia* shoot (PECTS) prompted us to carry out in silico studies to predict the possible mechanism of action of the detected compounds. In structure-based molecular docking, a small ligand molecule is aligned inside the binding cavity of the target protein, and the resulting docking pose is evaluated by a specific scoring function. The scoring function generates a score for each pose, and the resulting values are used to rank the different poses and ligands. In a methodological sense, there are two independent stages in the docking process: pose generation and the scoring. The first refers to the methods that are used to create different ligand and protein conformations and align different ligand conformations within the binding site of the protein. The latter, the scoring, is required in the docking process for the quantitative estimation of the pose quality. In our present study, 19 bioactive phytocompounds were identified from PECTS by GC–MS analysis, and these compounds were used for molecular docking studies. Two anti-inflammatory, one antioxidant, and two antidiabetic target proteins were selected for the molecular docking study. Ergosta-5,22-dien-3-ol, acetate, (3β,22E), was established, employing the iGEMDOCK tool, as the lead molecule among all of the identified compounds, and it exhibited the best anti-inflammatory, antioxidant, and antidiabetic activities by establishing the maximum binding interaction with respective to the target proteins at the cost of the lowest binding energy. When these phytocompounds were subjected to further docking study using Autodock 4.2, they were evaluated based on their free energy of binding (ΔG). The free energy of binding (ΔG), which is equal to the free energy of the protein–ligand complex minus the free energies of the protein and the ligand in their unbound states, is used to quantitatively measure the degree of spontaneity and strength of protein–ligand binding [55]. The binding interactions study conducted using the Autodock 4.2 programme revealed that the compound ergosta-5,22-dien-3-ol, acetate, (3β,22E), required total binding energy (−11.73 Kcal/mol) with aldo reductase, followed by α-amylase (−9.27 Kcal/mol), Cox-2 (−8.93 Kcal/mol), TNF-α (−8.21 Kcal/mol), and superoxide dismutase (−7.16 Kcal/mol). Enzymes are one of the most important groups of drug targets, and the identification of possible ligand–enzyme interactions is of major importance in many drug discovery processes [56]. Aldo reductase, an enzyme belonging to the aldo keto reductase superfamily, catalyses the rate-limiting step of the polyol pathway, an alternative path for glucose metabolism [57]. Reactive oxygen species (ROS) are produced when glucose is processed through the polyol pathway in hyperglycaemic situations [58]. These metabolic alterations cause osmotic and oxidative stresses, which in turn cause various degenerative diseases [59]. Additionally, the polyol pathway contributes to a number of biochemical alterations, including an increase in the generation of advanced glycation end-products and the activation of protein kinase C, both of which could be relevant to diabetes-induced vascular dysfunction [58]. Since aldo reductase is a central molecule and is known to control the rate-limiting step of the polyol pathway, its inhibition provides a possible strategy for preventing complications of chronic diabetes [60]. Experimental studies suggest that the inhibition of AR could be effective in the prevention of diabetic complications [61]. Thus, identifying potent AR inhibitors can pave the way for effective therapies against diabetes and related complications. In analogy with any spontaneous process, protein–ligand binding occurs only when the change in the Gibbs free energy (ΔG) of the system is negative at a constant pressure and temperature. Because the protein–ligand association extent is determined by the magnitude of the negative ΔG, ergosta-5,22-dien-3-ol, acetate, (3β,22E), could be a very potent inhibitor of aldo reductase. The van der Waals + hydrogen bonding + desolvation energy for ergosta-5,22-dien-3-ol, acetate, (3β,22E), and aldo reductase docked complex was −13.44 kcal/mol, much higher than their electrostatic energy of −0.08 kcal/mol, thus proposing that the hydrogen bonding and van der Waals interaction are the major forces stabilising the protein–ligand complex. Visualisation of the docked complex with the help of pyMol confirmed that the binding of ergosta-5,22-dien-3-ol, acetate, (3β,22E), with aldo reductase being stabilised by four bond interactions involving four amino acid residues (Thr-19, Lys-21, Ser-22, and Tyr-48) of the target protein. Our study recommends further in-depth investigation of this molecule to evaluate its bioactivities and toxicity. The proposed molecule should be subjected to the necessary clinical trials for broad-spectrum drug discovery.

To ensure the safety of plant products, systematic studies are required to predict the risks of toxicity and deliver scientific knowledge regarding the selection of safe dosages of the plant extracts [62]. In our experiment, the PECTS did not show any significant cell death when applied against Vero cells (ATCC-CCL-81) using the MTT assay, even at the highest dose of 250 mg/mL. After the clearance in the in vitro cytotoxicity assay, the PECTS was tested on mice for 14 days in an acute toxicity study. Here, we also did not observe any behavioural abnormalities, as well as deaths, of any mice (including both female and male) at very high doses (5000 mg/kg body weight). Thus, the use of this extract up to the highest administered dose was completely safe for the animals. According to the harmonised system for the classification of chemicals that cause acute toxicity, adopted by the Organization for Economic Co-operation and Development (OECD), this plant extract can strongly be recommended as a nontoxic and safe drug for humans and other animals. However, further repeated long-term chronic toxicological investigations are highly recommended to confirm its safety and effectiveness in humans.

4. Materials and Methods

4.1. Collection and Preparation of the Plant Materials

The fresh shoots of *Causonis trifolia* (L.) Mabb. & J.Wen were collected from roadside bushes in Santiniketan, Birbhum, West Bengal, India (11°03′15.46″ N, 076°32′23.58″ E) in October 2020. The collected plant species were identified with the help of the *Flora of West Bengal* [63] and was authenticated by an expert in plant taxonomy. The nomenclature of the identified plant species was updated following using a standard website, namely, "Plants of the World Online" [64]. The collected plant specimen was preserved as a herbarium specimen following standard herbarium techniques [65] and kept at the Department of Botany, Visva-Bharati, Santiniketan, India, for future references (voucher specimen number: INDIA, West Bengal, Birbhum district, Santiniketan, 10.10.2020, S Hazra 3 (VBH)). The collected plant materials were washed thoroughly under running tap water to remove the dust particles before they were cut into small pieces, shade dried, and uniformly ground using an electric grinder. The crude powder was stored in an airtight cellophane bag at 4 °C until further use.

4.2. Statistical Optimisation of the Extraction Process for the Enhanced Extraction of Phenols

In our present experiment, the response surface methodology (RSM) statistical tool was employed for the optimisation of the extraction of phenols from the shoots of *Causonis trifolia*. A three-factor and three-level Box–Behnken design was applied here for the RSM-based optimisation study (Table 4). A total of 17 experimental runs, including five central points, were generated using Design-expert 13 software (Stat-Ease Inc., Minneapolis, MN, USA) (Table 5). Here, the variables taken for this statistical design were the concentration of the extracted solvent (A, % *v/v*, methanol:water), temperature during the extraction (B, °C), and the duration of the extraction (C, hours). Each variable had two levels: −1 (i.e., lower limit) and +1 (i.e., higher limit). First, four different solvent systems, namely, methanol, ethyl acetate, chloroform, and petroleum ether, were examined to understand their extraction yield of phenols, and methanol was selected, as it exhibited a higher extraction potential. The extraction efficiency of the aqueous methanol was substantially higher than that of 100% methanol, as 100% methanol is not capable of extracting most of the bioactive phytochemicals. Thus, different proportions of methanol and water (ranging from 40% to 100%) (A, % *v/v*, methanol:water) were used as an independent variable for this investigation. The temperature range for this experiment was 20 °C to 80 °C (B, °C), while the extraction time ranged from 1 h to 24 h (C, hours). Meanwhile, the total phenolic content (TPC) was selected as the dependent or response variable. In Table 4, we have listed the ranges of all three parameters considered for the extraction of phenols during the RSM study. Multiple linear regression analysis was performed, and the experimental data were fitted to the second-order polynomial mode.

Table 4. Independent and dependent variables used in the Box–Behnken design for the optimisation of *Causonis trifolia* shoot extraction.

Symbols	Independent Variables	Unit	Coded Levels		
			−1 (Low)	0 (Medium)	+1 (High)
A	Methanol concentration	%	40	70	100
B	Temperature	°C	20	50	80
C	Extraction duration	Hours	1	12.5	24
	Dependent variable/ response variable			Goal	
Y1	Total phenolic content	mg GAE/g dry tissue		Maximise	

Table 5. Three-factorial RSM-based optimisation study using a Box–Behnken design.

Run	Solvent (%)	Temperature (°C)	Extraction Duration (Hours)	Total Phenolic Content (Mg GAE/g Dry Tissue)
1	70	20	24	100.24
2	70	20	1	108.45
3	100	50	24	103.88
4	70	50	12.5	154.35
5	70	50	12.5	151.25
6	70	50	12.5	155.84
7	40	50	24	108.68
8	70	80	24	105.34
9	70	80	1	100.23
10	40	50	1	105.22
11	100	80	12.5	100.55
12	70	50	12.5	150.44
13	70	50	12.5	151.33
14	40	80	12.5	116.45
15	40	20	12.5	111.22
16	100	20	12.5	107.45
17	100	50	1	102.22

For further detailed phytochemical analyses and bioactivity studies, the optimised phenol extract of the studied plant part was selected. As per the suggested extraction protocol of the RSM model, the unit mass (10 g) shoot powder was mixed with 150 mL of

67% methanol and kept at 49 °C for 12.53 h under constant agitation in a shaker incubator. The slurry obtained was then filtered through Whatman No. 1 filter paper, and the filtrate was collected. This procedure was repeated thrice to extract all phytochemicals from the powdered plant sample. All obtained filtrates were combined, and the methanol was evaporated in a rotary evaporator machine. The dried extract was stored in an air-tight container at 4 °C until further use.

4.3. *Quantitative Estimation of Certain Groups of Phytochemicals of the PECTS*

For the quantitative estimation of the extract, conventional methods were adopted. The total phenolic content using the Folin- Ciocaletu method [66], the total flavonoid content via the aluminium chloride colorimetric method [67], the total tannin content following the method of Afify et al. (2012) [68], and the total alkaloid content using the 1,10-phenanthroline method [69] were estimated.

4.4. *High-Performance Liquid Chromatography (HPLC) Analysis of the PECTS*

Agilent's 1260 Infinity II (Santa Clara, CA, USA) HPLC instrument was used, and the data were processed employing OpenLab software. The separation was achieved using a Luna C18 reversed-phase column (25 cm × 4.6 mm, 5 μm) (Phenomenex, Torrance, CA, USA). In our present study, at first, the stock solutions (1 mg/mL) of both the standard compounds and the plant extract were prepared. For this, a 1 mg sample was dissolved in 0.5 mL HPLC-grade methanol, followed by 10 min of sonication, and the resulting volume was 1 mL with the mobile phase solvent (acetonitrile and 1 percent aqueous acetic acid, 1:9). The mobile phase contained 1% aqueous acetic acid solution (Solvent A) and acetonitrile (Solvent B). The flow rate was set at 0.7 mL/min, the column was maintained at 28 °C, and the volume of the injected plant sample was fixed at 20 μL. A gradient elution was performed by changing the proportion of solvent B to solvent A. For a duration of 28 min, the gradient elution was changed from 10% to 40% B in a linear fashion, from 40% to 60% B in 39 min, and from 60% to 90% B in 50 min. It took 55 min for the composition of the mobile phase (solvent B: solvent A) to return to its original state. Then, it was allowed to run for another 10 min before injecting the next sample. The total duration of the analysis per sample was 65 min. According to the absorption maxima of the analysed compounds, the HPLC chromatograms were detected using a photo diode array UV detector at three distinct wavelengths (272, 280, and 310 nm). The retention period of each chemical was then determined by spiking with standards under the same conditions. The integrated peak area was used to quantify the phenolic and flavonoids present in the sample, and the content was estimated using a calibration curve by plotting the peak area against the concentration of the respective standard sample. The data are reported with a convergence limit in triplicate.

4.5. *Gas Chromatography-Mass Spectrometry (GC–MS) Analysis of the PECTS*

Five grams of optimised phenolic extract (PECTS) were dissolved in 100 mL of methanol. The solution was stirred continuously for 72 h before being filtered using Whatman No. 41 filter paper. Then, the extract was completely concentrated by evaporating the methanol in a rotary evaporator at 40 °C. Thereafter, 1 g of the dried concentrated extract was dissolved in 10 mL methanol. The stock solution of the sample extract was then transferred to an airtight container and stored at 4 °C until it was needed for the experiment.

The gas chromatography-mass spectrometry (GC-MS) analysis was carried out on a mass spectrometer (Agilent 6890-MS, Santa Clara, CA, USA). After filtering the sample solution via a sterilised Millipore filter (0.22 μm), it was used for the GC-MS analysis. One microliter of the sample was injected into a GC equipped with an MS and the nonpolar capillary column HP-5 (30 m × 0.25 mm; 0.25 μm). The oven program began at an initial temperature of 110 °C for 2 min, and the transfer line temperature was 200 °C. Then, the temperature was increased to 250 °C for 2 min at a rate of 5 °C/min. Finally, the temperature was raised to 280 °C at a rate of 6 °C/min for 7 min. The total run time was 50 min. The detector temperature was set at 250 °C. Helium (purity: 99.999%) was employed as the carrier gas with a flow rate of 1 mL/min. The sample injection was conducted following the split mode (1:10). In addition, 70 eV was set for the electron impact ionisation (EI). The mass range scanned was 50–550 m/z.

The identification of phytochemicals present in the optimised phenolic extract of *Causonis trifolia* was made by consulting the mass spectral library of NIST (NIST ver. 2.0, 2005). The NIST database was carefully checked, the spectra of the reference compounds were extracted, and they were matched with the unknown phytochemical compounds present in the plant sample.

4.6. *In Vitro Antioxidant Activity Study of the PECTS*

4.6.1. DPPH Radical Scavenging Activity

The DPPH radical scavenging activity was determined following the standard method [70] with slight modifications. The standard curve was prepared with ascorbic acid. The results are expressed as the % of the scavenging activity. The IC_{50} value was determined from the % of the inhibition vs. concentration of the different plant extracts and ascorbic acid by comparing the absorbance values of the control (A_0) and test compounds (A_t). The percentage of the DPPH radical scavenging activity was determined employing the following formula:

$$\text{Radical scavenging activity (\%)} = (A_0 - A_t/A_0) \times 100 \quad (1)$$

4.6.2. Hydrogen Peroxide Scavenging Assay

The ability of the optimised phenolic extract of *C. trifolia* shoots to scavenge hydrogen peroxide was determined according to the method of Ruch et al. (1989) [71]. The IC_{50} value was determined from the % of the inhibition vs. concentration of the plant extract and ascorbic acid by comparing the absorbance values of the control (A_0) and test compounds (A_t). The percentage of the H_20_2 radical scavenging activity was determined using the following formula:

$$\text{Radical scavenging activity (\%)} = (A_0 - A_t/A_0) \times 100 \quad (2)$$

4.6.3. ABTS Radical Scavenging Activity

This assay was carried out following the standard protocol of Thaipong et al. (2006) [70]. The standard curve was prepared using ascorbic acid. The results are expressed as the % of the scavenging activity. The IC_{50} value was determined from the % of the inhibition vs. concentration of the different plant extracts and ascorbic acid by comparing the absorbance values of the control (A_0) and test compounds (A_t). The percentage of the ABTS radical scavenging activity was determined using the following formula:

$$\text{Radical scavenging activity (\%)} = (A_0 - A_t/A_0) \times 100 \quad (3)$$

4.6.4. Phosphomolybdenum Assay

The total antioxidant activity of the extracts was evaluated using the phosphomolybdenum method [72]. Each test tube contained 0.3 mL of the optimised extract which was mixed with 3 mL of reagent solution (0.6 M sulfuric acid, 28 mM sodium phosphate, and 4 mM ammonium molybdate) and then incubated at 95 °C for 90 min. After the incubation, the absorbance of the test solution was measured at 695 nm using a UV-Vis spectrophotometer (Shimadzu UV1800 double beam spectrophotometer, Toshvin Analytical Pvt. Ltd., Kytoto, Japan) against a blank after cooling at room temperature. The total antioxidant activity is expressed as the milligram equivalent of ascorbic acid per gram sample.

4.7. *In Vitro Anti-Inflammatory Activity Study of the PECTS*

4.7.1. Inhibition of Albumin Denaturation Assay

First, 2 mL of 1% bovine serum albumin (BSA) was mixed with 400 μL of plant extract at different concentrations (100–500 μg/mL), and the pH of the reaction mixture was adjusted to 6.8 with 1N HCl. The reaction mixture was incubated at room temperature for 20 min and then heated to 55 °C for 20 min in a water bath. The mixture was cooled to room temperature, and the absorbance value was recorded at 660 nm. A BSA mixture with a 30% methanol solution was used as the control. An aqueous solution of diclofenac sodium at different concentrations was used as the standard [73]. The experiment was performed in triplicate. The percent inhibition was calculated using the following formula:

$$\text{Inhibition } (\%) = \frac{\text{Control O.D.} - \text{Sample O.D.}}{\text{Control O.D.}} \times 100 \tag{4}$$

where Control O.D. = optical density of the control; Sample O.D. = optical density of the test sample.

4.7.2. Antiprotease Activity Assay

First, 1 mL of 20 mM Tris HCl buffer (pH 7.4) was mixed with 0.06 mg of trypsin and 1 mL of extract at different concentrations (100–500 μg/mL). The reaction mixture was incubated at room temperature (37 °C) for 5 min. Then, 0.8% casein was added to the reaction mixture and further incubated for 20 additional minutes. A total of 2 mL of 70% perchloric acid was added to the mixture to terminate the reaction. The solution was centrifuged at 3000 rpm for 10 min. The absorbance of the supernatant was recorded at 210 nm. An aqueous solution of diclofenac sodium at different concentrations was used as the standard. The IC_{50} value was calculated [73]. The experiment was performed in triplicate. The percent inhibition was calculated using the following formula:

$$\text{Inhibition } (\%) = \frac{\text{Control O.D.} - \text{Sample O.D.}}{\text{Control O.D.}} \times 100 \tag{5}$$

4.7.3. Membrane Stabilisation Assay

A. Preparation of the R.B.C. Suspension

A blood sample was centrifuged at 3000 rpm for 10 min and washed thrice with normal isotonic saline. Then, 10% (*v/v*) normal saline was added to the tubes and stored at 4 °C for further experimental use.

B. Heat-Induced Haemolysis Assay

First, 1 mL of 10% (*v/v*) RBC suspension was mixed with 1 mL of the plant extract (100–500 μg/mL). The reaction mixture was incubated in a water bath at 56 °C for 30 min. The tubes were cooled to room temperature under running tap water. The tubes were centrifuged at 2500 rpm for 5 min, and the absorbance value of the supernatant was determined spectrophotometrically at 560 nm. Diclofenac sodium was used as a standard [73]. The percentage inhibition of the haemolysis was calculated as follows:

$$\text{Inhibition } (\%) = \frac{\text{Control O.D.} - \text{Sample O.D.}}{\text{Control O.D.}} \times 100 \tag{6}$$

4.8. In Vitro Antidiabetic Activity Study of the PECTS

4.8.1. α-Amylase Inhibitory Assay

The assay was carried out following the standard protocol [74]. Starch azure (2 mg) was suspended in 0.2 mL of 0.5 M Tris–HCl buffer (pH 6.9) containing 0.01 M $CaCl_2$ (i.e., substrate solution). The tubes containing the substrate solution were boiled for 5 min and then preincubated at 37 °C for 5 min. The plant extracts were dissolved in DMSO in order to obtain concentrations of 50, 100, 150, and 200 μg/mL. Then, 0.2 mL of plant extract of particular concentration was added to the tube containing the substrate solution. In addition, 0.1 mL of porcine pancreatic amylase in Tris–HCl buffer (2 units/mL) was added to the tube containing the plant extract and substrate solution. The reaction was carried out at 37 °C for 10 min, and it was stopped by adding 0.5 mL of 50% acetic acid to each tube. The reaction mixture was centrifuged at 3000 rpm for 5 min at 4 °C. The absorbance of the resulting supernatant was measured at 595 nm using a spectrophotometer (Shimadzu UV1800 double beam). Acarbose, a known α-amylase inhibitor, was used as the standard drug. The experiments were repeated thrice. The percentage of the α-amylase inhibitory activity was calculated using the following formula:

$$\text{Inhibition } (\%) = \frac{\text{Control O.D.} - \text{Sample O.D.}}{\text{Control O.D.}} \times 100 \quad (7)$$

The α-amylase inhibitory activities of the plant extracts and acarbose were calculated, and their IC_{50} values were determined.

4.8.2. Glucose Uptake Assay Using Yeast Cells

A glucose uptake assay was performed using yeast cells according to the method of Cirillo et al. (1963) [75]. Commercial baker's yeast suspended in distilled water was subjected to repeated centrifugation (3000 rpm for 5 min) until a clear supernatant was obtained, and 10% (*v*/*v*) of the suspension was prepared in distilled water. Various concentrations (50, 100, 150, and 200 μg/mL) of the plant extract were added to 1 mL of glucose solution (5 mM) and incubated together for 10 min at 37 °C. The reaction was started by adding 100 μL of the yeast suspension followed by vortexing and further incubation at 37 °C for 60 min. After the incubation, the tubes were centrifuged for 5 min at 3800 rpm, and the glucose was estimated using a spectrophotometer (Shimadzu UV1800 double-beam spectrophotometer) at 520 nm. Acarbose was used as the standard drug. All experiments were carried out in triplicate. The percentage increase in the glucose uptake by the yeast cells was calculated using the following formula:

$$\text{Glucose uptake } (\%) = \frac{\text{Control O.D.} - \text{Sample O.D.}}{\text{Control O.D.}} \times 100 \quad (8)$$

4.9. In Silico Molecular Docking Study

A docking study was performed to predict the possible anti-inflammatory, antioxidant, and antidiabetic properties of the phytocompounds identified from the GC-MS analysis of the shoot extract of the studied plant. The chemical structures of the compounds were retrieved as SDF files from the PubChem database "https:/pubchem.ncbi.nlm.nih.gov (accessed on 1 September 2021)". Then, the downloaded SDF files were converted into Mol, PDBQT, and PDB file formats using OPEN BABEL software "https://openbabel.org/wiki/Main Page (accessed on 1 September 2021)". The three-dimensional (3D) structures of the phytocompounds were optimised for molecular docking using Chimera 1.15 software [76]. Here, we employed a structure-based molecular docking approach to study the interactions of the phytocompounds with the proteins responsible for triggering inflammatory (e.g., tumour necrosis factor alpha (Tnf α) (PDB ID: 2AZ5) and cyclooxygenase-2 (Cox2) (PDB ID: 4PH9)); oxidative stress (superoxide dismutase (SOD) (PDB ID: 5YTO)); and diabetic (e.g., α-amylase (PDB ID: 1B2Y) and aldo reductase (PDB id: 4YS1)) conditions in the human body. The Research Collaboratory for Structural Bioinformatics (RCSB) Protein Data Bank's "https://www.rscb.org (accessed on 2 September 2021)" online server was exploited to obtain the structures of the selected target proteins. Finally, the molecular docking experiments were carried out using the iGEMDOCK 2.1 and AutoDock 4.2 programs [77,78].

Initially, hydrogen bonds, bond orders, charges, and flexible torsions were assigned to all of the target proteins and phytocompounds in order to prepare them for the docking study. The iGEMDOCK application was used to screen these compounds for protein–ligand interactions. The predefined parameters (population size: 200, generations: 60, and solutions: 2) were used in this docking study. The conformation with the lowest total binding energy among the different conformations generated was considered the best binding conformation of the phytocompounds against the target proteins. The identified phytocompounds were imported into the iGEMDOCK graphical user interface and sorted by the postdocking analysis based on their binding energies and compound fitness score measured using the iGEMDOCK docking algorithm [35]. The most potent phytocompound against the five target proteins was subjected to a further docking study using the Autodock 4.2 programme. Autodock 4.2 was employed to screen for the optimal binding position of the ligands along with an assessment of the relative intensities of the binding interactions. The interacting protein residues forming hydrogen bonds with phytocompounds were visible in the Pymol viewer [79].

4.10. Toxicity Analysis of the PECTS

4.10.1. In Vitro Cytotoxicity Assay

A cell toxicity assay was performed against the Vero cell line (ATCC-CCL-81) using the 3-(4,5-dimethylthiazol-2-yl)-2,5-diphenyltetrazolium bromide (MTT) assay following the standard protocol [80]. Approximately 5×10^3 cells/well were seeded in a 96-well plate and incubated at 37 °C with a 5% CO_2 atmospheric condition. After 24 h, the PECTS were added at concentrations ranging from 50 to 250 mg/L, and the incubation was conducted for 72 h. After the incubation, MTT was added at a 5 mg/L concentration to each well and the incubation was continued at 37 °C for a further 4 h. The residual medium was discarded, 0.1 mL of DMSO was added to solubilise the formazan crystals, and OD was taken at 540 nm for calculating the CC_{50}. The lowest concentration of a compound that resulted in a 50% reduction in the cell viability is called CC_{50}. The positive control was doxorubicin, and each experiment was repeated three times. The percentage of the cell viability was calculated using the following formula:

$$\text{Cell viability (\%)} = \frac{\text{Absorbence of treated cells } - \text{ Background absorbence}}{\text{Absorbence of untreated cells} - \text{background absorbence}} \times 100 \quad (9)$$

4.10.2. In Vivo Acute Toxicity Study

Animals and Ethical Use

An acute toxicity study was conducted on 6- to 8-week-old BALB/c mice acquired from the National Laboratory Animal Facility of the CSIR-Central Drug Research Institute, Lucknow. During the experiment, the Animal Ethical Committee's standards were strictly observed. Four experimental groups, each with 6 mice, were formed from a total of 24 male and female BALB/c mice weighing 20–22 g (3 males and 3 females per group). The animals were acclimatised for 1 week, maintained in standard laboratory conditions (24 ± 2 °C) with 12:12 h light and dark cycles in individual polypropylene cages, and they were fed with a pelleted standard rat diet (Lipton India Ltd., Bangalore, India) and water ad libitum. The experimental protocol and report were approved by the Institutional Animal Ethical Committee (IAEC) (IAEC/2019/114/Renew-2/Dated 25 May 2022).

Experiment Details

An acute toxicity assay was performed according to the Organization for Economic Cooperation and Development's (OECD) guideline 423 [81]. The acute toxicity study was performed on 6- to 8-week-old BALB/c. During the experiment, the Animal Ethical Committee's standards were strictly observed. A total of 24 male and female BALB/c mice weighing 20–22 g were randomly assigned to four experimental groups, each with six mice (3 males and 3 females per group). After fasting overnight, each treatment group was administered with PECTS at single doses of 1000, 2500, and 5000 mg/kg by oral gavage. The control groups received the same amount of distilled water. The mice were allowed free access to feed and drinking water. After dosing, all animals were observed individually for mortality and changes in general behaviour during the first 30 min and then at 2, 4, 6, 10, and 24 h following the treatment. Symptoms of toxicity, such as hypo-activity, breathing difficulty, tremors, piloerection, and convulsion, were observed after the administration of the various doses of the extract. During the remaining experimental period, animal observations were conducted at least once a day for a postdosing period of 14 days. The body weights were measured at the initiation of the treatment and on days 4, 7, 11, and 14 after the administration of the extract.

5. Conclusions

The present study explored the optimised phenolic extract of *Causonis trifolia* shoot regarding phytochemical analysis and biological activities. This extraction procedure was established using the response surface methodology (RSM), and it was employed as an effective tool for the extraction of the maximum amount of total phenolics from the minimum quantities of shoot biomass of *C. trifolia*. This technique is found to be essential for various purposes, such as extracting the highest number of intact, undistorted, and bioactive phenolic molecules in a cost-effective and rapid manner and facilitating the appearance of higher levels of antioxidant, anti-inflammatory, and antidiabetic properties by the extracted bioactive phenolics. The optimised extract was further analysed by HPLC and GC-MS analyses to identify and quantify the various bioactive compounds. An impressive and effective phytochemical fingerprint was explored in our study, which will further be employed to highlight the immense potential of the studied herb while curing various diseases and will also serve as a characteristic identifier of the studied plant extract. We have also experimentally demonstrated the antioxidant, anti-inflammatory, and antidiabetic potentials of the optimised phenolic shoot extract of the investigated plant. Although we focused on the bioactivities of the optimised phenolic extract, this study was not sufficient to detect specific active principle(s). In future various scientific projects, this can be undertaken to further fractionate this highly potential nexus of polyphenols. Despite the fact that the in vitro cell toxicity assay against the Vero cell line and the in vivo acute toxicity study on BALB/c mice showed that this extract is nontoxic, still, a detailed subacute and chronic toxicity study is required to comprehensively characterise the toxicity profile of this extract and to certify the hazardless use of the studied plant. From an in silico molecular docking study, we predicted that ergosta-5,22-dien-3-ol, acetate, (3β,22E), compound has the most promising binding affinity toward the binding sites of different target proteins associated with oxidative, inflammatory, and diabetic conditions in the human body. In the future, the identified phytocompound can be validated through molecular dynamics simulations of the protein model and experimental research on animal models to confirm its status as a novel compound. Our study extends the scope of scientific studies on this unexplored ethnomedicinal plant for the validation of its ethnomedicinal claims and development of a nontoxic drug candidate with maximum potential.

Supplementary Materials: The following supporting information can be downloaded at: https://www.mdpi.com/article/10.3390/plants12071495/s1, Table S1. Adequacy of the model tested; Table S2. The result of the ANOVA analysis; Table S3. Binding energy scores calculated by iGEMDOCK 2.1 for 19 phytocompounds against anti-inflammatory, anti-inflammatory (e.g., Tnf-α and Cox-2), antioxidant (e.g., SOD), and antidiabetic (α-amylase and aldo reductase) target proteins. Tnf-α: tumour necrosis factor alpha; Cox-2: Cy-clooxygenase 2; SOD: superoxide dismutase.

Author Contributions: S.H., A.S.R. and C.H.R. conceptualised and designed the present work; S.H. and A.S.R. performed the experiments and analysed and interpreted the data; S.D. performed the toxicity experiments and animal handling under the guidance of A.D.G.; S.H. and A.S.R. conducted the literature search and drafted the manuscript; S.H. prepared all of the figures after consulting with A.S.R. and C.H.R.; C.H.R. critically revised the manuscript and managed the correspondence. All authors have read and agreed to the published version of the manuscript.

Funding: This research received no external funding.

Data Availability Statement: All data are contained within the article.

Acknowledgments: The authors are thankful to the Department of Botany (UGC-DRS-SAP & DST-FIST sponsored), Visva-Bharati University, West Bengal, India, for the laboratory facilities to conduct the research. Two authors (S.H. and A.S.R.) acknowledge the University Grant Commission, India, for the financial assistance sanctioned in the forms of a non-NET research fellowship and a Departmental Research Support-Special Assistance Program (Phase II) research grant.

Conflicts of Interest: The authors declare no conflict of interest.

References

1. Kumar, D.; Kumar, S.; Gupta, J.; Arya, R.; Gupta, A. A review on chemical and biological properties of *Cayratia trifolia* Linn. (Vitaceae). *Pharmacogn. Rev.* **2011**, *5*, 184–188. [CrossRef]
2. Gupta, A.K.; Sharma, M. *Review on Indian Medical Plants;* ICMR: New Delhi, India, 2007; Volume 5, pp. 879–882.
3. Gaur, R.D.; Sharma, J.; Painuli, R.M. Plants used in traditional healthcare of livestock by Gujjar community of Sub-Himalayan tracts, Uttarakhand, India. *Indian J. Nat. Prod. Resour.* **2010**, *1*, 243–248.
4. Jasuja, N.D. Evaluation of Primary Metabolites and Antioxidant Potential Activity of *Cayratia trifolia* (Leaf and Stems). *J. Drug Deliv. Ther.* **2019**, *9*, 367–372. [CrossRef]
5. Khare, C.P. *Indian Medicinal Plants: An illustrated of Dictionary;* Springer: Berlin, Germany, 2007; p. 132.
6. Choudhary, K.; Singh, M.; Pillai, U. Ethnobotanical survey of Rajasthan-An update. *Am. Eurasian J. Bot.* **2008**, *1*, 38–45.
7. Swarnkar, S.; Katewa, S.S. Ethnobotanical observation on tuberous plants from tribal area of Rajasthan (India). *Ethnobot. Leafl.* **2008**, *1*, 87.
8. Yusuf, M.I.; Susanty, S.; Fawwaz, M. Antioxidant and antidiabetic potential of galing stem extract (*Cayratia trifolia* Domin). *Pharmacogn. J.* **2018**, *10*, 686–690. [CrossRef]
9. Meganathan, B.; Palanisamy, C.P.; Panagal, M. Antioxidant, antimicrobial and cytotoxicity potential of n-hexane extract of *Cayratia trifolia* L. *Bioinformation* **2021**, *17*, 452–459. [CrossRef]
10. Perumal, P.C.; Sophia, D.; Raj, C.A.; Ragavendran, P.; Starlin, T.; Gopalakrishnan, V.K. In vitro antioxidant activities and HPTLC analysis of ethanolic extract of *Cayratia trifolia* (L.). *Asian Pac. J. Trop. Dis.* **2012**, *2*, 952–956. [CrossRef]
11. Sowmya, S.; Perumal, P.C.; Gopalakrishnan, V.K. Chromatographic and spectrophotometric analysis of bioactive compounds from *Cayratia trifolia* (L.) stem. *Int. J. Pharm. Pharm. Sci.* **2016**, *8*, 56–64.
12. Rambo, D.F.; Biegelmeyer, R.; Toson, N.S.; Dresch, R.R.; Moreno, P.R.; Henriques, A.T. Box–Behnken experimental design for extraction optimization of alkaloids from *Erythrina verna* Vell. trunk barks and LC Method Validation. *Ind. Crop. Prod.* **2019**, *133*, 250–258. [CrossRef]
13. Kanthal, L.K.; Dey, A.; Satyavathi, K.; Bhojaraju, P. GC-MS analysis of bio-active compounds in methanolic extract of *Lactuca runcinata* DC. *Pharmacogn. J.* **2014**, *6*, 58–61. [CrossRef]
14. Fang, J.; Liu, C.; Wang, Q.; Lin, P.; Cheng, F. In silico polypharmacology of natural products. *Brief. Bioinform.* **2018**, *19*, 1153–1171. [CrossRef] [PubMed]
15. Ibrahim, M.B.; Sowemimo, A.A.; Sofidiya, M.O.; Badmos, K.B.; Fageyinbo, M.S.; Abdulkareem, F.B.; Odukoya, O.A. Sub-acute and chronic toxicity profiles of *Markhamia tomentosa* ethanolic leaf extract in rats. *J. Ethnopharmacol.* **2016**, *193*, 68–75. [CrossRef] [PubMed]
16. Heinrich, M.; Chan, J.; Wanke, S.; Neinhuis, C.; Simmonds, M.S. Local uses of Aristolochia species and content of nephrotoxic aristolochic acid 1 and 2—A global assessment based on bibliographic sources. *J. Ethnopharmacol.* **2009**, *125*, 108–144. [CrossRef] [PubMed]
17. Mlozi, S.H.; Mmongoyo, J.A.; Chacha, M. The in vivo toxicity evaluation of leaf and root methanolic extracts of *Tephrosia vogelii* Hook. f using animal model. *Clin. Phytosci.* **2020**, *6*, 73. [CrossRef]
18. Tauheed, A.M.; Mamman, M.; Ahmed, A.; Sani, N.I.A.; Suleiman, M.M.; Sulaiman, M.H.; Balogun, E.O. Acute, sub-acute, sub-chronic and chronic toxicity studies of four important Nigerian ethnomedicinal plants in rats. *Clin. Phytosci.* **2021**, *7*, 1. [CrossRef]
19. Mukherjee, N.; Mukherjee, S.; Saini, P.; Roy, P.; Babu, S.P.S. Antifilarial effects of polyphenol rich ethanolic extract from the leaves of *Azadirachta indica* through molecular and biochemical approaches describing reactive oxygen species (ROS) mediated apoptosis of *Setaria cervi*. *Exp. Parasitol.* **2014**, *136*, 41–58. [CrossRef]
20. Mukherjee, S.; Bandyopadhayay, B.; Basak, B.; Mandal, N.; Apurba, D.E.Y.; Mondal, B. An improved method of optimizing the extraction of polyphenol oxidase from potato (*Solanum tuberosum* L.) Peel. *Not. Sci. Biol.* **2012**, *4*, 98–107. [CrossRef]
21. Shaheena, S.; Chintagunta, A.D.; Dirisala, V.R.; Sampath Kumar, N.S. Extraction of bioactive compounds from *Psidium guajava* and their application in dentistry. *AMB Express.* **2019**, *9*, 208. [CrossRef]
22. Nonato, F.R.; Santana, D.G.; de Melo, F.M.; dos Santos, G.G.L.; Brustolim, D.; Camargo, E.A.; de Sousa, D.P.; Soares, M.B.P.; Villarreal, C.F. Anti-inflammatory properties of rose oxide. *Int. Immunopharmacol.* **2012**, *14*, 779–784. [CrossRef]
23. Hussein, H.R.; Hameed, I.H.; Ubaid, J.M. Analysis of the secondary metabolite products of *Ammi majus* and evaluation anti-insect activity. *Int. J. Pharmacogn. Phytochem. Res.* **2016**, *8*, 1403–1411.
24. Dahham, S.S.; Tabana, Y.M.; Iqbal, M.A.; Ahamed, M.B.; Ezzat, M.O.; Majid, A.S.; Majid, A.M. The anticancer, antioxidant and antimicrobial properties of the sesquiterpene β-caryophyllene from the essential oil of *Aquilaria crassna*. *Molecules* **2015**, *20*, 11808–11829. [CrossRef]
25. Noge, K.; Becerra, J.X. Germacrene D, a common sesquiterpene in the genus Bursera (Burseraceae). *Molecules* **2009**, *14*, 5289–5297. [CrossRef]

26. Kiran, S.R.; Devi, P.S. Evaluation of mosquitocidal activity of essential oil and sesquiterpenes from leaves of *Chloroxylon swietenia* DC. *Parasitol. Res.* **2007**, *101*, 413–418. [CrossRef] [PubMed]
27. Pérez Zamora, C.M.; Torres, C.A.; Nuñez, M.B. Antimicrobial activity and chemical composition of essential oils from Verbenaceae species growing in South America. *Molecules* **2018**, *23*, 544. [CrossRef] [PubMed]
28. Richter, G.; Hazzah, T.; Hartsel, J.A.; Eades, J.; Hickory, B. Cannabis sativa: An overview. In *Nutraceuticals*, 2nd ed.; Gupta, C.R., Lall, R., Srivastava, A., Eds.; Academic Press: Cambridge, MA, USA, 2021; Volume 38, pp. 603–624. [CrossRef]
29. Benelli, G.; Govindarajan, M.; AlSalhi, M.S.; Devanesan, S.; Maggi, F. High toxicity of camphene and γ-elemene from *Wedelia prostrata* essential oil against larvae of *Spodoptera litura* (Lepidoptera: Noctuidae). *Environ. Sci. Pollut. Res.* **2018**, *25*, 10383–10391. [CrossRef] [PubMed]
30. Perumal, R.; Dey, A.; Manavalan, R.; Prakasam, K.; Jayachandra, E.; Sreenivasa, G. Inhibition of albumin denaturation and antiinflammatory activity of furfuryl substituted pyrimidinoimidazolinones. *Int. J. Chem. Sci.* **2008**, *6*, 2016–2022.
31. Elisha, I.L.; Dzoyem, J.P.; McGaw, L.J.; Botha, F.S.; Eloff, J.N. The anti-arthritic, anti-inflammatory, antioxidant activity and relationships with total phenolics and total flavonoids of nine South African plants used traditionally to treat arthritis. *BMC Complement Altern. Med.* **2016**, *16*, 307. [CrossRef]
32. Naz, R.; Ayub, H.; Nawaz, S.; Islam, Z.U.; Yasmin, T.; Bano, A.; Wakeel, A.; Zia, S.; Roberts, T.H. Antimicrobial activity, toxicity and anti-inflammatory potential of methanolic extracts of four ethnomedicinal plant species from Punjab, Pakistan. *BMC Complement Altern. Med.* **2017**, *17*, 302. [CrossRef] [PubMed]
33. Anosike, C.A.; Obidoa, O.; Ezeanyika, L.U. Membrane stabilization as a mechanism of the anti-inflammatory activity of methanol extract of garden egg (*Solanum aethiopicum*). *DARU J. Pharm. Sci.* **2012**, *20*, 76. [CrossRef]
34. Puls, W.; Krause, H.P.; Müller, L.; Schutt, H.; Sitt, R.; Thomas, G. Inhibitors of the rate of carbohydrate and lipid absorption by the intestine. *Int. J. Obes.* **1984**, *8*, 181–190. [PubMed]
35. Konappa, N.; Udayashankar, A.C.; Krishnamurthy, S.; Pradeep, C.K.; Chowdappa, S.; Jogaiah, S. GC–MS analysis of phytoconstituents from *Amomum nilgiricum* and molecular docking interactions of bioactive serverogenin acetate with target proteins. *Sci. Rep.* **2020**, *10*, 16438. [CrossRef]
36. Hazra, S.; Ray, A.S.; Rahaman, C.H. Natural Phytocompounds from Common Indian Spices for Identification of Three Potential Inhibitors of Breast Cancer: A Molecular Modelling Approach. *Molecules* **2022**, *27*, 6590. [CrossRef] [PubMed]
37. Sharifi, A.; Mortazavi, S.A.; Maskooki, A.; Niakousari, M.; Elhamirad, A.H. Optimization of subcritical water extraction of bioactive compounds from barberry fruit (*Berberis vulgaris*) by using response surface methodology. *Int. J. Agri. Crop. Sci.* **2013**, *6*, 89–96.
38. Ilaiyaraja, N.; Likhith, K.R.; Babu, G.S.; Khanum, F. Optimisation of extraction of bioactive compounds from *Feronia limonia* (wood apple) fruit using response surface methodology (RSM). *Food Chem.* **2015**, *173*, 348–354. [CrossRef] [PubMed]
39. Begum, S.N.; Ray, A.S.; Rahaman, C.H. A comprehensive and systematic review on potential anticancer activities of eugenol: From pre-clinical evidence to molecular mechanisms of action. *Phytomedicine* **2022**, *107*, 154456. [CrossRef]
40. Paixao, N.; Perestrelo, R.; Marques, J.C.; Câmara, J.S. Relationship between antioxidant capacity and total phenolic content of red, rosé and white wines. *Food Chem.* **2007**, *105*, 204–214. [CrossRef]
41. Saha, S.; Verma, R.J. Antioxidant activity of polyphenolic extract of *Terminalia chebula* Retzius fruits. *J. Taibah Univ. Sci.* **2016**, *10*, 805–812. [CrossRef]
42. Ali, A.M.A.; El-Nour, M.E.M.; Yagi, S.M. Total phenolic and flavonoid contents and antioxidant activity of ginger (*Zingiber officinale* Rosc.) rhizome, callus and callus treated with some elicitors. *J. Genet. Eng. Biotechnol.* **2018**, *16*, 677–682. [CrossRef]
43. Liu, X.; Cui, C.; Zhao, M.; Wang, J.; Luo, W.; Yang, B.; Jiang, Y. Identification of phenolics in the fruit of emblica (*Phyllanthus emblica* L.) and their antioxidant activities. *Food chem.* **2008**, *109*, 909–915. [CrossRef] [PubMed]
44. Bandonien, D.; Murkovic, M. The detection of radical scavenging compounds in crude extract of borage (*Borago officinalis* L.) by using an on-line HPLC-DPPH method. *J. Biochem. Biophys. Methods* **2002**, *53*, 45–49. [CrossRef] [PubMed]
45. Banerjee, S.K.; Bonde, C.G. Total phenolic content and antioxidant activity of extracts of *Bridelia retusa* Spreng Bark: Impact of dielectric constant and geographical location. *J. Med. Plant. Res.* **2011**, *5*, 817–822.
46. Agata, I. Studies on caffeic acid derivatives in medicinal plants. *Yakugaku Zasshi J. Pharm. Soc. Jpn.* **1999**, *119*, 237–248. [CrossRef]
47. Tajner-Czopek, A.; Gertchen, M.; Rytel, E.; Kita, A.; Kucharska, A.Z.; Sokół-Łętowska, A. Study of antioxidant activity of some medicinal plants having high content of caffeic acid derivatives. *Antioxidants* **2020**, *9*, 412. [CrossRef] [PubMed]
48. Nićiforović, N.; Abramovič, H. Sinapic acid and its derivatives: Natural sources and bioactivity. *Compr. Rev. Food Sci. Food Saf.* **2014**, *13*, 34–51. [CrossRef]
49. Zhu, J.J.; Jiang, J.G. Pharmacological and nutritional effects of natural coumarins and their structure–activity relationships. *Mol. Nutr. Food Res.* **2018**, *62*, 1701073. [CrossRef]
50. Costa, E.V.; Dutra, L.M.; de Jesus, H.C.R.; de Lima Nogueira, P.C.; de Souza Moraes, V.R.; Salvador, M.J.; de Holanda Cavalcanti, S.C.; dos Santos, R.L.C.; do Nacimento Prata, A.P. Chemical composition and antioxidant, antimicrobial, and larvicidal activities of the essential oils of *Annona salzmannii* and *A. pickelii* (Annonaceae). *Nat. Prod. Commun.* **2011**, *6*, 1934578X1100600636. [CrossRef]
51. Salehi, B.; Quispe, C.; Sharifi-Rad, J.; Cruz-Martins, N.; Nigam, M.; Mishra, A.P.; Konovalov, D.A.; Orobinskaya, V.; Abu-Reidah, I.M.; Zam, W.; et al. Phytosterols: From preclinical evidence to potential clinical applications. *Front. Pharmacol.* **2021**, *11*, 1819. [CrossRef]

52. Byju, K.; Anuradha, V.; Vasundhara, G.; Nair, S.M.; Kumar, N.C. In vitro and in silico studies on the anticancer and apoptosis-inducing activities of the sterols identified from the soft coral, *Subergorgia reticulata*. *Pharmacogn. Mag.* **2014**, *10*, S65–S71. [CrossRef]
53. Mullard, A. New drugs cost US$2.6 billion to develop. *Nat. Rev. Drug. Discov.* **2014**, *13*, 877. [CrossRef]
54. Sun, D.; Gao, W.; Hu, H.; Zhou, S. Why 90% of clinical drug development fails and how to improve it? *Acta Pharm. Sin. B* **2022**, *12*, 3049–3062. [CrossRef]
55. Du, X.; Li, Y.; Xia, Y.-L.; Ai, S.-M.; Liang, J.; Sang, P.; Ji, X.-L.; Liu, S.-Q. Insights into Protein–Ligand Interactions: Mechanisms, Models, and Methods. *Int. J. Mol. Sci.* **2016**, *17*, 144. [CrossRef] [PubMed]
56. Konc, J.; Lešnik, S.; Janežič, D. Modeling enzyme-ligand binding in drug discovery. *J. Cheminform.* **2015**, *7*, 48. [CrossRef] [PubMed]
57. Yabe-Nishimura, C. Aldose reductase in glucose toxicity: A potential target for the prevention of diabetic complications. *Pharmacol. Rev.* **1998**, *50*, 21–34.
58. Tang, W.H.; Martin, K.A.; Hwa, J. Aldose reductase, oxidative stress, and diabetic mellitus. *Front. Pharmacol.* **2012**, *3*, 87. [CrossRef]
59. Qiu, L.; Lin, J.; Xu, F.; Gao, Y.; Zhang, C.; Liu, Y.; Luo, Y.; Yang, J.Y. Inhibition of aldose reductase activates hepatic peroxisome proliferator-activated receptor-α and ameliorates hepatosteatosis in diabetic db/db mice. *J. Diabetes Res.* **2012**, *2012*, 789730. [CrossRef] [PubMed]
60. Srivastava, S.K.; Ramana, K.V.; Bhatnagar, A. Role of aldose reductase and oxidative damage in diabetes and the consequent potential for therapeutic options. *Endocr. Rev.* **2006**, *26*, 380–392. [CrossRef]
61. Kador, P.F.; Kinoshita, J.H.; Sharpless, N.E. Aldose reductase inhibitors: A potential new class of agents for the pharmacological control of certain diabetic complications. *J. Med. Chem.* **1985**, *28*, 841–849. [CrossRef]
62. EL Moussaoui, A.; Bourhia, M.; Jawhari, F.Z.; Mechchate, H.; Slighoua, M.; Bari, A.; Ullah, R.; Mahmood, H.M.; Ali, S.S.; Ibenmoussa, S.; et al. Phytochemical Identification, Acute, and Sub-Acute Oral Toxicity Studies of the Foliar Extract of *Withania frutescens*. *Molecules* **2020**, *25*, 4528. [CrossRef]
63. Paul, T.K.; Lakshminarasimhan, P.; Chowdhery, H.J.; Singh, P. *Flora of West Bengal, Vol-II Leguminosae-Aizoaceae*; Botanical Survey of India: Kolkata, India, 2015.
64. POWO. Available online: https://powo.science.kew.org/ (accessed on 14 October 2020).
65. Jain, S.K.; Rao, R.R. *A Handbook of Field and Herbarium Methods*; Today and Tomorrows Publishers: New Delhi, India, 1977.
66. Swain, T.; Hillis, W.E. The phenolic constituents of *Prunus domestica*. I.—The quantitative analysis of phenolic constituents. *J. Sci. Food Agric.* **1959**, *10*, 63–68. [CrossRef]
67. Zhishen, J.; Mengcheng, T.; Jianming, W. The determination of flavonoid contents in mulberry and their scavenging effects on superoxide radicals. *Food Chem.* **1999**, *64*, 555–559. [CrossRef]
68. Afify, A.E.M.M.; El-Beltagi, H.S.; Abd El-Salam, S.M.; Omran, A.A. Biochemical changes in phenols, flavonoids, tannins, vitamin E, β–carotene and antioxidant activity during soaking of three white sorghum varieties. *Asian Pac. J. Trop. Biomed.* **2012**, *2*, 203–209. [CrossRef] [PubMed]
69. Ray, A.S.; Mandal, S.K.; Rahaman, C.H. Pharmacognostical fingerprinting and selective bioactivity studies of *Solanum glaucophyllum* Desf. *J. Tradit. Folk Pract.* **2018**, *6*, 3–25. [CrossRef]
70. Thaipong, K.; Boonprakob, U.; Crosby, K.; Cisneros-Zevallos, L.; Byrne, D.H. Comparison of ABTS, DPPH, FRAP, and ORAC assays for estimating antioxidant activity from guava fruit extracts. *J. Food Compos. Anal.* **2006**, *19*, 669–675. [CrossRef]
71. Ruch, R.J.; Cheng, S.J.; Klaunig, J.E. Prevention of cytotoxicity and inhibition of intercellular communication by antioxidant catechins isolated from Chinese green tea. *Carcinogenesis* **1989**, *10*, 1003–1008. [CrossRef]
72. Prieto, P.; Pineda, M.; Aguilar, M. Spectrophotometric Quantitation of Antioxidant Capacity through the Formation of a Phosphomolybdenum Complex: Specific Application to the Determination of Vitamin E. *Anal. Biochem.* **1999**, *269*, 337–341. [CrossRef] [PubMed]
73. Govindappa, M.; Channabasava, R.; Sowmya, D.V.; Meenakshi, J.; Shreevidya, M.R.; Lavanya, A.; Santoyo, G.; Sadananda, T.S. Phytochemical screening, antimicrobial and in vitro anti-inflammatory activity of endophytic extracts from *Loranthus* sp. *Pharmacogn. J.* **2011**, *3*, 82–90. [CrossRef]
74. Hansawasdi, C.; Kawabata, J.; Kasai, T. α-Amylase inhibitors from roselle (*Hibiscus sabdariffa* Linn.) tea. *Biosci. Biotechnol. Biochem.* **2000**, *64*, 1041–1043. [CrossRef] [PubMed]
75. Cirillo, V.P. Mechanism of glucose transport across the yeast cell membrane. *J. Bacteriol.* **1962**, *84*, 485–491. [CrossRef]
76. Pettersen, E.F.; Goddard, T.D.; Huang, C.C.; Couch, G.S.; Greenblatt, D.M.; Meng, E.C.; Ferrin, T.E. UCSF Chimera—A visualization system for exploratory research and analysis. *J. Comput. Chem.* **2004**, *25*, 1605–1612. [CrossRef]
77. Hsu, K.C.; Chen, Y.F.; Lin, S.R.; Yang, J.M. iGEMDOCK: A graphical environment of enhancing GEMDOCK using pharmacological interactions and post-screening analysis. *BMC Bioinform.* **2011**, *12*, S33. [CrossRef] [PubMed]
78. Morris, G.M.; Huey, R.; Lindstrom, W.; Sanner, M.F.; Belew, R.K.; Goodsell, D.S.; Olson, A.J. AutoDock4 and AutoDockTools4: Automated docking with selective receptor flexibility. *J. Comput. Chem.* **2009**, *30*, 2785–2791. [CrossRef] [PubMed]
79. Schrodinger, L.; DeLano, W. PyMol. 2020. Available online: http://www.pymol.org/pymol (accessed on 4 September 2021).

80. Singh, H.; Dhole, P.A.; Saravanan, R.; Baske, P.K. Ethnomedicinal plants used in sexual disorder in Balangir and Deogarh districts, Odisha, India. *Int. J. Curr. Sci.* **2017**, *20*, 57–62.
81. OECD. *OECD Guidelines for Testing of Chemicals: Acute Oral Toxicity—Acute Toxic Class Method. Test no. 423*; adopted 22 March 1996, and revised method adopted 17 December 2001; OECD: Paris, France, 2001.

plants

MDPI

Article

Keeping Healthy in Your Skin—Plants and Fungi Used by Indigenous Himalayan Communities to Treat Dermatological Ailments

Shiekh Marifatul Haq [1,*], Aadil Abdullah Khoja [2], Fayaz A. Lone [3], Muhammad Waheed [4], Rainer W. Bussmann [1,5], Ryan Casini [6], Eman A. Mahmoud [7] and Hosam O. Elansary [8,*]

1 Department of Ethnobotany, Institute of Botany, Ilia State University, Tbilisi 0162, Georgia
2 Department of Botany, Glocal University Saharanpur, Saharanpur 247121, India
3 Department of Botany, Government Degree College (Women), Kupwara 193222, India
4 Department of Botany, University of Okara, Okara 56300, Pakistan
5 Department of Botany, State Museum of Natural History, Erbprinzenstrasse 14, 76133 Karlsruhe, Germany
6 School of Public Health, University of California, Berkeley, 2121 Berkeley Way, Berkeley, CA 94704, USA
7 Department of Food Industries, Faculty of Agriculture, Damietta University, Damietta 34511, Egypt
8 Department of Plant Production, College of Food & Agriculture Sciences, King Saud University, P.O. Box 2460, Riyadh 11451, Saudi Arabia
* Correspondence: marifat.edu.17@gmail.com (S.M.H.); helansary@ksu.edu.sa (H.O.E.); Tel.: +966-581216322 (H.O.E.)

Abstract: Dermatological ailments are a major health problem, especially when related to human immune deficiency syndrome and acquired immune deficiency. The goal of this study was to identify the medicinal plants used by the indigenous peoples of the Northwestern Himalayas to treat dermatological diseases. Several field trips were conducted in the spring and summer seasons of 2020–2021 to collect the plants of dermatological value and information about their use through open-ended semi-structured interviews (n = 53) and group discussions (n = 33). The current investigation found 64 ethnomedicinal plants belonging to 34 families commonly used to treat a variety of dermatological ailments. The main growth form was herbs (80%), followed by trees (8%) and ferns (6%). It was found that leaves (51%) were the most commonly used plant part, followed by roots and the whole plant. Wound healing was the most dominant application, with 18 plant species used, followed by skin burns cured by 11 plant species and skin boils by eight plant species. Out of the total (18%) of medicinal plants with cosmetic uses, i.e., roots of *Jurinea dolomiaea*, *Rheum webbianum*, and *Rheum spiciforme* were crushed into powder and mixed with turmeric, and the paste is applied topically for glowing skin. Among the various preparation methods, paste (38%) was the most common way of preparation, followed by poultice (29%) and infusion (9%). Between ethnic groups, the maximum homogeneity was between Gujjar and Bakarwal ethnic groups (23 species, 36%), followed by Gujjars and Kashmiri (14 species, 22%). Bakarwals and Gujjar people live in the same geographical location, and they graze their animals in pastures, practice extensive transhumance pastoralism, and pass through different ecological landscapes, thus having sufficient experiences with certain plants and retaining more knowledge. The species identified with the highest utilization based on the number of citations and use value included *Ficus carica*, *Cichorium intybus*, *Euphorbia wallichii*, *Pinus wallichiana*, *Plantago major*, *Jurinea dolomiaea*, and *Artemisia absinthium*. The findings of this study demonstrate that people who reside in the Northwestern Himalayas region still rely on medicinal plants.

Keywords: dermatology; ethnomedicine; wound healing; traditional knowledge

1. Introduction

Ethnodermatology is a branch of Ethnobiology that deals with the identification and management of skin conditions and infections, as well as skincare and aesthetics in ethnically diverse populations [1]. The serious systemic diseases that develop as complex and severe damage from burns or generalized dermatitis demonstrate how important the skin is to the overall physiology of the body [1]. The skin barrier helps to prevent microbial infections, limit environmental contaminant absorption, and passive water loss from the body [2]. Natural residents derived from plants have a strong potential for cetaceous protection [3]. Despite the fact that plants have been used for skin health care management in many parts of the world since ancient times, herbal products were neglected for a long time after the development of modern synthetic soaps, creams, and other cosmetics [4,5]. However, as time progressed, the side effects of these chemical cosmetics became apparent, and people started turning back to herbal products for skin health care management. Traditional herbal medicine research became widely accepted around the world in the latter half of the 20th century as a result of the discovery of medicinal plants from native pharmacopeias that have been demonstrated to have significant healing power, either in their natural state or as a source of new pharmaceuticals, and the recognition that herbal formulations are less toxic than most pharmaceutical agents [6].

Traditional herbal practitioners play an important role in health care, especially in developing countries. Global estimates indicate that over 75% of the world's population cannot afford products of the Western pharmaceutical industry and depend on the use of traditional medicines mainly obtained from plants [7].

The skin is an external organ covering the body and serves many important functions, including percutaneous absorption, organ protection, fluid preservation, body shape maintenance, temperature regulation, and eliminating toxins from the body by sweat excretion [8]. The etiology of skin diseases displays a close connection between an individual's health and the sociocultural environment [9]. Skin diseases affect people of all age groups and gender. Ethno-medicinal studies showed that herbal medicine is an alternative therapy for the treatment and control of skin ailments. Skin ailments such as boils, itching, ringworm, skin disorders, leprosy, wound, dermatitis, eczema, scabies, skin allergy swelling, and psoriasis are caused by a variety of microorganisms [10]. Previous reports found that wound healing, eczema, dermatitis, fungal diseases, pyoderma, scabies, and skin allergies are the largest group of skin diseases that occur in most countries [7]. Most plants used for treating skin disorders may have other additional properties, such as anti-inflammatory, anti-microbial, anti-viral, cicatrizing, hemostatic, and analgesic effects requiring pharmacological confirmation [8].

Due to its remarkable biological and cultural diversity, the Northwestern Himalaya area is one of the most significant global repositories of medicinal plants and associated traditional knowledge. In terms of species composition and richness, the vegetation of the western and northern Himalayas exhibit increasing similarity to that of the Hindu Kush and the monsoon belt of the Karakorum mountains, respectively, due to comparable geological, physiographic, and climatic conditions [11]. At higher altitudes, especially in alpine and subalpine habitats, the limiting factor of altitude further reinforces the floristic similarities between the higher-elevation vegetation of the Hindu Kush and the Western Himalayas [12]. In the secluded valleys of the Himalayas, local communities rely on natural resources, including plants, for their livelihoods, often migrating with their livestock to summer pastures in accordance with the changing seasons. They possess valuable knowledge regarding ecosystem services [13]. However, the intensive use of plants for medicinal purposes, grazing, and fodder has resulted in heightened pressure on plant biodiversity, thereby posing significant implications for the long-term sustainability of the ecosystem. This continuous pressure has caused some species to become endangered, and they are at risk of extinction in the near future. The conflict between anthropogenic activities and biodiversity is apparent, as people tend to prioritize their own needs and place pressure on rare species [14].

In many areas where herbal medicine was developed and supported by oral tradition, its use has recently decreased. This could lead to the loss of important knowledge about the plants that will be used in the future [15]. Due to their lack of side effects, herbal medicines are becoming increasingly popular. Additionally, as healthcare costs rise, consumer interest in inexpensive treatments has encouraged them to reconsider the potential of alternatives [16]. The aim of the current research was to contribute to an improved understanding of wild ethnomedicinal plant uses to treat dermatological disorders from the remote area (administrative district "Kupwara") of Kashmir Himalayas. The main objectives of the study are (a) the collection of the medicinal plants used in the treatment of dermatological disorders; (b) to enlist the administration of different disease categories; (c) to record the medicinal plants used to treat dermatological disorders during the COVID-19 pandemic.

2. Results and Discussion

2.1. Demographic Status of Informants

A total of 86 informants were interviewed, all of whom followed Islam. Among them, male respondents had more knowledge about medicinal plants than female informants. Of the 86 informants, 58 were men, and the remaining 28 were women informants (Table 1). Most of the selected interviewees were illiterate, and a few had higher education. The informants were selected using snowball sampling based on their traditional knowledge about the use of medicinal plants to cure various diseases. In order to ensure the validity of traditional knowledge, an ongoing relationship with the local residents was maintained throughout the survey course. The informants above the age of 50–60 years were more knowledgeable. Other writers had similar findings [17–20]. Most of the participants also shared other experiences using medicinal plants. Dissemination of traditional knowledge was threatened because the younger generation had little interest in learning and had little knowledge of medicinal plants.

Table 1. Demographic status of informants in the study area.

Variable	Total Percentage	Gujjar	Bakarwal	Kashmiri
Respondents	86	31 (36.05%)	26 (30.23%)	29 (33.72%)
Gender				
Male	58 (67.44%)	22 (37.00%)	16 (27.00%)	20 (34.00%)
Female	28 (32.56%)	12 (42.90%)	10 (35.60%)	6 (21.50%)
Original Language		Gojeri	Gojeri	Kashmiri
Religion		Shia and Sunni Islam	Sunni Islam	Shia and Sunni Islam
Marriages		exogamous with other Muslims (Bakarwal)	exogamous with other Muslims (Gujjar)	exogamous with other Muslims (Pahari), endogamous (Sikh)
Livelihood source		Horticulture and pastoralism	Pastoralism	Horticulture and cattle rearing
Age groups				
28–40 years	19 (22.09%)	9 (47.37%)	6 (31.58%)	4 (21.05%)
41–50 years	22 (25.58%)	7 (31.82%)	9 (40.91%)	8 (36.36%)
51–60 years	34 (39.53%)	11 (32.35%)	13 (38.24%)	10 (29.41%)
61–70+ years	11 (12.79%)	5 (45.45%)	3 (27.27%)	3 (27.27%)
Education qualification				
Illiterate	48 (55.81%)	16 (33.33%)	18 (37.50%)	14 (29.17%)
Primary	23 (26.74%)	7 (30.43%)	5 (21.74%)	11 (47.83%)
Secondary	15 (17.44%)	5 (33.33%)	3 (20.00%)	7 (46.67%)
Occupation				

Table 1. *Cont.*

Variable	Total Percentage	Gujjar	Bakarwal	Kashmiri
Shepherds	21 (24.42%)	8 (38.10%)	10 (47.62%)	3 (14.29%)
Farmers	29 (33.72%)	10 (34.48%)	5 (17.24%)	14 (48.28%)
Hakeem's	12 (13.95%)	4 (33.33%)	5 (41.67%)	3 (25.00%)
Housewives	15 (17.44%)	5 (33.33%)	6 (40.00%)	4 (26.67%)
Unemployed	9 (10.47%)	2 (22.22%)	4 (44.44%)	3 (33.33%)
Religion				
Islam	86 (100%)			

2.2. Diversity of Medicinal Plants Used

The present study reported 64 species of medicinal plants used by local people to treat dermatological ailments. A detailed description of the therapeutic flora used to treat dermatological ailments is recorded in Table 2. These plant species were known for phytochemicals like alkaloids, phenols, terpenes, flavonoids, lactins, and saponins which might explain their efficacy in treating dermatological disorders [21–23]. In the study area, people mostly visit the high altitudes in the summers along with their cattle, where no medical facilities are available. This is why people use these medicinal plants to treat dermatological disorders. [24] reported 111 plant species from District Kayunga Uganda, and [25–27] reported 25 and 36 plant species from Jammu and Kashmir, respectively.

Table 2. Enumeration of plant species used by local people in district Kupwara of Jammu and Kashmir against dermatological ailments.

Family	Botanical Name and Voucher NO.	Common Name	Life Form	Part Used	Preparation	Administration	Skin Care	Total Citations	Use Value	Ethnic Groups			Native Range
										G	B	K	
Apiaceae	*Coriandrum sativum* L. (Cor.sat) 4081-KASH	Badyaan Souf	Herb	Leaves	Dried leaves are crushed into powder	Powder of leaves is mixed with mustard oil and applied topically	Hair tonic	19	0.22	—	—	19	Asia-Tropical Asia-Temperate,
Amaranthaceae	*Celosia argentea* L. (Cel.arg) 4217- KASH	Mawal, Lal jadi	Herb	Leaves	Dried leaves are crushed into powder	Powder of herb is mixed with water and applied topically	Anti-allergic	25	0.29	15	—	10	Africa
	Chenopodium album L. (Che.alb) 4085-KASH	Ganhaar	Herb	Seeds	Dried seeds are boiled in the water	Boiled seeds are mixed with dhesi ghee and applied topically	Dandruff and hair tonic	16	0.19	9	—	7	Asia, tropical Africa
Araceae	*Arisaema jacquemontii* Blume (Ari.jac)	Hapat Makai, Jadi booti	Herb	Tuber	Dried tuber is crushed into powder and mixed with water	A poultice of the tuber is applied topically for 2–5 days.	Skin burns	21	0.24	6	15	—	Himalayan region
	Arisaema propinquum Schott. (Ari.pro)	Sarfekal, Reach di jadi	Herb	Tuber	Dried tuber is crushed into powder and mixed with water	A poultice of the tuber is applied topically for 2–5 days.	Skin burns, blisters	25	0.29	—	25	—	Himalayan region
Asteraceae	*Achillea millefolium* L. (Ach.mil) 4097-KASH	Phelgass	Herb	Roots	Dried roots are crushed into powder	Dried root powder mixed with water is applied topically for 2–3 days	Insect sting	28	0.33	10	6	12	South America
	Anthemis cotula L. (Ant.cot)4244-KASH	Fackgass, Ganda	Herb	Leaves	Fresh leaves are crushed to make a paste	Paste is applied twice a day topically	Insect sting	15	0.17	—	—	15	Africa, Asia-Temperate, Europe
	Artemisia absinthium L. (Art.abs)	Tethwan, Tethyan	Herb	Leaves	Fresh leaves are soaked in water for 2 days	Infusion is taken orally early in the morning	Skin boils	31	0.36	20	11	—	Europe, Siberia, W. Himalaya

Table 2. *Cont.*

Family	Botanical Name and Voucher NO.	Common Name	Life Form	Part Used	Preparation	Administration	Skin Care	Total Citations	Use Value	Ethnic Groups			Native Range
										G	B	K	
	Centaurea iberica Trevir. and Spreng (Cen.ibe) 4084-KASH	Kreach	Herb	Leaves	Dried leaves are crushed into a paste and mixed with water	Paste of the leaves is applied topically	Skin burns and skin rashes	16	0.19	16	—	—	Europe, W. Himalaya
	Cichorium intybus L. (Cic.int)	Heand, Bagh hand	Herb	Leaves	Dried leaves are boiled in water, and the extract is taken orally after cooling	Extract is taken orally early in the morning	Skin irritation	36	0.42	16	8	12	Europe, Central Asia, W. Himalaya
	Erigeron annuus (L.) Desf. (Eri.ann) 4087-KASH	Dargass	Herb	Seeds	Dried seeds are crushed to extract oil	Oil is used topically for 3–5 days at bedtime	Hair tonic	14	0.16	14	—	-	Europe
	Jurinea dolomiaea Boiss. (Jur.dol) 4091-KASH	Gogalduup, Doop	Herb	Whole plant	The dried whole plant is crushed along with water to make a paste	Paste is applied topically for 2–3 days.	Eczema and burns	32	0.37	8	18	6	W. and C. Himalayas
	Saussurea costus (Kuth) (Sau.cos)	Kouth, Kuth	Herb	Roots	Dried roots are crushed into powder and mixed with water	A poultice of the roots is applied topically	Boils	16	0.19	6	10	-	Central Asia, W. Himalayas
	Senecio chrysanthemoides DC. (Sen.chr) 4101-KASH	Bouag	Herb	Leaves	Dried leaves are crushed into powder	Paste of the leaves is applied topically	Skin rashes and wound healing	24	0.30	—	24	-	W. Himalayas
	Taraxacum officinale (L.) Weber ex F.H. Wigg. (Tar.off)	Saz hand, Hand	Herb	Flowers	Fresh or dried flowers are soaked in water overnight	Infusion is taken twice a day orally	Skin rashes	30	0.35	10	5	15	Europe, NW. Africa
	Tussilago farfara L. (Tus.far) 4103-KASH	Wat pan, Pan pati	Herb	Leaves	Fresh leaves are crushed to make a poultice	A poultice of the leaves is applied topically	Wound healing	17	0.20	10	7	—	N. Africa, W. Himalayas

Table 2. *Cont.*

Family	Botanical Name and Voucher NO.	Common Name	Life Form	Part Used	Preparation	Administration	Skin Care	Total Citations	Use Value	Ethnic Groups			Native Range
										G	B	K	
Balsaminaceae	*Impatiens glandulifera* Royle (Imp.gla)	Goj-Gash, Truil	Herb	Whole plant	The whole plant is crushed into powder and mixed with water to make a paste	Paste is applied Topically for 2–3 days	Sunburns	12	0.14	7	5	—	W. Himalayas
	Impatiens edgeworthii Hook. f. (Imp.edg)	Truil, Tril fall	Herb	Leaves	Dried leaves are crushed into powder and mixed with water to make a paste	Paste is applied topically for 2–3 days	Skin burns	10	0.12	6	4	—	W. Himalayas
Berberidaceae	*Berberis lycium* Royle (Ber.lyc) 4102-KASH	Kawdach, Chex-machang	Shrub	Fruits	Dried fruits are crushed into powder and mixed with water to make a paste	Paste is applied topically	Healing of wounds	8	0.09	5	3	—	W. Himalayas
Brassicaceae	*Capsella bursa-pastoris* L. (Cap.bur) 4250-KASH	Kralmond	Herb	Leaves	Dried leaves are crushed into powder and mixed with mustard oil to make a paste	Paste is applied twice a day topically	Skin allergy	26	0.30	12	—	14	N. Africa
Boraginaceae	*Arnebia benthamii* Wall. ex G.Don (Arn.ben) 4096-KASH	Khazaban, Gowza-ban	Herb	Leaves	Dried leaves are crushed into powder and applied directly	Powder of the leaves is applied twice a day topically	Quick healing of wounds	31	0.36	8	18	5	W. Himalayas
	Cynoglossum glochidiatum Wall. ex Benth. (Cyn.glo) 4083-KASH	Nil toath	Herb	Roots	Fresh roots are crushed to make a paste	Paste is applied directly on infected portions	Pimples, Skin acne	26	0.30	16	—	10	Central Asia
	Cynoglossum nervosum Benth. (Cyn.ner)	Nil kin, Richola	Herb	Roots	Dried roots are crushed into powder and mixed with dhesi ghee to make a poultice	Poultice is applied to affected portions twice a day	Skin acne	19	0.22	9	10	—	W. Himalayas

Table 2. *Cont.*

Family	Botanical Name and Voucher NO.	Common Name	Life Form	Part Used	Preparation	Administration	Skin Care	Total Citations	Use Value	Ethnic Groups			Native Range
										G	B	K	
	Cynoglossum microglochin Benth. (Cyn.mic)	Cheur, Richola	Herb	Roots	Dried roots are crushed into powder and mixed with dhesi ghee to make a poultice	Poultice is applied to affected portions twice a day	Pimples	24	0.28	12	12	—	W. Himalayas
Cannabinaceae	*Cannabis sativa* L. (Can.sat)	Bang, Brand	Herb	Leaves	Dried leaves are crushed along with boiled water to make a paste	Paste is applied topically on infected portions	Skin Rashes	20	0.23	—	—	20	Central Asia, W. Himalayas
Caryophyllaceae	*Plantago major* L. (Pla.maj) 4118-KASH	Bead gull, Dada gulla	Herb	Whole plant	The fresh whole plant is crushed and applied directly	Poultice is applied on the infected portion for 2–3 days	Wound healing	34	0.40	18	16		Europe, N. and S. Africa, W. Himalayas
	Plantago lanceolata L. (Pla.lan)	Gull, Chota gulla	Herb	Leaves	Dried leaves are crushed along with dhesi ghee	Poultice is applied topically for 2–4 days twice a day	Skin boils	31	0.36	20	11		N. Africa, Eurasia
	Silene coronaria (L.) Clairv. (Sil.cor) 4229-KASH	Chock dawa	Herb	Leaves	Dried leaves are boiled in water for 10–15 min and applied after cooling	Boiled leaves are applied on affected portions for 2–3 days	Skin burns	24	0.28	18	-	6	E. Asia, Central Europe, W. Himalayas
	Stellaria media L. (Ste.med) 4249-KASH	Nick-haakh	Herb	Leaves	Dried leaves are crushed along with boiled water to make a paste	Paste is applied topically on infected portions	Skin irritation	11	0.13	-	11		N. Africa, Eurasia, W. Himalayas
Convolvulaceae	*Cascuta reflexa* Roxb. (Cas.ref) 4082-KASH	Kukliport	Herb	Whole plant	The dried whole plant is boiled in water for 10–20 min and taken after cooling	The decoction is taken twice a day orally	Hair tonic	17	0.20	7	-	10	Central Asia, W. Himalayas

Table 2. *Cont.*

Family	Botanical Name and Voucher NO.	Common Name	Life Form	Part Used	Preparation	Administration	Skin Care	Total Citations	Use Value	Ethnic Groups			Native Range
										G	B	K	
Dryopteridaceae	*Diplazium maximum* D. Don (Dip.max)	Longued, Kunji	Fern	Young frond	Boiled and dried Young fronds are cooked as a vegetable	The cooked vegetable is eaten along with rice	Anti-allergic	15	0.17	10	5		East Asia, W. Himalayas
Euphorbiaceae	*Euphorbia wallichii* Hook. F. (Eup.wil) 4216-KASH	Hirib	Herb	Latex	The fresh stem is cut to collect the latex	Latex is applied topically for 2–3 days	Warts, Acne	42	0.49	22	20		West and Central Asia, W. Himalayas
	Euphorbia hirta L. (Eup.hir)	Hearib	Herb	Latex	The fresh stem is cut to collect the latex	Latex is applied topically for 2–3 days	Acne, Eczema	15	0.17	10	5		Tropical and subtropical America
Geraniaceae	*Geranium wallichianum* Oliv. (Ger.wal)	Ratanjog, Rati bouti	Herb	Roots	Dried roots are crushed into powder and mixed with water to make a paste	Paste is applied twice a day topically	Skin rashes	20	0.23	12	8		W. Himalayas, Central Asia
Hamamelidaceae	*Parrotiopsis jacquemontiana* (Decne.) Rehder. (Par.jac)	Poah	Shrub	Leaves	Fresh leaves are crushed to make a paste	Paste is applied twice a day topically	Skin Boils	15	0.17	15	-		W. Himalayas, South Asia
Hypericaceae	*Hypericum perforatum* L. (Hyp.per) 4090-KASH	Chai kul, Julab ki jadi	Herb	Roots	Fresh roots are boiled in water for 10–15 min, and the extract is taken after cooling	The decoction is taken orally early in the morning for one week	Scabies	10	0.12	3	4	3	NW Africa, W. Himalayas, Europe
Lamiaceae	*Ajuga parviflora* Benth (Aju.par)	Jaind, Javind	Herb	Leaves	Dried leaves are crushed to make a powder	Powder is applied topically on affected portions	Wound healing	19	0.22	11	8		W. Himalayas, South Asia

Table 2. *Cont.*

Family	Botanical Name and Voucher NO.	Common Name	Life Form	Part Used	Preparation	Administration	Skin Care	Total Citations	Use Value	Ethnic Groups			Native Range
										G	B	K	
	Lamium album L. (Lam.alb) 4092-KASH	Zakhmedawa	Herb	Leaves	Fresh leaves are crushed to make a paste	Paste is applied topically on affected portions	Wound healing	25	0.29	—	25		Eurasia, W. Himalayas
	Nepeta cataria L. (Nep.cat) 4093-KASH	Gandsoi	Herb	Leaves	Dried leaves are crushed into a powder and mixed with mustard oil	Poultice is applied topically on affected portions	Skin irritation	15	0.17	9	—	6	E Europe, W. Himalayas
	Origanum vulgare L. (Ori.vul) 4100-KASH	Baber, Baberi	Herb	Seeds	Dried seeds are boiled in water for 15–30 min to make a decoction	The decoction is taken twice a day orally	Acne	28	0.33	8	-	20	Europe, Africa, W. Himalayas
	Prunella vulgaris L. (Pru.vul)	Kalyuth, Singrota	Herb	Whole plant	The fresh whole plant is soaked in water overnight to make an infusion	infusion is taken twice a day orally for 2–3 days	Skin allergy	31	0.36	—	31		Europe, Asia, W. Himalayas
Malvaceae	*Malva sylvestris* L. (Mal.syl)	Sochal, sucheal	Herb	Leaves	Fresh leaves are crushed to make a poultice	Poultice is applied topically on affected portions	Wound healing	24	0.28	8	—	18	Europe, C Asia, W. Himalayas
Morchellaceae	*Morchella esculenta* Fr. (Mor.esc) 4215-KASH	Gucchii	Fungi	Fruiting body	Dried fruiting body is crushed into powder and mixed with water or walnut oil to make a poultice	Poultice is applied topically on affected portions	Wound healing	26	0.30	11	9	6	W. Himalayas
Moraceae	*Ficus carica* L. (Fic.car) 4089-KASH	Anjeer	Tree	Latex	Leaves are cut to collect the latex	Latex is applied topically for 2–3 days on affected portions	Scabies, Eczema	35	0.41	7	5	23	Central Asia, W. Himalayas

Table 2. *Cont.*

Family	Botanical Name and Voucher NO.	Common Name	Life Form	Part Used	Preparation	Administration	Skin Care	Total Citations	Use Value	Ethnic Groups			Native Range
										G	B	K	
Pteridaceae	*Adinatum capillus-veneris* L. (Adi.cap) 4115-KASH	Guether	Fern	Leaves	dried leaves are crushed into a powder and mixed with mustard oil or dhesi ghee	The mixture is applied topically at bedtime for 2–3 days	Hair tonic and Dandruff	15	0.17	8	—	7	Europe
Pteridaceae	*Adiantum venustum* D. Don (Adi.ven) 4104-KASH	Guetheer	Fern	Leaves	dried leaves are crushed into a powder and mixed with mustard oil or dhesi ghee	The mixture is applied topically at bedtime for 2–3 days	Dandruff	18	0.21	8	3	7	W. Himalayas
Pinaceae	*Pinus wallichiana* A. B. Jacks (Pin.wal) 4227-KASH	Kayur, Fir	Tree	Latex	Latex is collected by making a deep cut into the stem of the tree using an axe	Latex is mixed with salt and applied topically on affected portions	Wound healing	37	0.43	15	15	7	W. Himalayas
Polygonaceae	*Rheum spiciforme* Royle (Rhe.spi)	Chithola, pambchalan	Herb	Roots	Dried roots are crushed into powder and mixed with water to make a poultice	Poultice is applied topically for 2–3 days	Wound healing and skin burns	24	0.28	5	19	—	W. Himalayas
Polygonaceae	*Rheum webbianum* Royle (Rhe.web) 4212-KASH	Chithola, pambchalan	Herb	Whole plant	Dried roots are crushed into powder and mixed with water to make a poultice	Poultice is applied topically for 2–3 days	Skin burns and wound healing	28	0.33	10	12	6	W. Himalayas
Ranunculaceae	*Aconitum heterophyllum* Wall. ex Royle (Aco.het) 4049-KASH	Patris, itees	Herb	Leaves	Fresh leaves are crushed into a paste	Paste is applied topically on affected portions	Scabies and healing of wounds.	14	0.16	—	14	—	W. Himalayas
Ranunculaceae	*Adonis aestivalis* L. (Ado.aes)	Tankbutton	Herb	Leaves	Fresh leaves are crushed into a paste	Paste is applied topically on affected portions	Skin boils and healing of wounds.	9	0.11	9	-	-	Europe

Table 2. *Cont.*

Family	Botanical Name and Voucher NO.	Common Name	Life Form	Part Used	Preparation	Administration	Skin Care	Total Citations	Use Value	Ethnic Groups G	B	K	Native Range
	Aconitum leave Royle (Aco.lea)	Mori	Herb	Leaves	Fresh leaves are boiled in water for 5–10 min	Boiled leaves are applied topically on affected portions	Dandruff	13	0.15	6	7	–	W. Himalayas
Rosaceae	*Cyndonia oblonga* Mill. (Cyn.obl)	Bumbchont	Tree	Seeds	Dried seeds are crushed into powder and mixed with dhesi ghee to make a poultice	Poultice is applied topically on affected portions for 2–3 days	Skin boils	16	0.17	5	—	11	Europe, Central Asia, W. Himalayas
	Fragaria nubicola Lindl. Ex Lacaita (Fra.nub) 4088-KASH	Ringrish, jungle gounch	Herb	Whole plant	The fresh whole plant is washed in water and crushed to make a paste	Paste is applied topically on affected portions	Wound healing	19	0.22	13	8	–	Central Asia, W. Himalayas
Rubiaceae	*Galium aparine* L. (Gal.apa)	Thapgass	Herb	Whole plant	The fresh whole plant is washed in water and crushed to make a paste	Paste is applied topically on affected portions twice a day	Wound healing	18	0.21	5	—	12	Eurasia
Saxifragaceae	*Bergenia ciliata* (Haw.) Sternb. (Ber.cil) 4213-KASH	Palfort, Butpewa	Herb	Roots	Dried leaves crushed into powder and mixed with water to make a paste	Paste is applied Topically on affected portions	Healing of wounds	35	0.41	8	22	5	W. Himalayas
Sapindaceae	*Aesculus indica* (Wall. ex Cambess.) Hook. (Aes.ind) 4111-KASH	Haandoon	Tree	Seeds	Seeds are crushed into powder and mixed with water to make a paste	Paste is applied Topically on affected portions	Skin burns and wounds	12	0.14	5	7	–	W. Himalayas
Salicaceae	*Salix acomphylla* Boiss. (Sal.aco)	Veer kul, veer	Tree	Leaves	Fresh leaves are crushed to make a paste	Paste is applied Topically on affected portions	Scabies	13	0.15	13	—	–	Europe, W. Himalayas

Table 2. *Cont.*

Family	Botanical Name and Voucher NO.	Common Name	Life Form	Part Used	Preparation	Administration	Skin Care	Total Citations	Use Value	Ethnic Groups			Native Range
										G	B	K	
Scrophulariaceae	*Verbascum thapsus* L. (Ver.tha) 4242-KASH	Sarfe Makai, Bunder tund	Herb	Leaves	Fresh leaves are soaked in water for 2 days to make an infusion	Infusion is taken orally for 2–4 days twice a day	Skin burns	18	0.21	7	3	8	Europe, W. Himalayas
Solanaceae	*Datura stramonium* L. (Dat.str) 4086-KASH	Datur, Datuer	Herb	Seeds	Dried seeds are crushed into powder and mixed with water to make a paste	Paste is applied topically on affected portions	Skin boils	16	0.19	9	—	7	C. America
	Solanum nigrum L. (Sol.nig)	Kambailkul	Herb	Leaves	Dried leaves are crushed into powder and mixed with water to make a paste	Paste is applied topically on affected portions	Skin boils	21	0.24	—	—	21	Eurasia, N. and E. Africa, W. Himalayas
Urticaceae	*Urtica dioica* L. (Urt.dio) 4219-KASH	Soi	Herb	Leaves	Dried leaves are crushed into powder and mixed with water to make a paste	Paste is applied topically on affected portions	Wound healing	19	0.22	10	9	—	Europe, N. and W. Africa, W. Himalayas
Valerianaceae	*Valeriana jatamansi* Jones (Val.jat) 4237-KASH	Mushkbalay	Herb	Whole plant	The fresh whole plant is crushed to make a paste	Paste is applied topically on affected portions	Pimples	22	0.26	10	8	4	Central Asia, W. Himalayas
Violaceae	*Viola odorata* L. (Vio.odo)	Banposh, Vanpoash	Herb	Roots	Dried roots are crushed to make a poultice	Poultice is applied topically on affected portions	Pimples	14	0.16	8	6	—	Europe, N. and W. Africa

The main growth habit of the used species was herbs (80%), followed by trees (8%), ferns (6%), shrubs (4%), and fungi (2%) (Figure 1). The reason behind the use of herbs might be the presence of a high content of bio-active compounds [26] and a higher efficacy than other plant parts [27]. Herbs are the primary component of forest ecosystems and are easily available in nature [28]. The plants belonged to 34 families, with Asteraceae (11 species) being the most dominant, followed by Lamiaceae (five species), Boraginaceae, and Caryophyllaceae (four species each). Other important families in the study area were Ranunculaceae (three species), Araceae, Amaranthaceae, Balsaminaceae, Pteridaceae, Euphorbiaceae, Polygonaceae, and Rosaceae (two species each), while the remaining 21 families were monotypic. The present study results aligned with previous studies where Asteraceae has been reported as the dominant medicinal plant family in various other study regions [29–31].

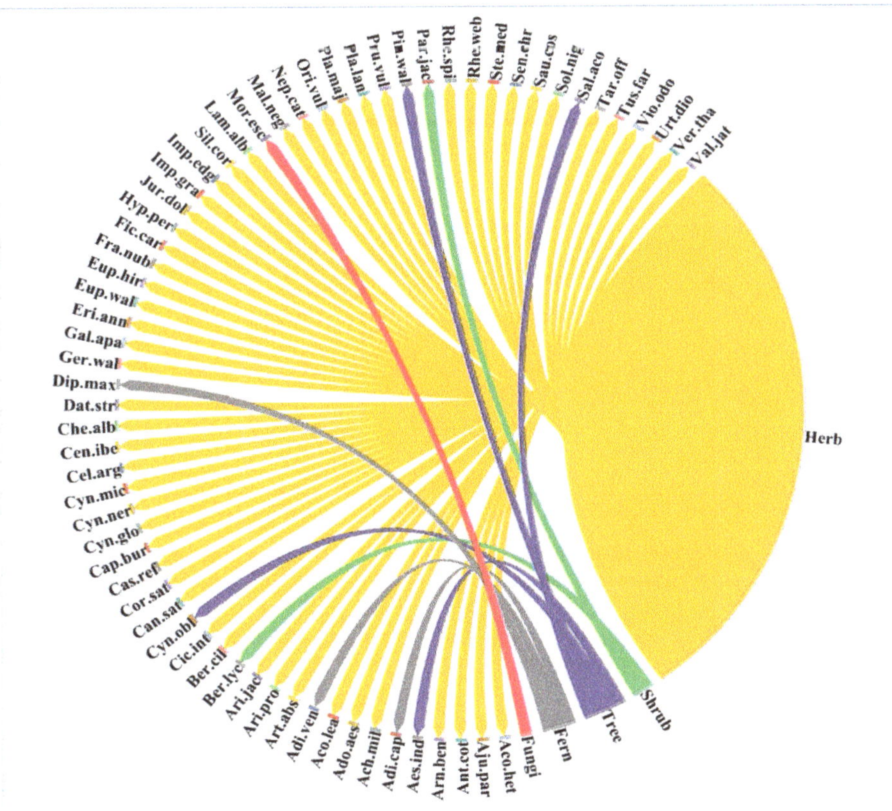

Figure 1. Percentage of species found among the various types of life in the study area. The direction of the lines indicates which species belong to which category of life forms, and the thickness of each bar indicates how many species are present in each category of life forms. In Table 2, each species' full name is displayed.

2.3. *Plant Parts Used as Medicine*

It was found that leaves (n = 32, 51%) were the most commonly used plant parts, followed by roots and the whole plant (n = 12, 19%), the whole plant (n = 6, 9%), seeds (n = 5, 8%), latex (n = 4, 6%), tuber (n = 2, 3%), and fruits and flowers (n = 1, 2% each) (Figure 2). Consistent with current research, it is reported that leaves were the main part used to cure ailments in various previous studies [32,33]. The reason behind the frequent use of leaves may be that the leaves are the center of photosynthesis and other metabolic activities, so most of the metabolites are formed in the leaves [34,35]. In addition, it is easy to collect and prepare medicines from leaves instead of other parts of the plant. It has been reported that the consumption of leaves is a sustainable collection of medicinal plants because there is no need to uproot the whole plant, so this use increases the protection of medicinal plants [36,37]. As a result of the indigenous population's reliance on foraged plants for their daily diet, various plant sections are favored depending on how they can be used. Employing leaves and aerial components is also considered sustainable and safe [37]. Local shepherds, herbalists, and herbal drug dealers, as well as other ethnic groups, frequently prefer to utilize or trade the roots of plants for medicinal purposes [38]. Roots are also known to possess a good concentration of bioactive substances [39]. Overharvesting subsurface components or entire plants should be discouraged, especially for vulnerable species, since this practice results in eradication and a decline in the plant's position in the wild [40,41]. Roots are often the most favored part of plants as they often comprise a higher concentration of bioactive constituents [42–44].

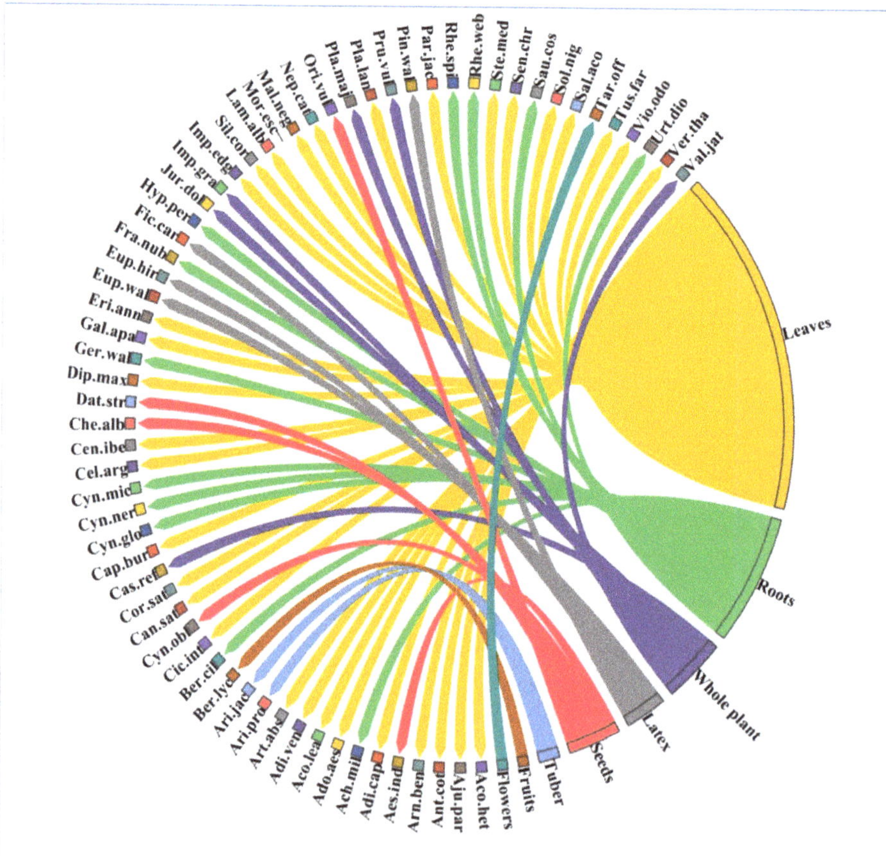

Figure 2. The proportion of each species found in the various plant parts used in the study area. The thickness of each bar indicates the number of species in each type of plant part used, and the direction of the lines identifies which species are related to which types of plant parts used. Table 2 displays each species' complete name.

2.4. Mode of Preparation

The method of administration of herbal medicines used to treat dermatological ailments includes cooking, decoction, infusion, juice, paste, powder poultice, and oil. Among the various preparation methods, paste (n = 25, 38%) is the most common way of preparation, followed by poultice (n = 19, 29%), infusion (n = 6, 9%), cooked (n = 5, 8%), juice (n = 4, 6%), powder (n = 3, 5%), decoction (n = 2, 3%), and oil (n = 1, 2%) (Figure 3). Similarly, [24,45] reported paste as the most dominant preparation, consistent with our findings.

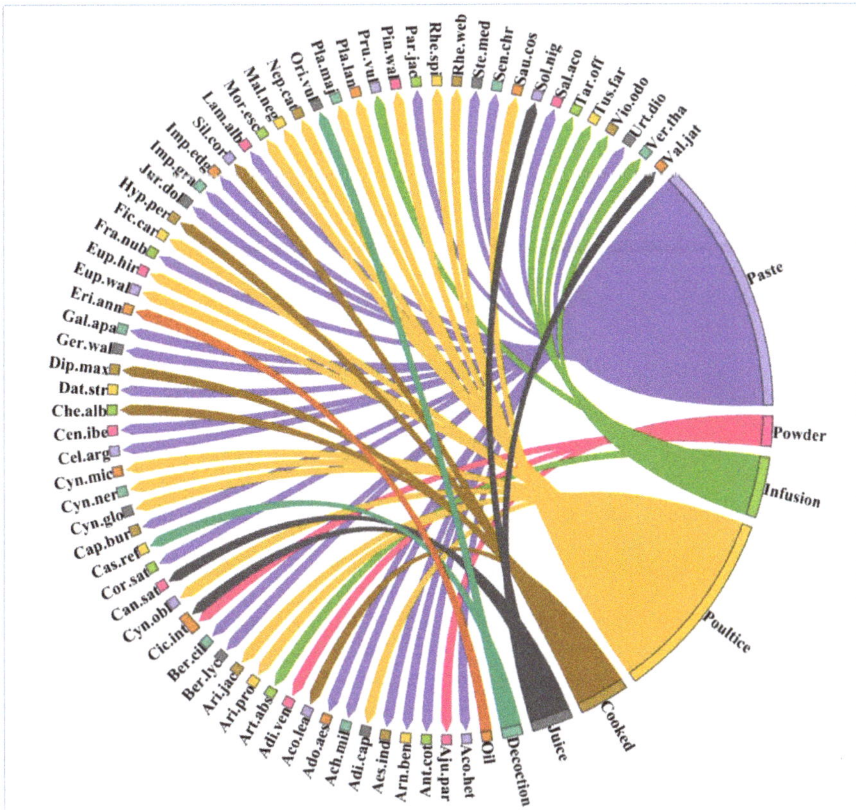

Figure 3. Percent of the species present in various plant preparations in the study area. The thickness of each bar indicates the number of species in each plant preparation, and the direction of the lines indicates which species are related to which type of plant preparation. Table 2 displays each species' complete name.

In the triplot, PC1, PC2, and PC3 explained 51.1% of the mode of preparations. Eight clusters of plant preparations based on species presence/absence can be seen here: paste, poultice, infusion, cooked, juice, powder, decoction, and oil (Figure 4). PC1 and PC2 showed a variation of 2.0, while PC1 and PC3 showed a variation of 4.5, and PC2 and PC3 showed a variation of 2.5, respectively. The plant preparations were often stored in glass bottles or other containers and used during off-seasons or during heavy snowfall in winter. According to [46], grinding, boiling, and smashing were the most common ways of extracting active ingredients in major parts of the world. Most of the plant preparations were applied topically (n = 54, 84%), and only (n = 10, 14%) were taken orally (Table 2). Our results align with [47], who reported topical application as dominant compared to oral.

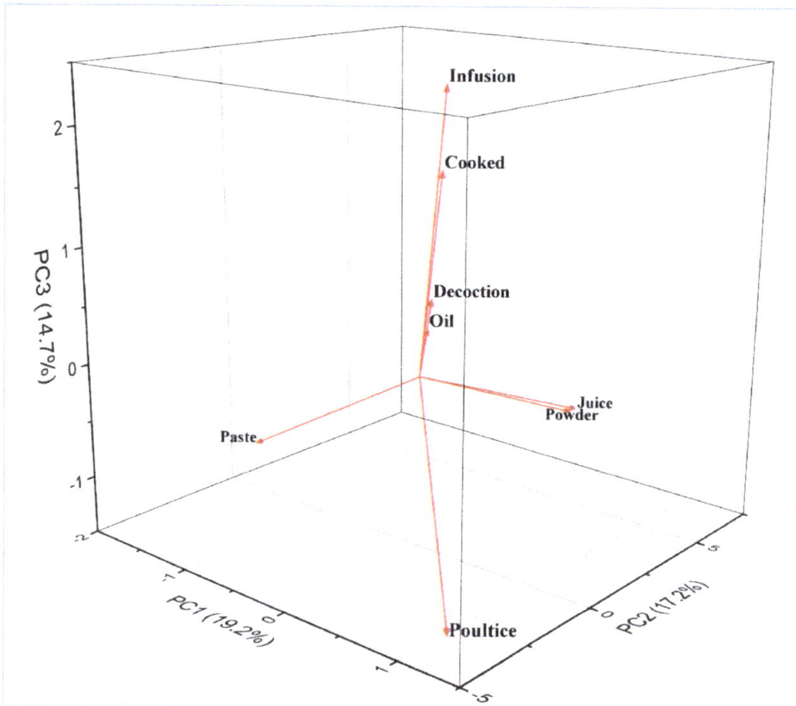

Figure 4. Principal component analysis (PCA) triplot of different plant preparations in district Kupwara of Jammu and Kashmir, India.

2.5. Disease Categories in Skin Diseases and Wound Healing

In this study, wound healing was the most dominant disease as it was cured by 18 different plant species, followed by skin burns cured by 11 plant species, skin boils (eight species), skin rashes (six species), skin acne, and hair tonic (five species each), scabies, dandruff, pimples and skin allergy (four species each), eczema (three species), insect sting and skin irritations (two species each), and warts and blisters (one species each) (Figure 5). [45–47] reported wound healing as the dominant disease category, which aligns with our findings. Wound healing consists of minor cuts to major injuries as most people in the study area are aligned with agricultural practices, which is the major reason for wound healing to be the dominant disease category.

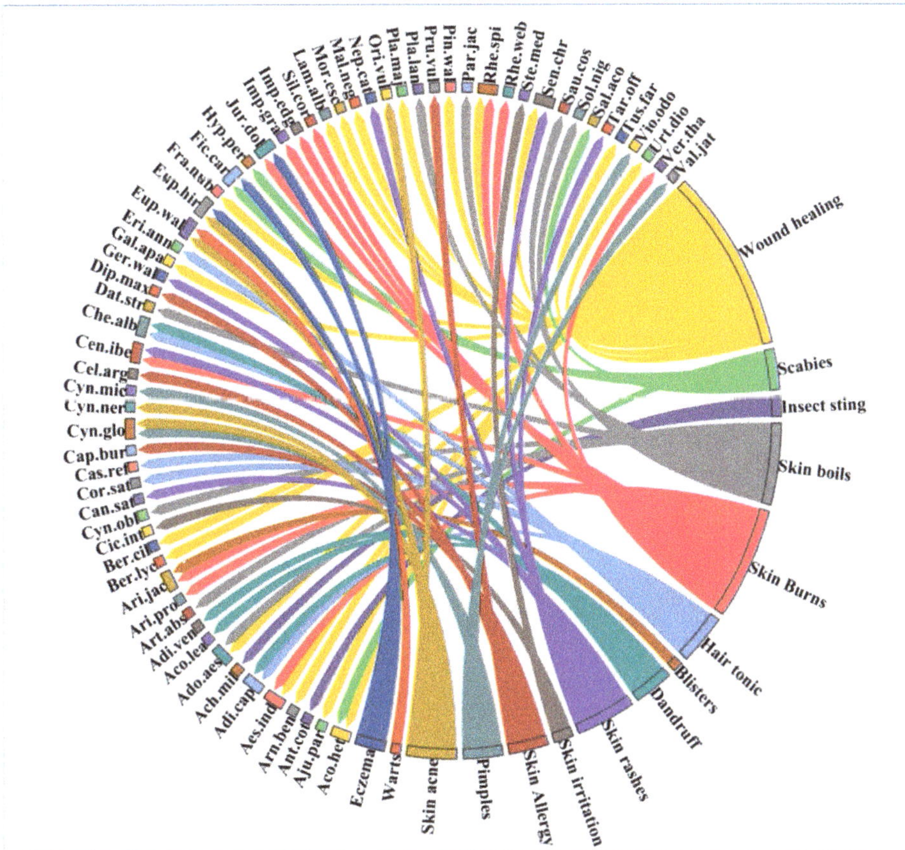

Figure 5. The percentage of species in the study area that fall into various disease categories. The direction of the lines indicates which species are related to which disease categories, and the thickness of each bar indicates how many species fall into each disease category. Table 2 displays each species' complete name.

In the triplot, PC1, PC2, and PC3 explained 30.8% of the disease categories. Fifteen clusters of disease categories based on species presence/absence can be seen there: wound healing, scabies, skin boils, skin burns, skin rashes, blister, hair tonic, skin acne, warts, insect sting, dandruff, skin allergy, eczema, pimples, and skin irritation (Figure 6). PC2 and PC3 showed a small variation of 0.3, while PC1 and PC2 showed a variation of 1.6, and PC1 and PC3 showed a variation of 1.3 (Figure 6). The research focuses heavily on influencing the best way to categorize diseases and treatments based on ethnomedical data, and translation from an emic to an etic outlook is necessary for the best biomedical screening of traditional treatments [48]. [49] reported wound healing as the most dominant disease category in Uttarakhand, India, which aligns with our research.

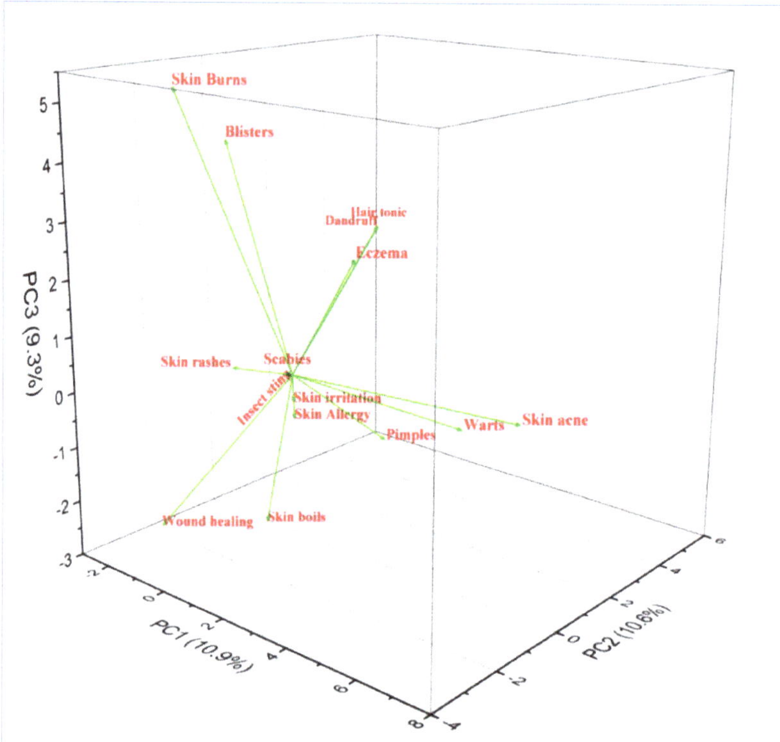

Figure 6. Principal component analysis (PCA) triplot of different disease categories in district Kupwara of Jammu and Kashmir, India.

2.6. Medicinal Plants Used during the COVID-19 Pandemic

A total of (n = 20, 31.25%) plant species were used to treat different dermatological disorders during the COVID-19 pandemic. Species like *Ficus carica, Cichorium intybus, Euphorbia wallichii, Arisaem ajacquemontiana, Achillea millefolium, Anthemis cotula, Jurinea dolomiaea, Arnebia benthamii, Cynoglossum glochidiatum, Silene coronaria, Cascuta reflexa, Geranium wallichianum, Lamium alba, Prunella vulgaris, Morchella esculanta, Adiantum venustum, Pinus wallichiana, Rheum webbianum, Bergenia ciliate,* and *Verbascum thapsus* were used by different ethnic communities especially Gujjars, Bakarwals and Phari. *Ficus carica, Jurinea dolomiaea, Cichorium intybus, Euphorbia wallichii, Cynoglossum glochidiatum, Pinus wallichiana, Rheum webbianum,* and *Bergenia ciliata* were the few commonly used medicinal plants to treat different dermatological disorders.

2.7. Cosmetic Use of Plants

Out of 64 medicinal plants (n = 12, 18.75%) have cosmetic uses. Roots of *Jurinea dolomiaea, Rheum webbianum,* and *Rheum spiciforme* were crushed into powder and mixed with turmeric, and the paste applied topically for glowing skin. Similarly, the latex of *Ficuscarica* and *Euphorbia wallichii* were applied topically to the skin. *Coriandrum sativum, Adinatum capillus-veneris, Adiantum venustum,* and *Chenopodium album* leaves and seeds were used as a hair tonic and anti-dandruff.

2.8. Cross-Cultural Analysis

A cross-cultural comparison of the medicinal plants used to treat dermatological ailments by the three ethnic groups has revealed a high degree of heterogeneity, and only a small proportion (14 species, 22%) of medicinal uses of certain recorded taxa were commonly shared among the three ethnic groups (Figure 7A). Between the three groups (Figure 7B), maximum homogeneity was between Gujjar and Bakarwal ethnic groups (23 species, 36%), followed by Gujjars and Kashmiri (14 species, 22%). Bakarwals and Gujjars live in the same geographical area and graze their animals in pastures, engage in extensive transhumance pastoralism, and travel through various ecological landscapes, gaining sufficient experience with certain plants and retaining more knowledge. The Bakarwal and Gujjars raise animals and have extensive traditional ecological knowledge of natural resources. They are particularly closely connected to nature due to their greater economic disadvantage and reliance on medicinal plants. The Bakarwal and Kashmiri ethnic groups show an overlap, and the dissimilarity in medicinal plants in terms of use reports may indicate certain sociocultural gaps, which in turn have prevented the sharing of traditional knowledge among the respective ethnic groups, especially since they do not intermarry (even though they share the same faith). All three groups live in the same region; therefore, geography has also played an important role in keeping the idiosyncrasy of the recorded medicinal usage among the considered groups. Comparatively, the Bakarwal ethnic and Gujjar reported maximum uses (five and eight each); the reason behind Bakarwal and Gujjar ethnic group usage is related to cattle rearing and that they mostly depend upon forest products as well as possessing more information on the medicinal uses of high mountain plant species, which is why they exclusively prefer and use these plants. The Kashmiris reported fewer plants (3, 4%), probably as a result of their urban settings and increased exposure to medical facilities and drug stores. It is significant to note that despite living in urban settings, Kashmiris reported using *Solanum nigrum*, *Anthemis cotula*, and *Cannabis sativa* in peculiar ways. This may also be related to the fact that, in contrast to the Gujjar and Bakarwal populations, which frequently intermarry, the Kashmiri community is strongly endogamous due to its geographical location. Similar results were reported by [8] from the North-West Frontier Province, Pakistan, [33,50,51] and from the Western Himalayas of Jammu and Kashmir.

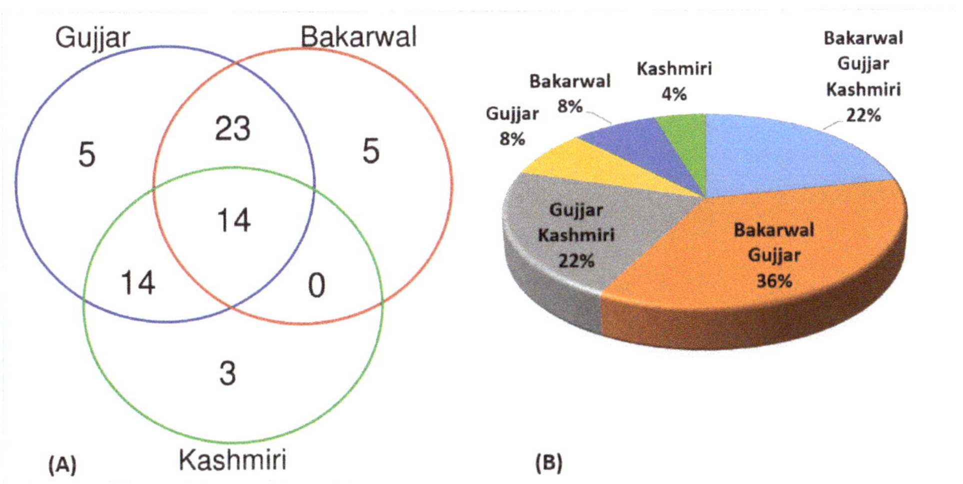

Figure 7. Venn diagrams (**A**), (**B**) showing a cross-culture comparison of medicinal plants used to treat different dermatological ailments among different ethnic groups.

2.9. Use Value (UV)

The use value index is used to determine the relative importance of medicinal plants in the study area. The value ranges from 0–1. Medicinal plants with the highest reports of use have high use value, while medicinal plants with few reports have the lowest use value. In the current research, the highest UV was calculated for *Ficus carica* (0.49) and the lowest UV of (0.11) for *Adonis aestivalis* (Table 2). In addition, it is not true that medicinal plants with low use values are less important, but it indicates that the knowledge of these medicinal plants is at risk or availability of the particular medicinal plant is less [23,51,52]. The high UV of medicinal plants in the study region is attributed to their common distribution in the area, and the local people were familiar with their medicinal uses [52]. The second highest value of UV is (0.42) of *Cichorium intybus* and *Euphorbia wallichii* (0.41). The highest use value of *Ficuscarica* is because it is commonly available and easy to use. Along with its dermatological uses, its fruits were edible and used to treat piles.

Bakarwal, followed by Gujjar, cites the highest number of plants for dermatological disorders, and the lowest number of plants is cited by the Kashmiri ethnic group (Figure 8). The reason behind this is the lack of modern medical facilities in this particular ethnic group; they commonly use traditional medicine to treat different illnesses. This particular ethnic group spent most of their time in forests (nomadic life) and transferred from one place to another, so living in a remote forest and alpine environment where modern facilities are not available means that this ethnic group is heavily dependent on traditional medicine.

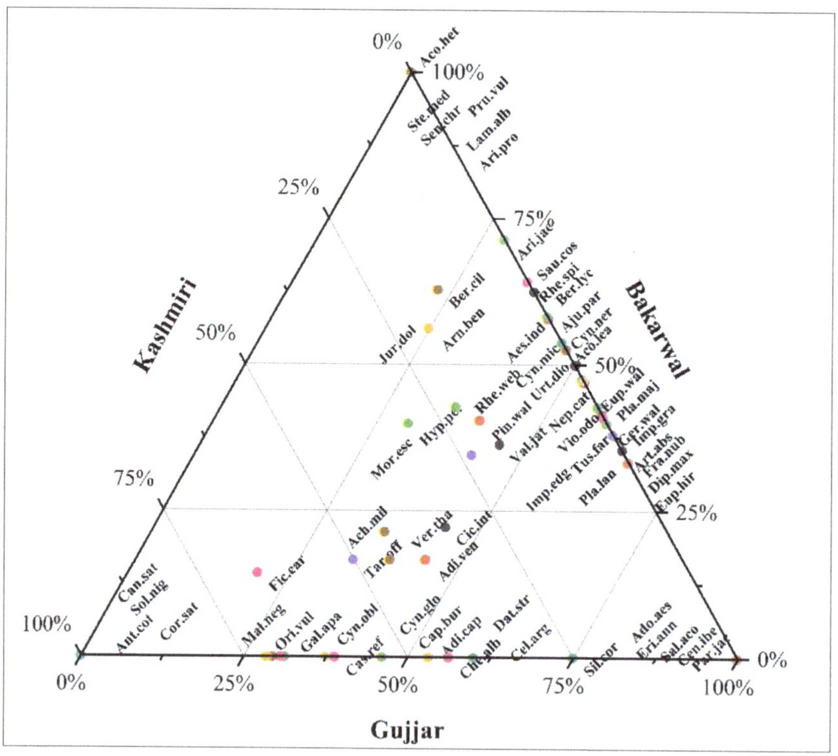

Figure 8. Ternary plot showing the distribution of plants and fungi on the basis of citation in different ethnic groups.

2.10. Biological Activity and Phytochemistry of Highly Cited Plants

As a result of phytochemical studies on *Ficus carica*, phytosterols, phenolic components, organic acids, anthocyanins, amino acids, fatty acids, hydrocarbons, aliphatic alcohols, volatile components, and a few more secondary metabolites have been isolated. Other compounds exhibited antioxidant, cytotoxic, anti-cancer anti-inflammatory, and hypolipidemic actions [53]. *Ficus carica* leaf extracts in petroleum ether, chloroform, and ethanol have been shown to have anti-inflammatory properties against carrageenan-induced rat paw edema [54]. Numerous studies have focused on exploring the pharmacological properties and bioactive constituents of *F. carica* latex. The phenolic compounds present in *F. carica* latex are primarily responsible for its pharmacological activity. One such compound is mangiferin, which is abundant in *F. carica* latex extract and has several therapeutic benefits, such as anti-diabetic, anti-microbial, antioxidant, and anti-cancer effects. Additionally, *F. carica* latex is considered a valuable source of proteins, such as proteases, which are utilized to treat skin conditions such as warts and wounds [53]. Ali et al. [55] studied the antibacterial activity of *Euphorbia wallichii* against six bacteria viz., *Bacillus subtilis, Escherichia coli, Staphylococcus aureus, Shigella flexenari, Pseudomonas aeruginosa,* and *Salmonella typhi* and proved that the chloroform extract and ethyl acetate were found to be most effective exhibiting high activity. Similarly, when tested against *Staphylococcus aureus, Pseudomonas aeruginosa, Escherichia coli, and Candida albicans*, the crude aqueous extract, root extract, and organic seed extract of *Cichorium intybus* were found to have antibacterial activity [56]. *Euphorbia* latex has antibacterial, antioxidant, anti-inflammatory, anti-angiogenic, wound healing, cytotoxic, hemostatic, genotoxic/mutagenic, and insecticidal activities [57]. *Cichorium intybus* was studied in vivo and in vitro in mice to determine its anti-allergic activity. The results demonstrated that the aqueous extract quickly suppressed mast cell-mediated allergic responses in a dose-dependent way. In rats, it also prevented the passive cutaneous anaphylactic reaction caused by anti-dinitrophenyl IgE [58].

A wide variety of compounds, including hydrocarbons, terpene acids, organic acids, flavonoids, flavonoid glycosides, and terpene alcohols, were present in the alcoholic extracts of *Pinus wallichiana* [59], and the species showed antifungal qualities, given that the growth of *Microsporium canis* was suppressed by the n-hexane fraction of the ethanolic extract of needles at a minimum inhibitory dose of 25 L/mL [60]. *Plantago major* has shown laxative, anti-inflammatory, antipyretic, astringent, antibacterial, and immune-boosting properties. It has an immunostimulatory effect by increasing lymphocyte proliferation and interferon-gamma production [61]. *Artemisia absinthium* is regarded as a raw material for oil extraction [62]. The most frequently listed compounds are thujyl alcohol esters, α-thujone, β-thujone, camphene, α-cadinene, guaiazulene (Z)-epoxyocimene, (E)-sabinyl acetate, (Z)-chrysantenyl acetate [63]. *A. absinthium* has potent antifungal and antibacterial properties and is used in cosmetics for scalp, face, and hair care. The use of *Artemisia absinthium* in five forms is permitted by CosIng (Cosmetic Ingredient database, a European database that collects data on cosmetic ingredients). There are skincare items, perfumes, and antimicrobial compounds among them. The plant's raw elements are applied in cosmetic items such as shampoos, face serums, masks, essences, tonics, SPF-filtered moisturizing creams, and under-eye patches. These types of cosmetics are intended to protect, cleanse, moisturize, and erase skin blemishes [64].

3. Materials and Methods

3.1. Study Area

The Kashmir valley of J and K (UT) has 10 districts under its jurisdiction, of which Kupwara is one of the isolated border areas, located in the northern part of Kashmir valley, at 34°01′60.00″ N and 74°15′60.00″ E. The total geographic area of the region is 2379 $km^{2,}$ with 368 villages. According to the 2011 census, the population density is 366 persons per $km^{2,}$ and the total population is 870,354 (Figure 9). The scheduled caste and scheduled tribe population of the area is 7.97%. The most common language spoken is Kashmiri, followed by Phari and Gojree. It is the home to many ethnic communities, such as Bakarwals (nomads occupying the high-altitude regions of Kupwara), Kashmiri (living in the main valleys and are in the majority), and Gujjars (nomad group that surrounds the area and migrates seasonally) [33,65].

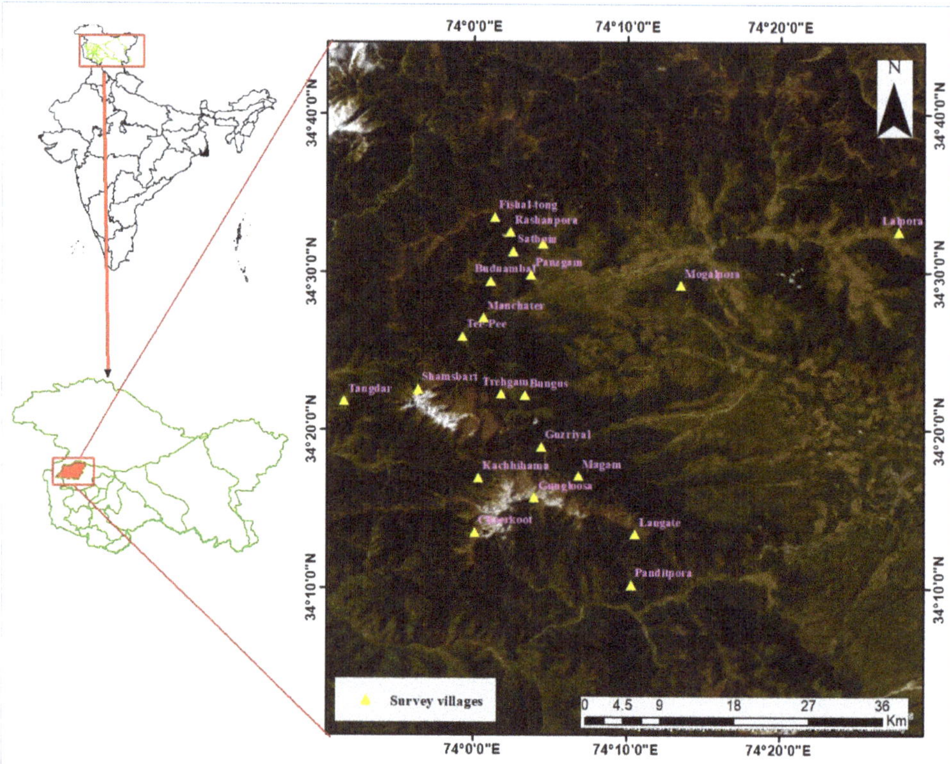

Figure 9. Map of the study area and points showing the surveyed villages.

The typical mountainous vegetation varies from temperate, alpine to alpine meadows, interspersed with unique taxa of shrubs and herbs. *Cedrus deodara* (Roxb. ex D.Don) G. Don, *Pinus willichiana* A.B Jacks., *Abies pindrow* (Royle ex D. Don) Royle are the main tree components of the western belts of the Himalaya. The District Kupwara includes high-altitude mountain ranges like Sadnatop (3200 m), Keran sector (1530 m–3353 m), InchasFalmarg (3048 m), RashanporaDutt (3200 m), Bungus valley (3800 m), Budnamal (2800), Raja-Ram-Ki-ladi (3500 m), while as deep valley includes Tee_Pee (2100 m), Lolab valley (1737 m). During winter, the study area faces severe cold but pleasant weather during summer. The temperature ranges between–4 °C minimum in winter and up to 32 °C maximum in summer [66].

3.2. Field Survey

Field trips (n = 35) were conducted in the spring and summer seasons of 2020–2021 to collect medicinal plants' medicinal value and record the indigenous knowledge associated with their use. Information was gathered through open-ended semi-structured interviews (n = 53) and group discussions (n = 33) following [67]. Based on mutual agreement and in accordance with the Nagoya Protocol, neither the ethnicity of the participants nor the linguistic information provided is released. We adhered to the International Society of Ethnobiology's code of ethics [68]. The information on ethnic medicinal plants used to treat various skin diseases like scabies (Darder), pimples (fefad), skin burns (Dazun), Acne (Dane), etc., and treatment of wounds were collected from local communities, hakims, and tribal people (Gujjar and Bakarwals). The photograph representation from the collection to the application of plants is shown in figure (Figure 10). In this study, skin diseases and wound healing were divided into 15 groups of dermatological ailments, including wound healing, dandruff, skin burns, skin boils, hair tonic, scabies, pimples, skin allergy, skin rashes, insect sting, skin acne, blisters, warts, eczema, and skin irritation. Informants were asked questions in their local language. A quantitative analysis was conducted to evaluate and organize records during the field survey. The obtained specimens were cross-checked with the assistance of a taxonomist at the University of Kashmir, Srinagar (J and K), where the specimens were also deposited. To authenticate plant names, POWO 2019 (http://www.plantsoftheworldonline.org/ accessed on 1 March 2022) was used.

Figure 10. The photograph representation of *Rheum spiciforme* from (**a**) collection to (**d**) application and (**f**) application of latex from *Ficuscarica* on affected scabies.

3.3. *Quantitative Data Analysis*

Overall trends in the total citations and Used Value were expressed illustratively through Generalized Linear Regression Models (GLM) through Graph Pad Prism version 9 (Graph Pad Software, San Diego, CA, USA). Using the package "vegan" [69] and R software 4.0.0 [70], principal component analysis (PCA) was carried out to display the provisioning services and plant part utilized. A ternary plot was created using Origin Pro software. The circlize package in [71] R software 3.6.1 (R Core Team 2020) was used to create chord diagrams displaying the species' percentage contributions to the study parameters (life form, plant parts used, plant preparations, and disease categories) that were applied in the study [72].

3.4. *Use Value (UV)*

The Use Value index is used to calculate the relative value of each medicinal plant species used by the local population. In the current study use value was calculated using the following formula:

$$\mathbf{UV = Ui/N}$$

where **Ui** is the total number of use reports by each informant, and **N** indicates the total number of informants participating in the study [73]. Use Value is high when there are many usage reports for a given medicinal plant species, and use value is low when very few reports are associated with its use.

4. Conclusions

Medicinal plants are the backbone of our traditional healthcare system. In most developing countries, a large part of the population still depends on traditional medicine. The study attempts to highlight potential medicinal plants for the treatment of various dermatological ailments and activities of plants in the Kupwara district of Jammu and Kashmir. During the study period, 64 different medicinal plants belonging to 34 families were reported from the study area. It can be concluded from the current research that the people in the study area have a wealth of traditional knowledge inherited from their ancestors, and the record of this valuable knowledge provides new knowledge for the area. Indigenous people still depend on medicinal plants for primary healthcare but, at the same time, are alarmed by the degradation of wild flora. In order to verify this indigenous knowledge, we recommend that species with high use value (**UV**) be used for feature phytochemical and pharmacological analysis.

Author Contributions: Conceptualization and Study Design S.M.H.; Data collection A.A.K.; Data analysis, S.M.H. and M.W.; Original draft S.M.H., H.O.E., R.C., E.A.M. and A.A.K. thorough review and editing S.M.H., F.A.L., R.W.B., H.O.E., R.C., E.A.M. and A.A.K. All authors have read and agreed to the published version of the manuscript.

Funding: Researchers Supporting Project number (RSP2023R118), King Saud University.

Institutional Review Board Statement: Not applicable.

Informed Consent Statement: Present paper does not contain any individual's personal data; therefore, this section does not apply to our study.

Data Availability Statement: All data have already been included in the manuscript.

Acknowledgments: The authors would like to thank the Researchers Supporting Project number (RSP2023R118), King Saud University, Riyadh, Saudi Arabia.

Conflicts of Interest: All the authors declare no conflict of interest.

References

1. Idu, M.; Erhabor, J.O.; Timothy, O.; Osazuwa, E.S. Ethnodermatological study among the Itsekiri people of Warri South local government area of delta state, Nigeria. *J. Plant Dev. Sci.* **2011**, *3*, 67–73.
2. Vaughn, A.R.; Clark, A.K.; Sivamani, R.K.; Shi, V.Y. Natural oils for skin-barrier repair: Ancient compounds now backed by modern science. *Am. J. Clin. Ddermatol.* **2019**, *19*, 103–117. [CrossRef] [PubMed]
3. Ayantunde, A.A.; Briejer, M.; Hiernaux, P.; Udo, H.M.; Tabo, R. Botanical knowledge and its differentiation by age, gender and ethnicity in Southwestern Niger. *Hum. Ecol.* **2008**, *36*, 881–889. [CrossRef]
4. Kumar, A.; Pandey, V.C.; Singh, A.G.; Tewari, D.D. Traditional uses of medicinal plants for dermatological healthcare management practices by the Tharu tribal community of Uttar Pradesh, India. *Gen. Res. Crop. Evol.* **2013**, *60*, 203–224. [CrossRef]
5. Pieroni, A.; Quave, C.L.; Villanelli, M.L.; Mangino, P.; Sabbatini, G.; Santini, L.; Tomasi, M. Ethnopharmacognostic survey on the natural ingredients used in folk cosmetics, cosmeceuticals and remedies for healing skin diseases in the inland Marches, Central-Eastern Italy. *J. Eethnopharmacol.* **2004**, *91*, 331–344. [CrossRef]
6. Kumar, B.; Vijayakumar, M.; Govindarajan, R.; Pushpangadan, P. Ethnopharmacological approaches to wound healing—Exploring medicinal plants of India. *J. Ethnopharmacol.* **2007**, *114*, 103–113. [CrossRef]
7. Sharma, J.; Gairola, S.; Sharma, Y.P.; Gaur, R.D. Ethnomedicinal plants used to treat skin diseases by Tharu community of district Udham Singh Nagar, Uttarakhand, India. *J. Ethnopharmacol.* **2014**, *158*, 140–206. [CrossRef]
8. Abbasi, A.M.; Khan, M.A.; Ahmad, M.; Zafar, M.; Jahan, S.; Sultana, S. Ethnopharmacological application of medicinal plants to cure skin diseases and in folk cosmetics among the tribal communities of North-West Frontier Province, Pakistan. *J. Ethnopharmacol.* **2010**, *128*, 322–335. [CrossRef]
9. Madikizela, B.; Ndhlala, A.R.; Finnie, J.F.; Van Staden, J. Study of plants from Pondoland used against diarrhoea. *J. Ethnopharmacol.* **2012**, *141*, 61–71. [CrossRef]
10. Gonzalez-Aspajo, G.; Belkhelfa, H.; Haddioui-Hbabi, L.; Bourdy, G.; Deharo, E. Sacha Inchi Oil (*Plukenetiavolubilis* L.), effect on adherence of *Staphylococus aureus* to human skin explant and keratinocytes in vitro. *J. Ethnopharmacol.* **2015**, *171*, 330–334. [CrossRef]
11. Arshad, F.; Waheed, M.; Harun, N.; Fatima, K.; Khan, B.A.; Fatima, K.; Abbas, Z.; Jabeen, S.; Majeed, M. Indigenous farmer's perception about fodder and foraging species of Semi-arid lowlands of Pakistan: A case study of District Kasur, Pakistan. *Taiwania* **2022**, *67*, 510–523.

12. Arshad, F.; Waheed, M.; Iqbal, M.; Fatima, K.; Fatima, K. Ethnobotanical assessment of woody flora of district Kasur (Punjab), Pakistan. *Ethnobot. Res. Appl.* **2020**, *20*, 1–13.
13. Waheed, M.; Haq, S.M.; Arshad, F.; Bussmann, R.W.; Pieroni, A.; Mahmoud, E.A.; Casini, R.; Yessoufou, K.; Elansary, H.O. Traditional Wild Food Plants Gathered by Ethnic Groups Living in Semi-Arid Region of Punjab, Pakistan. *Biology* **2023**, *12*, 269. [CrossRef]
14. Khan, S.M.; Page, S.; Ahmad, H.; Shaheen, H.; Ullah, Z.; Ahmad, M.; Harper, D.M. Medicinal flora and ethnoecological knowledge in the Naran Valley, Western Himalaya, Pakistan. *J. Ethnobiol. Ethnomed.* **2013**, *9*, 4. [CrossRef]
15. Gilca, M.; Tiplica, G.S.; Salavastru, C.M. Traditional and ethnobotanical dermatology practices in Romania and other Eastern European countries. *Clin. Dermatol.* **2018**, *36*, 338–352. [CrossRef]
16. Saikia, A.P.; Ryakala, V.K.; Sharma, P.; Goswami, P.; Bora, U. Ethnobotany of medicinal plants used by Assamese people for various skin ailments and cosmetics. *J. Ethnopharmacol.* **2006**, *106*, 149–157. [CrossRef]
17. Ahmad, L.; Semotiuk, A.; Zafar, M.; Ahmad, M.; Sultana, S.; Liu, Q.R.; Yaseen, G. Ethnopharmacological documentation of medicinal plants used for hypertension among the local communities of DIR Lower, Pakistan. *J. Ethnopharmacol.* **2015**, *175*, 138–146. [CrossRef]
18. Khoja, A.A.; Andrabi, S.A.H.; Mir, R.A. Traditional medicine in the treatment of gastrointestinal diseases in northern part of Kashmir Himalayas. *Ethnobot. Res. Appl.* **2022**, *23*, 1–17.
19. Farooq, S.; Gulzar, A.; Haq, S.M.; Hassan, M.; Bussmann, R.W. Millennial traditions-Exploration of ethno-medicinal knowledge among different ethnic groups of Northern part Kashmir Himalaya, India. *Ethnobot. Res. Appl.* **2022**, *23*, 1–18. [CrossRef]
20. Haq, S.M.; Yaqoob, U.; Majeed, M.; Amjad, M.; Hassan, M.; Ahmad, R.; la Nuez Morales-de, A. Quantitative ethnoveterinary study on plant resource utilization by indigenous communities in high-altitude regions. *Front. Vet. Sci.* **2022**, *9*, 1–14. [CrossRef]
21. Khoja, A.A.; Andrabi, S.A.H.; Mir, R.A.; Bussmann, R.W. Ethnobiological uses of plant species among three ethnic communities in the administrative (Kupwara) of Jammu and Kashmir-India: A cross cultural Analysis. *Ethnobot. Res. Appl.* **2022**, *24*, 1–22. [CrossRef]
22. Tabassum, N.; Hamdani, M. Plants used to treat skin diseases. *Pharmacogrevi* **2014**, *8*, 52. [CrossRef] [PubMed]
23. Martínez, G.J.; Barboza, G.E. Natural pharmacopoeia used in traditional Toba medicine for the treatment of parasitosis and skin disorders (Central Chaco, Argentina). *J. Ethnopharmacol.* **2010**, *132*, 86–100. [CrossRef] [PubMed]
24. Namukobe, J.; Sekandi, P.; Byamukama, R.; Murungi, M.; Nambooze, J.; Ekyibetenga, Y.; Nagawa, C.B.; Asiimwe, S. Antibacterial, antioxidant, and sun protection potential of selected ethno medicinal plants used for skin infections in Uganda. *Trop Med. Health* **2021**, *15*, 49. [CrossRef] [PubMed]
25. Mandal, U.; Mallick, S.K.; Mahalik, G. Ethnomedicinal plants used for the treatment and healing of skin diseases in Odisha, India: A review. *Shodh Sanchar Bull.* **2020**, *10*, 100–108.
26. Shrestha, P.M.; Dhillion, S.S. Medicinal plant diversity and use in the highlands of Dolakha district, Nepal. *J. Eethnopharmacol.* **2003**, *86*, 81–96. [CrossRef]
27. Kayani, S.; Ahmad, M.; Zafar, M.; Sultana, S.; Khan, M.P.Z.; Ashraf, M.A.; Yaseen, G. Ethnobotanical uses of medicinal plants for respiratory disorders among the inhabitants of Gallies–Abbottabad, Northern Pakistan. *J. Ethnopharmacol.* **2014**, *156*, 47–60. [CrossRef]
28. Heinrich, M.; Edwards, S.; Moerman, D.E.; Leonti, M. Ethnopharmacological field studies: A critical assessment of their conceptual basis and methods. *J. Eethnopharmacol.* **2009**, *124*, 1–17. [CrossRef]
29. Giday, M.; Asfaw, Z.; Woldu, Z. Medicinal plants of the Meinit ethnic group of Ethiopia: An ethnobotanical study. *J. Ethnopharmacol.* **2009**, *124*, 513–521. [CrossRef]
30. Tamang, S.; Singh, A.; Bussmann, R.W.; Shukla, V.; Nautiyal, M.C. Ethno-medicinal plants of tribal people: A case study in Pakyong subdivision of East Sikkim, India. *Acta Ecol. Sin.* **2021**, *43*, 34–46. [CrossRef]
31. Mesfin, F.; Demissew, S.; Teklehaymanot, T. An ethnobotanical study of medicinal plants in Wonago Woreda, SNNPR, Ethiopia. *J. Ethnobiol. Ethnomed.* **2009**, *5*, 28. [CrossRef]
32. Ghorbani, A. Studies on pharmaceutical ethnobotany in the region of Turkmen Sahra, north of Iran:(Part 1): General results. *J. Ethnopharmacol.* **2005**, *102*, 58–68. [CrossRef]
33. Abdelouahab, N.; Heard, C.M. Dermal and transcutaneous delivery of the major glycoside constituents of *Harpagophytum procumbens* (Devil's Claw) in vitro. *Planta Med.* **2008**, *74*, 527–531. [CrossRef]
34. Kadhirve, K.; Ramya, S.; Sudha, T.S.; Ravi, A.V.; Rajasekaran, C.; Selvi, R.V.; Jayakumararaj, R. Ethnomedicinal survey on plants used by tribals in Chitteri Hills. *Environ. We. Int. J. Sci. Tech.* **2010**, *5*, 35–46.
35. Pandey, M.M.; Rastogi, S.; Rawat, A.K.S. *Saussurea costus*: Botanical, chemical and pharmacological review of an ayurvedic medicinal plant. *J. Ethnopharmacol.* **2007**, *110*, 379–390. [CrossRef]
36. Adhikari, M.; Thapa, R.; Kunwar, R.M.; Devkota, H.P.; Poudel, P. Ethnomedicinal uses of plant resources in the Machhapuchchhre Rural Municipality of Kaski District, Nepal. *Medicines* **2019**, *6*, 69. [CrossRef]
37. Yousuf, S.; Haq, S.M.; Rasool, A.; Zulfajri, M.; Hanafiah, M.M.; Nafees, H.; Mahboob, M. Evaluation of antidepressant activity of methanolic and hydroalcoholic extracts of Acorus calamus L. rhizome through tail suspension test and forced swimming test of mice. *J. Tradit. Chin. Med. Sci.* **2020**, *7*, 301–307. [CrossRef]
38. Rathore, S.; Debnath, P.; Kumar, R. KuthSaussureacostus (Falc.) Lipsch.: A critically endangered medicinal plant from Himalaya. *J. Appl. Res. Med. Aromat. Plants* **2021**, *20*, 177.

39. Uprety, Y.; Asselin, H.; Boon, E.K.; Yadav, S.; Shrestha, K.K. Indigenous use and bio-efficacy of medicinal plants in the Rasuwa District, Central Nepal. *J. Ethnoboil. Ethnomed.* **2010**, *6*, 3. [CrossRef]
40. Hassan, M.; Abdullah, A.; Haq, S.M.; Yaqoob, U.; Bussmann, R.W.; Waheed, M. Cross-ethnic use of ethnoveterinary medicine in the Kashmir Himalaya-A Northwestern Himalayan region. *Acta Ecol. Sin.* 2022; in press. [CrossRef]
41. Haq, S.M.; Hamid, M.; Lone, F.A.; Singh, B. Himalayan hotspot with Alien Weeds: A case study of biological spectrum, phenology, and diversity of weedy plants of high-altitude mountains in District Kupwara of J&K Himalaya, India. *Proc. Natl. Acad. Sci. India Sect. B Biol. Sci.* **2021**, *91*, 139–152.
42. Papp, N.; Czégényi, D.; Tóth, M.; Dénes, T.; Bartha, S.G.; Csepregi, R.; Filep, R. Ethnomedicine survey on folk dermatology in Transylvania, Romania. *Clin. Dermatol.* **2020**, *40*, 651–664. [CrossRef] [PubMed]
43. Aadil, A.; Andrabi, S.A.H. Wild edible plants and fungi used by locals in Kupwara district of Jammu and Kashmir, India. *Pleione* **2021**, *15*, 179–189.
44. Haq, S.M.; Calixto, E.S.; Yaqoob, U.; Ahmed, R.; Mahmoud, A.H.; Bussmann, R.W.; Abbasi, A.M. Traditional usage of wild fauna among the local inhabitants of Ladakh, Trans-Himalayan Region. *Animals* **2020**, *10*, 2317. [CrossRef] [PubMed]
45. Cakilcioglu, U.; Turkoglu, I. An ethnobotanical survey of medicinal plants in Sivrice (Elazığ-Turkey). *J. Ethnopharmacol.* **2010**, *132*, 165–175. [CrossRef]
46. Ishtiyak, P.; Hussain, S.A. Traditional use of medicinal plants among tribal communities of Bangus Valley, Kashmir Himalaya, India. *Stud. Ethno-Med.* **2017**, *11*, 318–331. [CrossRef]
47. Shah, S.A.; Shah, N.A.; Ullah, S.; Alam, M.M.; Badshah, H.; Ullah, S.; Mumtaz, A.S. Documenting the indigenous knowledge on medicinal flora from communities residing near Swat River (Suvastu) and in high mountainous areas in Swat-Pakistan. *J. Ethnopharmacol.* **2016**, *182*, 67–79. [CrossRef]
48. Zari, S.T.; Zari, T.A. A review of four common medicinal plants used to treat eczema. *J. Med. Plants Res.* **2021**, *9*, 702–711.
49. Staub, P.O.; Geck, M.S.; Weckerle, C.S.; Casu, L.; Leonti, M. Classifying diseases and remedies in ethnomedicine and ethnopharmacology. *J. Ethnopharmacol.* **2015**, *174*, 514–519. [CrossRef]
50. Zaid, N.A.; Sekar, M.; Bonam, S.R.; Gan, S.H.; Lum, P.T.; Begum, M.Y.; Fuloria, S. Promising natural products in new drug design, development, and therapy for skin disorders: An overview of scientific evidence and understanding their mechanism of action. *Drug Des. Dev. Ther.* **2022**, *16*, 23–66. [CrossRef]
51. Kadir, M.F.; Sayeed, M.S.B.; Setu, N.I.; Mostafa, A.; Mia, M.M.K. Ethnopharmacological survey of medicinal plants used by traditional health practitioners in Thanchi, Bandarban Hill Tracts, Bangladesh. *J. Ethnopharmacol.* **2014**, *155*, 495–508. [CrossRef]
52. Muhammad, A.; Shaheen, B.; Abdul, L.K.; Abdul, M.T.; In-Jung, L. Medicinal plants and their uses in selected temperate zones of Pakistani Hindukush-Himalaya. *J. Med. Plants Res.* **2012**, *6*, 4113–4127. [CrossRef]
53. Badgujar, S.B.; Patel, V.V.; Bandivdekar, A.H.; Mahajan, R.T. Traditional uses, phytochemistry and pharmacology of Ficuscarica: A review. *Pharm. Biol.* **2014**, *52*, 1487–1503. [CrossRef]
54. Amol, P.P.; Vikas, V.P.; Vijay, R.P.; Rajesh, Y.C. Anthelmintic and preliminary phytochemical screening of leaves of *Ficus carica* Linn against intestinal helminthiasis. *Int. J. Res. Ayurveda Pharm. (IJRAP)* **2010**, *1*, 601–605.
55. Ali, I.; Naz, R.; Khan, W.N.; Gul, R.; Choudhary, M.I. Biological screening of different root extracts of *Euphorbia wallichii*. *Pak. J. Bot.* **2009**, *41*, 1737–1741.
56. Mares, D.; Romagnoli, C.; Tosi, B.; Andreotti, E.; Chillemi, G.; Poli, F. Chicory extracts from *Cichorium intybus* L. as potential antifungals. *Mycopathologia* **2005**, *160*, 85–91. [CrossRef]
57. Choudhary, S.; Kaurav, H.; Chaudhary, G. Kasanibeej (*Cichorium intybus*): Ayurvedic view, folk view, phytochemistry and modern therapeutic uses. *Int. J. Res. Appl. Sci. Biotechnol.* **2021**, *8*, 14–125.
58. Benjamaa, R.; Moujanni, A.; Kaushik, N.; Choi, E.H.; Essamadi, A.K.; Kaushik, N.K. *Euphorbia* species latex: A comprehensive review on phytochemistry and biological activities. *Front. Plant Sci.* **2022**, *13*, 1008881. [CrossRef]
59. Sinha, D. A review on ethnobotanical, phytochemical and pharmacological profile of *Pinus wallichiana* AB Jacks. *Pharmacogn. J.* **2019**, *11*, 624–631. [CrossRef]
60. Dambolena, J.S.; Gallucci, M.N.; Luna, A.; Gonzalez, S.B.; Guerra, P.E.; Zunino, M.P. Composition, antifungal and antifumonisin activity of *Pinus wallichiana*, *Pinus monticola* and *Pinus strobus* essential oils from Patagonia Argentina. *J. Essent. Oil Bear. Plants* **2016**, *19*, 1769–1775. [CrossRef]
61. Chiang, L.C.; Chiang, W.; Chang, M.Y.; Lin, C.C. In vitro cytotoxic, antiviral and immunomodulatory effects of Plantago major and Plantago asiatica. *Am. J. Chin. Med.* **2003**, *31*, 225–234. [CrossRef]
62. Ahamad, J.; Mir, S.R.; Amin, S.A. Pharmacognostic review on *Artemisia absinthium*. *Int. Res. J. Pharm.* **2019**, *10*, 25–31. [CrossRef]
63. Beigh, Y.A.; Ganai, A.M. Potential of wormwood (*Artemisia absinthium* Linn.) herb for use as additive in livestock feeding: A review. *Pharma Innov.* **2017**, *6*, 176–187.
64. Szopa, A.; Pajor, J.; Klin, P.; Rzepiela, A.; Elansary, H.O.; Al-Mana, F.A.; Mattar, M.A.; Ekiert, H. *Artemisia absinthium* L.—Importance in the History of Medicine, the Latest Advances in Phytochemistry and Therapeutical, Cosmetological and Culinary Uses. *Plants* **2020**, *9*, 1063. [CrossRef] [PubMed]
65. Abdullah, A.; Andrabi, S.A.H. An approach to the study of traditional medicinal plants used by locals of block Kralpora Kupwara Jammu and Kashmir India. *Int. J. Bot. Stud.* **2021**, *6*, 1433–1448.
66. Haq, S.M.; Hassan, M.; Bussmann, R.W.; Calixto, E.S.; Rahman, I.U.; Sakhi, S.; Ali, N. A Cross-Cultural Analysis of Plant Resources among Five Ethnic Groups in the Western Himalayan Region of Jammu and Kashmir. *Biology* **2022**, *11*, 491. [CrossRef]

67. Khoja, A.A.; Haq, S.M.; Majeed, M.; Hassan, M.; Waheed, M.; Yaqoob, U.; Zaman, W. Diversity, Ecological and traditional knowledge of pteridophytes in the western Himalayas. *Diversity* **2022**, *14*, 28. [CrossRef]
68. Policepatel, S.S.; Manikrao, V.G. Ethnomedicinal plants used in the treatment of skin diseases in Hyderabad Karnataka region, Karnataka, Indian. *Asian Pac. J. Trop. Biomed.* **2013**, *3*, 882–886. [CrossRef]
69. Haq, S.M.; Hassan, M.; Jan, H.A.; Al-Ghamdi, A.A.; Ahmad, K.; Abbasi, A.M. Traditions for Future Cross-National Food Security—Food and Foraging Practices among Different Native Communities in the Western Himalayas. *Biology* **2022**, *11*, 455. [CrossRef]
70. Haq, S.M.; Yaqoob, U.; Calixto, E.S.; Rahman, I.U.; Hashem, A.; Abd_Allah, E.F.; Ijaz, F. Plant resources utilization among different ethnic groups of Ladakh in Trans-Himalayan Region. *Biology* **2021**, *10*, 827. [CrossRef]
71. Waheed, M.; Haq, S.M.; Arshad, F.; Bussmann, R.W.; Iqbal, M.; Bukhari, N.A.; Hatamleh, A.A. Grasses in Semi-Arid Lowlands—Community Composition and Spatial Dynamics with Special Regard to the Influence of Edaphic Factors. *Sustainability* **2022**, *14*, 14964. [CrossRef]
72. Haq, S.M.; Calixto, E.S.; Rashid, I.; Srivastava, G.; Khuroo, A.A. Tree diversity, distribution and regeneration in major forest types among an extensive elevational gradient in Indian Himalaya: Implications for sustainable forest management. *For. Ecol. Manag.* **2022**, *506*, 119968. [CrossRef]
73. Phillips, O.; Gentry, A.H. The useful plants of Tambopata, Peru: I. Statistical hypotheses tests with a new quantitative technique. *Econ. Bot.* **1993**, *47*, 15–32. [CrossRef]

plants

MDPI

Article

Green Synthesis of Zinc Oxide Nanoparticles Using *Viscum album* Extracts: Unveiling Bioactive Compounds, Antibacterial Potential, and Antioxidant Activities

Waheeda Mushtaq [1], Muhammad Ishtiaq [1,*], Mehwish Maqbool [1], Muhammad Waqas Mazhar [1], Ryan Casini [2], Ahmed M. Abd-ElGawad [3] and Hosam O. Elansary [3,*]

1 Department of Botany, Mirpur University of Science & Technology (MUST), Mirpur 10250, Pakistan
2 School of Public Health, University of California, Berkeley, 2121 Berkeley Way, Berkeley, CA 94704, USA
3 Department of Plant Production, College of Food & Agriculture Sciences, King Saud University, P.O. Box 2460, Riyadh 11451, Saudi Arabia
* Correspondence: drishtiaq.bot@must.edu.pk (M.I.); helansary@ksu.edu.sa (H.O.E.)

Abstract: The current study explored the antioxidant and antibacterial capabilities of zinc oxide nanoparticles (ZnONPs) synthetized using methanolic leaf extracts of the medicinal herb *Viscum album*. Through TEM investigation and UV–Vis analysis, which peaked at 406 nm, the synthesis of ZnONPs was verified. TEM analyses showed that the synthesized ZnONPs had a size distribution with an average of 13.5 nm and a quasi-spherical shape. Forty-four phytoconstituents were found in the methanolic leaf extracts of *V. album*. Additionally, a comparison of the antibacterial effectiveness and antioxidant capacity of aqueous and methanolic extracts of wild-grown *V. album* phytomedicine and green-manufactured ZnONPs was conducted. The green-generated ZnONPs were examined against *Escherichia coli*, *Staphylococcus aureus*, and *Pseudomonas aeruginosa* and shown to have superior antibacterial activity by 22%, 66%, and 44%, respectively, as compared to wild herbal medicinal extracts. Since the ZnONPs' aqueous extracts had higher concentrations of DNA gyrase-B inhibitory components, they were shown to be more effective in limiting bacterial growth. In contrast to the percentages of 49% and 57% for a wild plant extract, the aqueous- and methanolic-extract-mediated green ZnONPs, with a 100 g/mL concentration, showed 94% and 98% scavenging capacity for DPPH free radicals, respectively. However, methanolic extracts were more effective than aqueous extracts in terms of the antioxidant analyses. This study establishes that greenly produced ZnONPs have the potential to be used in nanomedicine to treat bacteria that are resistant to a variety of drugs, as well as those with reactive oxygen species toxicity.

Keywords: green synthesis; zinc oxide nanoparticles; *Viscum album*; antimicrobial activities; phytomedicines

1. Introduction

Since the beginning of time, humankind has relied on plants to meet their basic needs. The main uses of plants are food and medicine. Botanic or herbal pharmaceuticals made from plants are thought to be more affordable, less harmful, and effective in treating a variety of disorders. Multi-drug resistance (MDR) strains of bacteria are becoming increasingly difficult to treat or minimize due to evolutionary trends and resistance built into microorganism genomes. According to recent literature, the increased resistance that different microorganisms have developed to various medications, including antibiotics, is posing a serious threat to world health [1]. The current heightened resistance that microorganisms used against many multiple drugs are displaying to a variety of antibiotics is becoming a critical global health concern. The current situation makes it more important than ever to create innovative antibiotics and antimicrobial medications [1]. One method uses botanicals, which are bioactive substances derived from plants. Researchers from all over the world have documented and continue to document the roles of botanicals as antimicrobial drugs due to their effects on microbes, such as ATP loss, protein synthesis inhibition, pH disturbance, intracytoplasmic damage, and DNA damage [2]. A novel addition to these phytomedicines are nanomedicines consisting of nanomaterials, often derived from the medicinal plants, which are proving to be promising medications in biomedical research [3].

Free radicals in the body create oxidative stress, which is thought to be the primary factor in a number of diseases and conditions, including some types of tumors, ageing, and inflammatory conditions such as glomerulonephritis, vasculitis, and arthritis [4]. The body produces free radicals as a by-product of ordinary human metabolism. They can also develop as a result of exposure to environmental toxins including tobacco smoke and industrial air pollutants. In order to promote free-radical-scavenging activity and avoid cellular damage, the human body needs antioxidants to provide electrons to free radicals. In addition to being edible, cultural medicinal plants are a significant source of antioxidants, which helps them to prevent a number of ailments [5]. Therefore, there is certain need to develop promising antioxidant medications.

The environmentally friendly and cost-effective green synthesis method uses plant extracts to create nanoparticles. The synthesis fluid contains certain phytochemicals and botanicals that may improve the biocompatibility and immunomodulatory abilities of the nanoparticles. Additionally, the stability provided by the green manufacturing technique enhances the nanoparticles' medicinal capabilities [6]. Nanotechnology is progressing quickly in the present day due to its numerous applications in biomedical research and agriculture [7]. Zinc oxide nanoparticles (ZnONPs), in particular, have applications in agriculture, sunscreens, the cosmetics industry, and food packaging [8]. Due to their capacity to penetrate the cell envelops of microbes, ZnONPs have also been found to possess antibacterial capabilities [9]. The green synthesis of nanomaterials is a potentially environmentally friendly and cost-effective way to mitigate environmental pollution. Several studies have reported the green synthesis of nanomaterials. For instance, one study [10] focused on transforming industrial and organic waste into a composite material that can effectively remove organic pollutants. The researchers synthesized the composite material by incorporating titanium-doped activated carbon and cellulose nanofibers into a paper matrix. They characterized the material using various techniques, such as SEM, XRD, and BET, and evaluated its performance in removing organic pollutants from water. Another study [11] explored the possibility of using waste materials for the green synthesis of TiO_2 nanoparticles as a photocatalytic additive for indoor lime-based finishings. The researchers synthesized the nanoparticles using a sol–gel method with waste materials and evaluated their photocatalytic activity by incorporating them into lime-based finishings and testing their methylene blue degradation capability.

Today, there exists approximately 4500 species of angiosperm parasitic plants, yet they only account for 1% of all flowering plants [12]. *Viscum album* L. belongs to the parasitic plant family Santalaceae. In Pakistan, the usage of *V. album* for treating conditions including nausea, vomiting, cancer, diarrhea, diabetes, and hypertension is significant. It is also regarded as a crucial component of many herbal preparations, and their popularity is continuously rising [13]. Additionally, it is frequently utilized as holiday décor. It is believed to possess magical properties and is beneficial for treating a variety of urinogenital issues, fractured limbs, skin conditions, and chest conditions [14].

V. album is a widely distributed plant species in the AJK area, and its abundance and accessibility make it a convenient and cost-effective source for obtaining leaf extracts for nanoparticle synthesis. Furthermore, green synthesis aims to minimize the use of hazardous chemicals and energy-intensive processes. *V. album* leaf extracts offer a sustainable and environmentally friendly alternative to traditional chemical methods for synthesizing ZnONPs. Additionally, *V. album* is known to contain a rich diversity of bioactive compounds. such as phenolic compounds, flavonoids, alkaloids, and polysaccharides. These compounds possess various therapeutic properties, including antibacterial and antioxidant activities. By using *V. album* leaf extracts, a wide range of these bioactive compounds can be incorporated into the green synthesis of ZnONPs, potentially enhancing their therapeutic potential. Moreover, the combination of different bioactive compounds present in the extracts may exhibit enhanced antibacterial and antioxidant activities compared to individual compounds. Utilizing the whole extract in the synthesis process allows for the preservation of these synergistic effects, potentially resulting in nanoparticles with improved biological properties.

Little work has been reported in the literature on the use of *V. album* in the green synthesis of ZnONPs, but the literature reports a few studies utilizing *V. album* for a green synthesis approach. For instance [15], researchers used *V. album* leaf extract for the green synthesis of ZnONPs. They characterized the nanoparticles using various techniques, such as UV–Vis spectroscopy, XRD, SEM, and TEM. They also evaluated the antimicrobial activity of the synthesized nanoparticles against various bacterial strains. The study found that the *V. album* leaf extract was an effective reducing and capping agent for the synthesis of ZnONPs. The synthesized nanoparticles were spherical and had a mean particle size of around 20 nm. The antimicrobial activity assessment showed that the ZnONPs had a significant inhibitory effect on the growth of the tested bacterial strains. In another study [16], the researchers used *V. album* extract as a reducing and capping agent to synthesize ZnONPs. They characterized the nanoparticles using various techniques and evaluated the cytotoxicity of the nanoparticles using the MTT assay on the MCF-7 breast cancer cell line. The study concluded that *V. album* extract was an effective green reducing agent for the synthesis of ZnONPs. The synthesized nanoparticles were found to be crystalline and had a hexagonal wurtzite structure. The cytotoxicity assessment showed that the ZnONPs had a dose-dependent cytotoxic effect on the MCF-7 cell line. The above studies provide convincing evidence that plant extracts of *V. album* might be used in the green synthesis of ZnONPs.

The current work is based on the GCMS analysis of some volatile compounds found in the *V. album* leaf extracts. The aim of the study was to investigate and compare the chemical compositions and antibacterial and antioxidant properties of *V. album* leaf extracts with green-synthesized zinc oxide nanoparticles. Furthermore, the study aimed to assess the potential of green-synthesized zinc oxide nanoparticles as novel antimicrobial and antioxidant agents. The objectives of the study included an analysis of the chemical composition of *V. album* leaf extracts using GCMS, aiming to identify the major compounds present in the extracts and to determine the antibacterial activity of *V. album* leaf extracts and green-synthesized zinc oxide nanoparticles against a range of pathogenic bacteria. Furthermore, the antioxidant activity of *V. album* leaf extracts and green-synthesized zinc oxide nanoparticles was evaluated using in vitro assays, and the antibacterial and antioxidant activities of *V. album* leaf extracts with green-synthesized zinc oxide nanoparticles were compared. The authors seek to provide an insight into the potential use of *V. album* leaf extracts and green-synthesized zinc oxide nanoparticles in the development of new therapeutic agents for the treatment of infectious diseases and oxidative-stress-related disorders. The leaf extracts of *V. album* might serve as a promising medium for green nano-zinc oxide synthesis. The roles of ZnONPs and methanolic leaf extracts of *V. album* are compared for their antimicrobial efficacy and antioxidant activities. We believe that green-synthesized ZnONPs might serve as better antimicrobial agents and might be useful in addressing the challenge of antimicrobial resistance (AMR). The effectiveness of the plant as a potential source of green medicines against microbes is shown through this investigation. The comprehensive research aim is to understand the green synthesis of ZnONPs using *V. album* extracts and explore their antibacterial and antioxidant properties. This study aims to contribute to the development of sustainable and eco-friendly nanomaterials with potential biomedical applications.

2. Results

2.1. GC-MS Analysis of Methanolic Extracts

The identification was performed by comparing the linear retention index to the hydrocarbons and available MS literature. A detailed summary of the compounds identified is presented in Table 1.

Table 1. Phytochemical constituents of *Viscum album* plant methanolic extract derived via GC-MS.

S.NO.	Systematic Name	Molecular Formula	CAS #	Mol. Mass g/mol	Retention Time Min.	Calculated Retention Indices	Documented Retention Indices
1	*n*-Tetradecane	$C_{14}H_{30}$	629-59	198.3	4.940	837	239.35
2	*n*-Decane	$CH_3(CH_2)_8CH_3$	124-18-5	142.2	4.940	837	163.22
3	2-Methylundecane	$C_{12}H_{26}$	7045-71-8	170.3	4.940	837	1165.1
4	5-Ethyl-2-methyloctane	$C_{11}H_{24}$	62016-18-6	156.31	5.368	858	-
5	*n*-Hexadecane	$C_{16}H_{34}$	544-76-3	226.4	5.368	858	272.02
6	2,3-Dimethylundecane	$C_{13}H_{28}$	17312-77-5	184.36	5.368	858	-
7	*n*-Dodecane	$C_{12}H_{26}$	112-40-3	170.3	6.839	927	202.83
8	*n*-Tridecane	$CH_3(CH_2)_{11}CH_3$	629-50-5	184.37	6.839	927	221.55
9	2,6-Dimethylundecane	$C_{13}H_{28}$	17301-23-4	184.36	7.069	937	1210
10	Benzene, 1,3-bis(1,1-dimethylethyl)-	$C_{14}H_{22}$	1014-60-4	190.32	7.920	974	-
11	*n*-Pentadecane	$C_{15}H_{32}$	629-62-9	212.41	8.481	999	256.12
12	*n*-Eicosane	$C_{20}H_{42}$	112-95-8	282.5	8.481	999	333.60
13	2-Methyloctane	C_9H_{20}	3221-61-2	128.2	9.743	1054	864.90
14	1-Iodo-2-methylnonane	$C_{10}H_{21}I$	101-47-9	268.17	9.743	1054	
15	*n*-Heptadecane	$C_{17}H_{36}$	629-78-7	240.47	11.557	1135	282.99
16	*n*-Nonadecane	$C_{19}H_{40}$	629-92-5	268.52	11.851	1148	
17	2-Methyltricosane	$C_{24}H_{50}$	1928-30-9	338.66	11.851	1148	2365
18	1-Iodotetradecane	$C_{14}H_{29}I$	19218-94-1	324.28	11.851	1148	
19	2-Methyldodecane	$C_{13}H_{28}$	1560-97-0	184.36	12.610	1183	1261
20	2,6,10,14-Tetramethylheptadecane	$C_{21}H_{44}$	018344-37-1	296.574	12.701	1187	
21	5-Propyldecane	$C_{13}H_{28}$	17312-62-8	184.36	12.701	1187	
22	1-Iodododecane	$C_{12}H_{25}I$	4292-19-7	296.23	13.102	1206	
23	*n*-Octacosane	$C_{28}H_{58}$	630-02-4	394.77	13.177	1210	442.42
24	Methoxyacetic acid, 2-tetradecyl ester	$C_{17}H_{34}O_3$	282-04-8	286.4	13.316	1216	
25	10-Methylnonadecane	$C_{20}H_{42}$	56862-62-5	282.55	13.316	1216	
26	*n*-Heneicosane	$C_{21}H_{44}$	629-94-7	296.58	13.349	1218	348.41
27	*n*-Tetradecane	$C_{14}H_{30}$	629-59-4	198.39	13.349	1218	239.35
28	2,6-Di-tert-butyl-4-methylphenol	$C_{15}H_{24}O$	128-37-0	220.35	13.488	1225	1514.
29	2,4-Di-tert-butylphenol	$C_{14}H_{22}O$	96-76-4	206.32	13.536	1227	1512
30	Methoxyacetic acid, 2-tridecyl ester	$C_{16}H_{32}O_3$	282-04-5	272.42	13.664	1233	
31	Octadecane-1-sulfonyl chloride	$C_{18}H_{37}ClO_2S$	342-70-4	353.0	13.664	1233	
32	Benzene, 1-propylheptyl	$C_{16}H_{26}$	4537-12-6	218.38	14.162	1257	
33	Benzene, 1-ethyloctyl	$C_{16}H_{26}$	4621-36-7	218.37	14.509	1274	
34	Benzene, 1-pentylhexyl	$C_{17}H_{28}$	4537-14-8	232.4	15.632	1330	
35	Benzene, 1-butylheptyl	$C_{17}H_{28}$	4537-15-9	232.4	15.691	1333	
36	Benzene, 1-propyloctyl	$C_{17}H_{28}$	4536-86-1	232.4	15.878	1343	
37	Tetradecane, 4-ethyl	$C_{16}H_{34}$	55045-14-2	226.44	16.750	1388	
38	Dodecane, 2-methyl-6-propyl	$C_{16}H_{34}$	55045-08-4	226.44	16.750	1388	
39	Docosane, 11-butyl	$C_{26}H_{54}$	013475-76-8	366.71	16.814	1391	
40	Undecane, 2-phenyl	$C_{17}H_{28}$	4536-88-3	232.41	16.911	1396	
41	Dodecane, 6-phenyl	$C_{18}H_{30}$	2719-62-2	246.4	17.290	1417	
42	Dodecane, 5-phenyl	$C_{18}H_{30}$	2719-63-3	246.4	17.371	1421	
43	Tetradecane, 2-bromo	$C_{14}H_{29}Br$	74036-95-6	277.28	17.510	1429	
44	Benzene, 1-ethyldecyl	$C_{18}H_{30}$	2400-00-2	246.4	17.964	1454	

2.2. Total Phenolics and Free Proline Contents

The assayed total phenolics and leaf free proline contents are presented in Table 2. The presence of phenolics and free proline contents in both aqueous and methanolic extracts varied slightly. The proline contents were assayed to demonstrate that the source plant was not under stress. Usually, elevated levels of free proline indicate that a plant is exposed to drought stress or some oxidative damage. Our analysis shows that the plant is a rich source of botanicals.

Table 2. Free proline and total phenolics assayed in the leaves of *V. album*.

Sample	Leaf Free Proline ($mg \cdot g^{-1} \cdot FW$)	Total Phenolics (mg GAE/g Extract)
Aqueous extracts of *V. album*	10.23 ± 0.12	6.43 ± 0.56
Methanolic extracts of *V. album*	9.28 ± 0.11	8.22 ± 0.43

2.3. Green Synthesis and Characterization

The XRD patterns of ZnONPs synthesized from different origins, such as different solvent extracts, differ due to differences in the synthesis conditions and resulting particle sizes and crystal structures. As shown in Figure 1A, the methanolic extracts produced ZnONPs with smaller particle sizes, which resulted in broader diffraction peaks and changes in the peak positions and intensities compared to those obtained from larger particles synthesized using aqueous extracts, as shown in Figure 1B. The (100), (002), (101), (102), (110), (103), (112), and (201) reflection lines of the ZnONPs were identified at 2θ angles of 31.83, 34.48, 36.32, 47.60, 56.66, 62.97, 68.01, and 69.09 based on the peak positions observed in the XRD patterns. The Debye–Scherrer equation was employed to determine the size of the nanoparticles, as mentioned below.

$$D = K \times \lambda / \beta \times \cos(\theta)$$

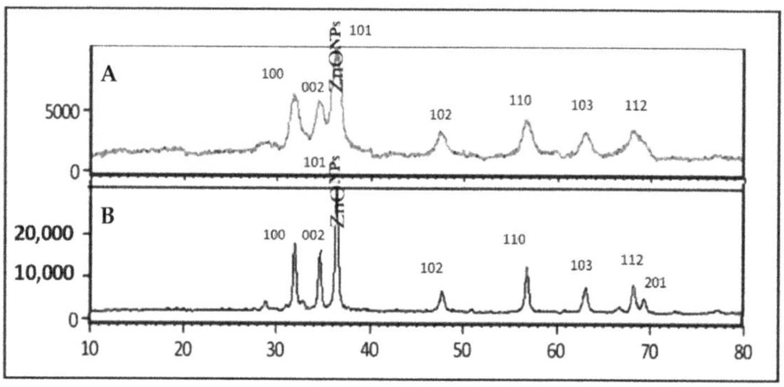

Figure 1. XRD patterns of ZnONPs. (**A**) XRD pattern of ZnONPs synthesized using methanolic extracts. (**B**) XRD pattern of ZnONPs synthesized using aqueous extracts. XRD patterns were recorded in a range from 10° to 80° 2θ.

Above, D refers to the average crystallite size of the nanoparticles. It represents the characteristic dimension of the nanoparticles being studied. K represents a dimensionless shape factor, which depends on the geometry of the sample and the X-ray diffraction setup. It is typically a constant specific to the experimental setup (in our case, we used 0.9 as an approximation). Λ denotes the wavelength of the X-ray used in the experiment (0.1542 nm). It is usually expressed in units of length, such as meters or angstroms. Β represents the full width at half maximum (FWHM) of the diffraction peak in the X-ray diffraction pattern. It is a measure of the broadening of the peaks due to various factors, such as the size and strain of the crystalline domains. θ denotes the Bragg angle, which is the angle between the incident X-ray beam and the crystal lattice planes in the sample. It is typically measured in degrees. By plugging in the appropriate values for these variables, the Debye–Scherrer equation allows for the determination of the average crystallite size of the nanoparticles based on their X-ray diffraction patterns. The average crystalline size of the formed nanoparticles was estimated using Scherrer's formula, resulting in values of 39.82 nm and 8.14 nm for the ZnONPs green-synthesized using aqueous and methanolic extracts of *V. album*, respectively.

The methanolic-extract-mediated ZnONPs' UV–visible peak was noted at 406 nm, indicating that the particles were synthesized (Figure 2A). TEM pictures of these ZnONPs are shown in Figure 2B. A copper grid with a distinctive carbon layer was employed for imaging purposes. The 98.7% pure ZnONPs green-synthesized using methanolic extracts of *V. album* employed in the experimental trial had a surface area of 178 $m^2 \cdot g^{-1}$ (BET Surface Area and Porosity Analyser, Model: Tristar II 3020, manufactured by Micromeritics Instrument Corporation based in Norcross, GA, USA) and a 5.1 $Kg \cdot L^{-1}$ density and were green in color. The ZnONPs with quasi-spherical forms are shown to have an average size distribution of 13.5 nm in the TEM images (Figure 2B).

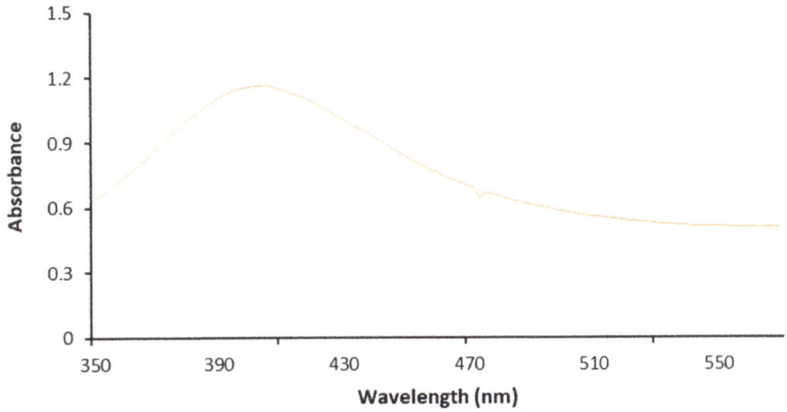

(**A**)

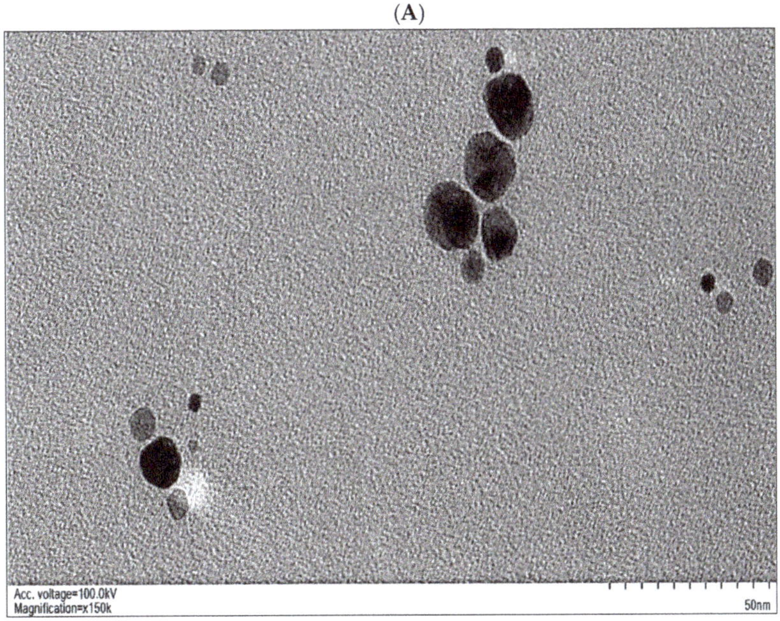

(**B**)

Figure 2. *Cont.*

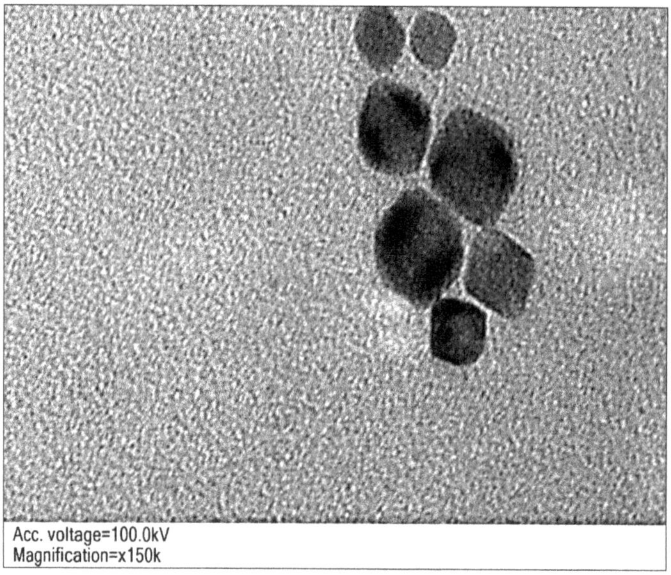

(C)

Figure 2. (**A**). Absorbance spectra of ZnONPs green-synthesized from the methanolic extracts of *V. album* peaking at 406 nm. (**B**). TEM image of ZnONPs synthesized from methanolic extract showing a quasi-spherical to spherical shape and size distribution averaging 13.5 nm. (**C**). TEM micrograph of the ZnONPs green-synthesized using aqueous extracts of *V. album*. The particles were larger in size and had smaller surface areas.

The TEM micrographs for the ZnONPs synthesized using aqueous plant extract are shown in Figure 2C. These particles were quasi-spherical, having a larger particle size averaging 50 nm with a smaller surface area of around 13 m^2/g.

2.4. Antibacterial Activities

The antimicrobial efficacy of nano-zinc oxide derived from *V. album* and the source plant extracts, in terms of the zone of inhibition, is shown in Table 3 and was compared to that of the standard drug, moxifloxacin (25 μg/mL). The maximum zone of inhibition was recorded for the ZnONPs prepared from aqueous extracts of *V. album*. The green-synthesized ZnONPs showed better antibacterial behavior compared to the source plant extract.

Table 3. Zone of inhibition under the influence of ZnONPs and source plant extracts of *V. album*.

Test Medicine	*E. coli*		*S. aureus*		*P. aeruginosa*		*B. subtilis*	
	Aq.	Met.	Aq.	Met.	Aq.	Met.	Aq.	Met.
Source plant extract of *V. album*	36 ± 0.9	34 ± 0.4	24 ± 0.8	23 ± 0.8	25 ± 0.2	22 ± 0.2	22 ± 0.9	21 ± 0.3
Gree-synthesized ZnONPs	44 ± 0.2	43 ± 0.2	40 ± 0.3	39 ± 0.3	36 ± 0.9	33 ± 0.6	21 ± 0.7	20 ± 0.7
Standard drug	45 ± 0.4		41 ± 0.6		35 ± 0.5		22 ± 0.5	

n = 3 measurements, diameter of well = 6 mm.

2.5. DNA Gyrase-B Inhibition

Observations were recorded regarding the invitro DNA Gyrase-B (Gyr-B) inhibition for the test extracts of the green-synthesized ZnONPs and source plant *V. album*. The results were compared to the standard drug, novobiocin (Figure 3). The results show that green-synthesized ZnONPs are potent inhibitors of Gyr-B activity, and their effects are comparable to the effects produced by novobiocin. Aqueous extracts of *V. album* were found to be least effective in limiting Gyr-B activity.

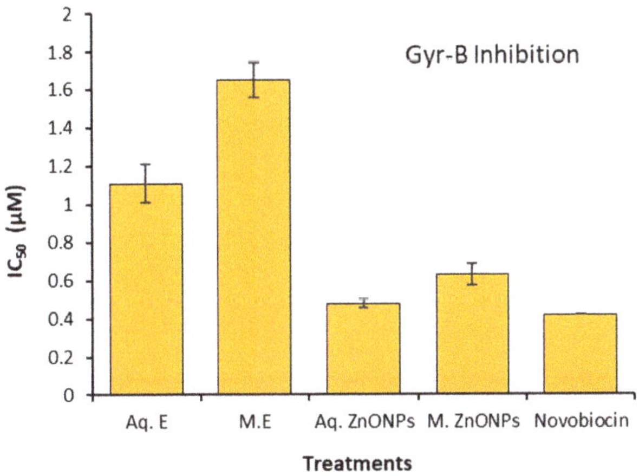

Figure 3. Invitro Gyr-B activity as demonstrated through the application of test extracts from *V. album* and green-synthesized ZnONPs. Aq: aqueous; E: extracts; M: methanolic; ZnONPs: zinc oxide nanoparticles (Mean ± S.E; n = 3).

2.6. Antioxidant Potential

Figure 4 provides an illustration of the antioxidant activity of the green-produced ZNONPs and *V. album* leaf extracts. Compared to aqueous extracts, methanolic extract has stronger antioxidant activity. According to the study's findings, at a dosage of 750 µg/mL, the plant extracts can scavenge 80% of DPPH radicals, whereas ascorbic acid, a common antioxidant, can do so at a concentration of 15.0 µg/mL. The effects of the methanolic ZnONPs were the best of all the test extracts, and they were able to demonstrate promising antioxidant potential. The aqueous- and methanolic-extract-mediated green ZnONPs, at a concentration of 100 µg/mL, were able scavenge 94% and 98% of DPPH free radicals.

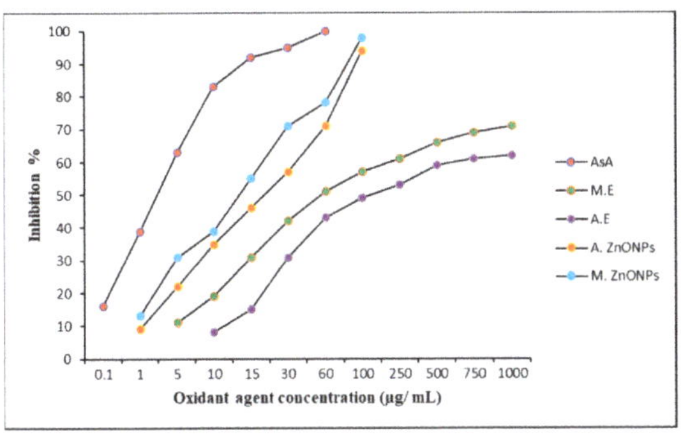

Figure 4. Antioxidant potential of green-synthesized ZnONPs and source plant extracts. As A: ascorbic acid; M.E.: methanolic extract of *V. album*; A.E.: aqueous extract of *V. album*; A. ZnONPs: zinc oxide nanoparticles green-synthesized using aqueous extracts of *V. album*; and M. ZnONPs: ZnONPs green-synthesized using methanolic extracts of *V. album*.

3. Materials and Methods

3.1. Plant Collection and Extraction

The local peasant association of Mirpur Azad Jammu and Kashmir (AJK), Pakistan, provided the tree species of Quercus dilatata from which the *V. album* plants were harvested. The sampling process was finished in November 2021. A sonicator produced by ISO LAB Germany was used to sonicate 40 grammes of the crushed herb in 1 L of methanol for 30 min at 40 °C, after which it was kept for 24 h in the dark. Using medium-grade filter paper, the mixture was purified, and methanol was added to bring the level to 1 L. The same procedure was used to create the aqueous extracts, which required 1 L of distilled water [17].

3.2. GC-MS Analysis of the Methanolic Extracts

A Shimadzu QP2020 GC-MS (Shimadzu Corporation, Kyoto, Japan) equipped with a split–splitless-type injector was used to analyze the methanol extracts following the method described by Naik et al. [18]. The mass spectrum of each original component was examined and compared to the literature data from the libraries of NIST 2017 and ADAMS-2007.

3.3. Determination of the Total Phenolic and Proline Contents of Extracts from the Sample Plant

The technique described by [19] was used to determine the total free proline concentration. Using the recommended reagents, frozen leaves were extracted through a process that involved centrifugation, boiling, and cooling. Readings were taken at 520 nm, and toluene was used as the reference. Using a calibration curve, the amount of proline was calculated and expressed as mg proline g^{-1} FW.

Total phenolics were assayed following [20]. Table 4 outlines the reagent values used in the method.

Table 4. Composition of different solutions used for the total phenolic content (TPC) assay.

Chemicals Used	Conc. Used	Final Amount Used (500 mL)
Folin–Ciocalteu reagent	2 N	0.1 mL
Na_2CO_3	10%	2.8 mL
Plant Extract	0.5 mg/mL	100 µg/mL

3.4. Green Synthesis and Characterization of Zinc Oxide Nanoparticles

The use of either methanolic or aqueous extracts of *V. album* may have an impact on the size and shape of the synthesized ZnONPs; thus, the choice of extract was considered in relation to the intended application of the nanoparticles. Additionally, to ensure reproducibility, the experimental conditions (such as pH, temperature, and stirring speed) were kept consistent between different batches of synthesized ZnONPs. The choice of solvent, whether aqueous or methanolic extracts of *V. album* are investigated, can affect the size and shape of the synthesized ZnONPs. This is because the solvent used can affect the reaction kinetics and, thus, the nucleation and growth rate of the nanoparticles.

A liter of 0.01 M zinc acetate dihydrate solution was prepared in distilled water. Using a magnetic stirrer, the zinc acetate dihydrate solution was vigorously agitated as 25 mL of herbal extract (either methanolic or aqueous extract) was added dropwise. The mixtures were heated to 70 °C for 30 min under constant stirring, and the pH of the reaction mixture was adjusted to 9 using sodium hydroxide (NaOH) for the purpose of the experiment. The combination abruptly changed color to a deep black hue. Subsequently, the resulting suspension was centrifuged for 15 min at 10,000 rpm using a Hermle Z326 centrifuge (Labtechnik GmbH, Wehingen, Germany), and the synthesized ZnONPs were washed three times with deionized water to remove any unreacted reagents or impurities. The NPs were collected and dried for further analysis [18].

Utilizing transmission electron microscopy and UV–visible spectroscopy, the characterization of NPs was carried out at the Department of Science, University of Central Punjab

(UCP), Lahore, Punjab, Pakistan. Using an ELICO SL-159 Spectrophotometer (Suzhou, China) in the 350–570 nm region, a UV–Vis spectral analysis was performed. These samples' transmission electron microscopy (TEM) pictures were captured using an FEI Tecnai 12 from Gatan, Germering, Germany that was accessible at the Institute of Space Technology (IST), Islamabad [21,22]. The crystal structure of the nanoparticles was investigated through powder X-ray diffraction (XRD) using a Phillips X Pert Model X-ray diffractometer, PANalytical, Almelo, Netherlands. CuKα radiation with a wavelength of 0.1542 nm was used, and the XRD pattern was recorded in a range from 10° to 80° 2θ.

3.5. *Antibacterial Activities*

The test bacterial strains included Pseudomonas aeruginosa (ATCC 10145), Escherichia coli (ATCC 10799), Staphylococcus aureus (ATCC 29213), and Bacillus subtilis (ATCC 11774). We determined whether the extract, silver nanoparticles, and the common antibiotic moxifloxacin could all limit bacterial growth. The inhibition zones produced by the inoculum of the studied bacteria were measured on 9 cm plates filled with sterile nutrient agar. Using a 6 mm puncher that had been sanitized, a 6 mm well was created. Then, 40 L of test solution (25 g/mL) and the moxifloxacin standard solution was poured into the wells. The control solution was 5% DMSO. At 37 C, the plates were incubated for 24 h. The outcomes of every experiment were replicated [23]. Gyr-B activity was assayed following [24].

3.6. *Determination of Antioxidant Potential*

The strategy described in [25] was applied with a few changes. In brief, 0.15 mL of 0.2 mM DPPH solution (dissolved in MeOH) was added to 0.15 mL of each extract solution. A Synergy HTC multimode reader was used to assess the solutions' absorbance at 517 nm in comparison to a methanol blank after 30 min of dark incubation at room temperature. The ability to scavenge the DPPH radical was calculated using the equation below:

DPPH Scavenging activity (%) = (Absorbance of blank-absorbance of extract)/(absorbance of blank) × 100.

4. Discussion

The hemi-parasitic herb *V. album* has been identified as a traditional ethnomedicine in several rural and hilly areas of AJK, Pakistan. Ethnobotanical and folklore research offered the original, fundamental foundation for selecting this plant for further phytochemical and analytical exploration or the isolation of single components with some efficacy and reactivity for managing different bacterial diseases and strong antioxidant potential. The methanolic extract of *V. album* was analyzed using GC-MS, revealing the presence of nonadecane and 2-methyltricosane, which are naturally occurring compounds in *Astragalus mongholicus* and *Achillea millefolium* [26]. These compounds have been identified as natural pest killers and growth regulators in plants. Additionally, the presence of potential antioxidants, di-tert-butyl-4-methylphenol and 2,4-di-tert-butylphenol, was detected in *V. album* [27]. Octacosane, a hydrocarbon compound with 28 carbon atoms, was identified as a natural product in Mikania cordifolia and is found in various plant species [28]. GC-MS analysis also revealed the presence of n-tetradecane, which is commonly used in organic synthesis. The methanolic extracts showed the presence of liquid alkanes, including 2-methylundecane. Tridecane, a natural product of *Juglan nigra*, was also noted [29]. Heneicosane, on its own, is primarily considered an inert hydrocarbon with minimal biological activity. However, it is important to note that heneicosane can contribute to the physical properties of substances or organisms in which it is found, for example, by acting as a component of the cuticle in plants and providing protection against moisture loss and physical damage. The presence of eicosane, a major antifungal metabolite, was observed in the extracts, and it exhibits antimicrobial properties. Furthermore, the methanolic extracts of *V. album* were found to contain heneicosane [30], which is reported as a potential antimicrobial alkane produced by plants. Additionally, the GCMS of the extract revealed the presence of pentadecane and heptadecane, which are major hydrocarbons and have been reported for their antimicrobial activity [31]. The compound is also a potential antigingivitic. The compounds 6-phenyldodecane and 5-phenyldodecane have been reported in *Abrus precatorius*, with no potential activity identified thus far [32]. The results of GC-MS are in accordance with the work reported in [33]. In the GC-MS analyses, a total ion chromatogram (TIC) was produced, which is a plot of the sum of the intensities for all ions in a predetermined range as a function of time. It is often helpful to take the average of the mass spectra gathered across a peak to identify an unknown analyte. To enhance chemical identification with library searching, background ions can also be removed. Furthermore, for the current study, methanolic extracts of *V. album* were preferred, since methanol has a high solubility power, which makes it a good choice for extracting a wide range of compounds from plants, including bioactive molecules with antimicrobial properties. Methanol is a non-polar solvent, which means that it can extract a broad range of compounds, including lipids and non-polar secondary metabolites that are often associated with antimicrobial activity. Moreover, methanol is a relatively stable solvent and is less prone to oxidation or degradation than other solvents, making it an ideal solvent for the long-term storage of plant extracts. Additionally, methanol is a safe and inexpensive solvent compared to other solvents such as chloroform, acetone, or ethanol, which are toxic or flammable [34].

Plants store proline and soluble carbohydrates to regulate their osmotic potential. The proline content of leaves is understood to be one of the earliest indicators of stress [35]. Variables including the kind of stress and length of exposure may play a role in the low proline concentration of *V. album* leaves. The proline content of the leaves suggests that under the conditions used in this study, stress was not a factor for *V. album*. Proline catabolism may be successfully stimulated in leaves that are in the last stages of senescence, during drought stress, and during changes in nutritional status [36]. According to our findings, *V. album* is a rich source of biocomponents, and its extracts, when grown under typical conditions, have the ability to reduce zinc ions. Overall, the finding of this research support the contribution of phenolics to this plant's antioxidant and anti-inflammatory properties [37].

In the current study, we used a green synthesis method to create ZnONPs. It has been demonstrated that between 330 and 460 nm, zinc oxide nanoparticles exhibit a characteristic broad absorption peak [38]. The production of zinc nanoparticles from *V. album* leaf extracts is thus successfully demonstrated by an optical absorption band peaking at 406 nm. This quick extracellular precipitation process has the potential to become a straightforward method for producing nanoparticles [39]. To understand the crystalline characteristics and size of the nanoparticles, a TEM study was carried out. The particles were nearly quasi-spherical in form, with only a slight thickness fluctuation, as shown in the ZnO TEM images. This research supports the findings of [22,39,40]. The methanolic extracts produced ZnONPs with smaller particle sizes, which resulted in broader diffraction peaks and changes in the peak positions and intensities compared to those obtained from larger particles synthesized using aqueous extracts. When a material is analyzed with XRD, a beam of X-rays is directed towards the sample, and the resulting diffracted X-rays are detected and measured. The XRD pattern produced by a material is a plot of the intensity of the diffracted X-rays versus the angle of diffraction. The width of the diffraction peak reflects the crystalline size of the material, with broader peaks indicating smaller crystallites. When a material has a larger surface area, this often means that it has smaller crystallites, which can result in broader XRD peaks. This is because smaller crystallites have fewer atoms in the crystal lattice, which results in a wider distribution of scattering angles and broader peaks in the XRD pattern. Additionally, it was noticed that smaller particles, produced via methanolic-extract-mediated green synthesis, had a larger surface area compared to the-aqueous extract-mediated green-synthesized ZnONPs, which had a smaller surface area. This smaller particle size and higher surface area of ZnONPs produced using methanolic extracts might be due to the higher solubility of Zn precursors in organic solvents. The above findings are in good agreement with the pool of previous studies [41,42].

The measured particle sizes of the ZnONPs differed for both TEM and XRD. Significant discrepancies in particle sizes, as measured through TEM and XRD, are not uncommon and can be attributed to various factors. TEM provides a direct visualization of individual particles, offering a more precise measurement of particle size by capturing the actual morphology and dimensions. On the other hand, XRD calculates the average crystallite size based on diffraction patterns, which can be influenced by factors such as crystallite shape, strain, and preferred orientation. The measurement principles of these techniques differ, leading to variations in the reported sizes. Additionally, the sample preparation techniques for TEM and XRD can introduce artifacts and affect the size measurements due to factors such as aggregation, drying effects, or uneven distribution. The limitations and calibration of the instruments used, as well as the heterogeneity of the particle samples themselves, can also contribute to the observed discrepancies [41,42].

The green production method is currently preferred for reducing the harmful problems and dangerous effects of nanoparticles. The green synthesis method ensures the controlled and targeted distribution of nanomedicine, and it is supported by the scientific community at large. Additionally, there are fewer adverse effects associated with the development of nanoparticles using a green synthesis method [43]. The controlled release of the nanomedicine is made possible by the plants' bioactive substances and natural botanicals [44]. According to the proposed mechanism, combinations of compounds, as extracts, were utilized in the green synthesis of ZnONPs. Based upon the previous literature [15,16], we used a combination of extracts in green synthesis. The use of plant extracts as reducing agents, stabilizers, or capping agents for nanoparticle formation is crucial in the green synthesis approach.

The behavior of the green-produced ZnONPs, in terms of their antibacterial properties, was improved compared to the source plant extracts. These results could be attributed to the green synthesis' non-hazardous and safe methodology, which uses plant extracts as capping and reducing agents [45]. The synthesized ZnONPs may not necessarily contain the exact same bioactive compounds as those found in the extracts. The green synthesis process may selectively incorporate certain compounds from the extracts or lead to the formation of new compounds during the nanoparticle synthesis. This might be a possible reason for the better antibacterial activities of ZnONPs compared to source plant extracts. The antibacterial properties of ZnONPs were further enhanced through the use of aqueous extracts in green synthesis. This can be explained by the greater negative charge and more homogeneous size distribution of aqueous extract nanoparticles compared to methanolic extract nanoparticles. The synthesized ZnONPs are expected to exhibit higher stability compared to the extracts, which may degrade over time due to various factors, such as exposure to light, heat, or enzymatic degradation. Thus, presumably, the better antibacterial properties of green-synthesized ZnONPs, as compared to plant extracts, are due to their stability. Several researchers have reported on the function of *V. album* leaf extracts as antibacterial agents [46,47]. The ZnONPs are known to interfere with bacterial DNA replication and drastically change how virulence-causing genes are expressed in the bacteria, which limits the pathogenic potential of the bacteria [48]. In the green synthesis method, phenolics and flavonoids found in plants serve as the primary reducing and capping agents. By blocking the activity of DNA gyrase, these flavonoids have been found to induce the negative supercoiling of bacterial DNA [49]. Gyr-B structural dynamics indicate that the enzyme is composed of two subunits (alpha and beta subunits). Antibiotics target the alpha subunit to interfere with bacterial DNA amplification. According to Singh et al. [50], the application of ZnONPs causes an upregulation of genes such as those for gyrase and single-stranded binding protein, which prevents bacteria from replicating their DNA and encourages bacterial mortality and stagnant development. These results provide proof of our current findings. Additionally, according to Slade and Radman [51], the application of ZnONPs causes a downregulation of genes involved in DNA repair mechanisms, such as base excision and mismatch repair.

The antioxidant potential of the ZnONPs was higher compared to the source plant extract, and the methanolic extracts were better at enhancing antioxidant potential. For Tunisian brachychiton, the methanolic extract had better antioxidant activity than the aqueous extract of the leaves [52]. The existence of alpha-tocopherol and other undiscovered phytochemicals in plant extracts, as well as the presence of phenolics and flavonoids, may contribute to the antioxidant action. Additionally, the synergistic effects of the extract's numerous biocomponents, each of which has a unique structure, polarity, and capacity to dissolve in a particular solvent, will affect the antioxidant capacity [53,54].

According to our results, ZnONPs proved effective on most occasions, as compared to the source plant extracts. ZnONPs have gained significant attention in antibacterial and antioxidant studies due to their unique properties and superior performance compared to plant extracts. ZnONPs possess a large surface-area-to-volume ratio, which allows for a larger contact area with bacterial cells or reactive oxygen species (ROS) in antioxidant studies. This increased contact enhances the interaction between the nanoparticles and the target, leading to enhanced antibacterial and antioxidant effects [16,21]. In contrast, the source plant extracts contained a mixture of compounds, as verified by the GCMS study, and their activity may vary depending on the extraction method and plant source, making them less reliable and consistent. ZnONPs exhibit potent antibacterial properties by inducing oxidative stress in bacteria. They generate reactive oxygen species (ROS) such as hydrogen peroxide (H_2O_2), superoxide ions (O_2), and hydroxyl radicals (OH-) that can damage bacterial cell membranes, proteins, and DNA, leading to bacterial death. In comparison, *V. album* extracts may contain different active compounds that can have varying degrees of antibacterial activity, and their mechanisms of action can be more complex and less well-understood. Additionally, ZnONPs demonstrate long-lasting stability and the sustained release of antibacterial and antioxidant agents, which can provide extended protection against bacteria and oxidative damage. Their stable structure allows them to retain their activity over time, whereas plant extracts can degrade or lose their potency due to environmental factors, storage conditions, or interactions with other components [9,16,21].

5. Conclusions

A GC-MS analysis of *V. album* methanolic extracts was conducted. The analysis showed that the plant is enriched with botanicals and volatile compounds useful for various biomedical applications. The green synthesis of nano-zinc oxide was accomplished using both the aqueous extracts and methanolic extracts of *V. album*, and the method was confirmed in its success through characterization techniques. The XR diffractograms revealed that methanolic extracts were associated with smaller particle sizes, which resulted in broader diffraction peaks and changes in the peak positions and intensities as compared to those obtained from larger particles synthesized using aqueous extracts. A comparative study on the antimicrobial efficacy and antioxidant potential of the synthesized ZnONPs and their source extracts was conducted, which revealed that the green-synthesized ZnONPs might be a potential nanomedicine for treating microbial infections. The aqueous extracts of the ZnONPs were found to more reliably limit bacterial growth. Both aqueous- and methanolic-extract-mediated green ZnONPs showed significant antioxidant activity, with methanolic extracts showing better efficacy. From the perspective of antimicrobial resistance, we recommend the use of green-synthesized nanoparticles as substitutes for antibiotic drugs.

Author Contributions: Conceptualization, methodology, M.I., M.M., M.W.M., R.C., W.M. and A.M.A.-E.; software, validation, formal analysis, A.M.A.-E. and H.O.E.; investigation, resources, data curation, M.W.M., M.I., M.M., R.C. and A.M.A.-E.; writing—original draft preparation, writing—review, editing, visualization, supervision, project administration, funding acquisition, W.M., M.I., M.M., M.W.M., R.C., A.M.A.-E. and H.O.E. All authors have read and agreed to the published version of the manuscript.

Funding: Researchers Supporting Project (number RSPD2023R676), King Saud University.

Data Availability Statement: All data are available within this publication.

Acknowledgments: The authors would like to thank the Researchers Supporting Project (number RSPD2023R676), King Saud University, Riyadh, Saudi Arabia.

Conflicts of Interest: The authors declare no conflict of interest.

References

1. Frieri, M.; Kumar, K.; Boutin, A. Antibiotic resistance. *J. Infect. Public Health* **2017**, *10*, 369–378. [CrossRef]
2. AlSheikh, H.M.; Sultan, I.; Kumar, V.; Rather, I.A.; Al-Sheikh, H.; Tasleem Jan, A.; Haq, Q.M. Plant-based phyto-chemicals as possible alternative to antibiotics in combating bacterial drug resistance. *Antibiotics* **2020**, *9*, 480. [CrossRef]
3. Ishtiaq, M.; Maqbool, M.; Ajaib, M.; Ahmed, M.; Hussain, I.; Khanam, H.; Mushtaq, W.; Hussain, T.; Azam, S.; Hayat Bhatti, K.; et al. Ethnomedicinal and folklore inventory of wild plants used by rural communities of valley Samahni, District Bhimber Azad Jammu and Kashmir, Pakistan. *PLoS ONE* **2021**, *16*, e0243151. [CrossRef] [PubMed]
4. Lobo, V.; Patil, A.; Phatak, A.; Chandra, N. Free radicals, antioxidants and functional foods: Impact on human health. *Pharmacogn. Rev.* **2010**, *4*, 118. [CrossRef] [PubMed]
5. Aranaz, M.; Costas-Rodríguez, M.; Lobo, L.; García, M.; González-Iglesias, H.; Pereiro, R.; Vanhaecke, F. Homeostatic alterations related to total antioxidant capacity, elemental concentrations and isotopic compositions in aqueous humor of glaucoma patients. *Anal. Bioanal. Chem.* **2022**, *414*, 515–524. [CrossRef] [PubMed]
6. Cai, F.; Li, S.; Huang, H.; Iqbal, J.; Wang, C.; Jiang, X. Green synthesis of gold nanoparticles for immune response regulation: Mechanisms, applications, and perspectives. *J. Biomed. Mater. Res. Part A* **2022**, *110*, 424–442. [CrossRef]
7. Moratin, H.; Ickrath, P.; Scherzad, A.; Meyer, T.J.; Naczenski, S.; Hagen, R.; Hackenberg, S. Investigation of the immune modulatory potential of zinc oxide nanoparticles in human lymphocytes. *Nanomaterials* **2021**, *11*, 629. [CrossRef]
8. Siddiqi, K.S.; Husen, A. Properties of zinc oxide nanoparticles and their activity against microbes. *Nanoscale Res. Lett.* **2018**, *13*, 141. [CrossRef]
9. Rambabu, K.; Bharath, G.; Banat, F.; Show, P.L. Green synthesis of zinc oxide nanoparticles using *Phoenix dactylifera* waste as bioreductant for effective dye degradation and antibacterial performance in wastewater treatment. *J. Hazard. Mater.* **2021**, *402*, 123560. [CrossRef]
10. Maqbool, Q.; Czerwinska, N.; Giosue, C.; Sabbatini, S.; Ruello, M.L.; Tittarelli, F. New waste-derived TiO_2 nanoparticles as a potential photocatalytic additive for lime based indoor finishings. *J. Clean. Prod.* **2022**, *373*, 133853. [CrossRef]
11. Maqbool, Q.; Barucca, G.; Sabbatini, S.; Parlapiano, M.; Ruello, M.L.; Tittarelli, F. Transformation of industrial and organic waste into titanium doped activated carbon–cellulose nanocomposite for rapid removal of organic pollutants. *J. Hazard. Mater.* **2022**, *423*, 126958. [CrossRef]
12. Nickrent, D.L. Parasitic angiosperms: How often and how many? *Taxon* **2020**, *69*, 5–27. [CrossRef]
13. Ajaib, M.; Ishtiaq, M.; Bhatti, K.H.; Hussain, I.; Maqbool, M.; Hussain, T.; Mushtaq, W.; Ghani, A.; Azeem, M.; Khan, S.M.; et al. Inventorization of traditional ethnobotanical uses of wild plants of Dawarian and Ratti Gali areas of District Neelum, Azad Jammu and Kashmir, Pakistan. *PLoS ONE* **2021**, *16*, e0255010. [CrossRef]
14. Valle, A.C.V.; de Carvalho, A.C.; Andrade, R.V. Viscum Album-Literature Review. *Int. J. Sci. Res.* **2021**, *10*, 63–71.
15. Pujar, P.R.; Giri, R.S.; Malashetty, B.B.; Nandibewoor, S.T. Green synthesis of Zinc oxide nanoparticles using aqueous leaf extract of Viscum album and their characterization. *J. Taibah Univ. Sci.* **2017**, *11*, 117–126. [CrossRef]
16. Ahmad, M.Z.; Fatima, S.; Alzohairy, M.S.; Alam, M.F.; Al-Khedhairy, A.A.; Al-Qahtani, M.A. Green synthesis of Zinc oxide nanoparticles using Viscum album extract: Characterization and cytotoxicity assessment. *J. Nanomater.* **2015**, *2015*, 626142.
17. Nsor, B.; Paul, M.P.; Modibbo, A.A.; Eneji, A.A. Effects of methanolic extract of mistletoe (*Viscum album*) on alloxan-induced diabetic albino rats. *J. Clin. Med. Images Case Rep.* **2022**, *2*, 1162. [CrossRef]
18. Naik Rathod, H.; Mallappa, B.; Malenahalli Sidramappa, P.; Reddy Vennapusa, C.S.; Kamin, P.; Revanasiddappa Nidoni, U.; Desai, B.R.; Rao, S.N.; Mariappan, P. Determination of 77 multiclass pesticides and their metabolites in capsicum and tomato using GC-MS/MS and LC-MS/MS. *Molecules* **2021**, *26*, 1837. [CrossRef] [PubMed]
19. Bates, L.S.; Waldren, R.P.; Teare, I.D. Rapid determination of free proline for water-stress studies. *Plant Soil* **1973**, *39*, 205–207. [CrossRef]
20. Singleton, V.L.; Orthofer, R.; Lamuela-Raventós, R.M. Analysis of total phenols and other oxidation substrates and antioxidants by means of folin-ciocalteu reagent. *Methods Enzymol.* **1999**, *299*, 152–178.
21. Mazhar, M.W.; Ishtiaq, M.; Maqbool, M.; Akram, R. Seed priming with zinc oxide nanoparticles improves growth, osmolyte accumulation, antioxidant defense, and yield quality of water-stressed mung bean plants. *Arid. Land Res. Manag.* **2022**, *37*, 222–246. [CrossRef]
22. Waqas Mazhar, M.; Ishtiaq, M.; Hussain, I.; Parveen, A.; Hayat Bhatti, K.; Azeem, M.; Thind, S.; Ajaib, M.; Maqbool, M.; Sardar, T.; et al. Seed nano-priming with Zinc Oxide nanoparticles in rice mitigates drought and enhances agronomic profile. *PLoS ONE* **2022**, *17*, e0264967. [CrossRef] [PubMed]
23. Al-Ramamneh, E.A.; Ghrair, A.M.; Shakya, A.K.; Alsharafa, K.Y.; Al-Ismail, K.; Al-Qaraleh, S.Y.; Mojski, J.; Naik, R.R. Efficacy of Sterculia diversifolia Leaf Extracts: Volatile Compounds, Antioxidant and Anti-Inflammatory Activity, and Green Synthesis of Potential Antibacterial Silver Nanoparticles. *Plants* **2022**, *11*, 2492. [CrossRef] [PubMed]

24. Durcik, M.; Tammela, P.; Barančoková, M.; Tomašič, T.; Ilaš, J.; Kikelj, D.; Zidar, N. Synthesis and evaluation of N-phenylpyrrolamides as DNA gyrase B Inhibitors. *Chem. Med. Chem.* **2018**, *13*, 186–198. [CrossRef] [PubMed]
25. Al-Jaber, H.I.; Shakya, A.K.; Elagbar, Z.A. HPLC Profiling of Selected Phenolic Acids and Flavonoids in *Salvia eigii*, *Salvia hierosolymitana* and *Salvia viridis* growing wild in Jordan and their in vitro antioxidant activity. *PeerJ.* **2020**, *8*, e9769. [CrossRef]
26. Miyazawa, M.; Kameoka, H. Volatile flavor components of *Astragali radix* (*Astragalus membranaceus* Bunge). *Agric. Biol. Chem.* **1987**, *51*, 3153–3154.
27. Jin, X.L.; Wang, K.; Li, Q.Q.; Tian, W.L.; Xue, X.F.; Wu, L.M.; Hu, F.L. Antioxidant and Anti-Inflammatory Effects of Chinese propolis during palmitic acid-induced lipotoxicity in cultured hepatocytes. *J. Funct. Foods* **2017**, *34*, 216–223. [CrossRef]
28. Mohammad, N.; Majumder, M.S.; Rashed-al-Qayum, K.M.; Bhattacharjee, S.H.; Kar, A.U. Phytochemical Screening of Medicinal Plant-Mikania cordifolia and Determination of Its Characteristics. *Mintage J. Pharm. Med. Sci.* **2013**, *2*, 14–17.
29. Islam, M.T.; Ali, E.S.; Uddin, S.J.; Shaw, S.; Islam, M.A.; Ahmed, M.I.; Shill, M.C.; Karmakar, U.K.; Yarla, N.S.; Khan, I.N.; et al. Phytol: A Review of biomedical activities. *Food Chem. Toxicol.* **2018**, *121*, 82–94. [CrossRef]
30. Kaur, N.; Chaudhary, J.; Jain, A.; Kishore, L. Stigmasterol: A Comprehensive Review. *Int. J. Pharm. Sci. Res.* **2011**, *2*, 2259.
31. Tundis, R.; Peruzzi, L.; Menichini, F. Phytochemical and biological studies of *Stachys* species in relation to chemotaxonomy: A review. *Phytochemistry* **2014**, *102*, 7–39. [CrossRef]
32. Lebri, M.; Tilaoui, M.; Bahi, C.; Achibat, H.; Akhramez, S.; Fofie, Y.B.; Gnahoué, G.; Lagou, S.M.; Zirihi, G.N.; Coulibaly, A.; et al. Phytochemical analysis and in vitro anticancer effect of aqueous extract of *Abrus precatorius* Linn. *Der Pharma Chem.* **2015**, *7*, 112–117.
33. Song, C.; Wei, X.Y.; Qiu, Z.D.; Gong, L.; Chen, Z.Y.; Ma, Y.; Shen, Y.; Zhao, Y.J.; Wang, W.H.; Yang, B. Exploring the resources of the genus *Viscum* for potential therapeutic applications. *J. Ethnopharmacol.* **2021**, *277*, 114233. [CrossRef] [PubMed]
34. Pál, M.; Tajti, J.; Szalai, G.; Peeva, V.; Végh, B.; Janda, T. Interaction of polyamines, abscisic acid and proline under osmotic stress in the leaves of wheat plants. *Sci. Rep.* **2018**, *8*, 12839. [CrossRef]
35. Wang, W.; Li, J.; Zhang, H.; Wang, X.; Fan, J.; Zhang, X. Phenolic compounds and bioactivity evaluation of aqueous and methanol extracts of *Allium mongolicum* Regel. *Food Sci. Nutr.* **2019**, *7*, 779–787. [CrossRef]
36. Siddique, A.; Kandpal, G.; Kumar, P. Proline accumulation and its defensive role under diverse stress conditions in plants: An overview. *J. Pure Appl. Microbiol.* **2018**, *12*, 1655–1659. [CrossRef]
37. de Oliveira Melo, M.N.; Oliveira, A.P.; Wiecikowski, A.F.; Carvalho, R.S.; de Lima Castro, J.; de Oliveira, F.A.; Pereira, H.M.; da Veiga, V.F.; Capella, M.M.; Rocha, L.; et al. Phenolic compounds from Viscum album tinctures enhanced antitumor activity in melanoma murine cancer cells. *Saudi Pharm. J.* **2018**, *26*, 311–322. [CrossRef]
38. Dutta, S.; Jaiswal, K.K.; Verma, R.; Basavaraju, D.M.; Ramaswamy, A.P. Green synthesis of zinc oxide catalyst under microwave irradiation using banana (*Musa* spp.) corm (rhizome) extract for biodiesel synthesis from fish waste lipid. *Biocatal. Agric. Biotechnol.* **2019**, *22*, 101390. [CrossRef]
39. Sadhasivam, S.; Shanmugam, M.; Umamaheswaran, P.D.; Venkattappan, A.; Shanmugam, A. Zinc oxide nanoparticles: Green synthesis and biomedical applications. *J. Clust. Sci.* **2021**, *32*, 1441–1455. [CrossRef]
40. Parveen, A.; Siddiqui, Z.A. Zinc oxide nanoparticles affect growth, photosynthetic pigments, proline content and bacterial and fungal diseases of tomato. *Arch. Phytopathol. Plant Prot.* **2021**, *54*, 1519–1538. [CrossRef]
41. Kumar, D.; Thota, S.; Dixit, S.; Kumari, P. Synthesis and characterization of zinc oxide nanoparticles using *Plectranthus amboinicus* leaf extract: Antibacterial and anticancer activities. *J. Nanomater.* **2018**, *2018*, 4528642.
42. Moghimi, A.; Fathinia, M.; Erfani-Moghadam, V.; Salavati-Niasari, M. Synthesis, characterization and application of ZnO nanoparticles using methanolic and aqueous extract of *Cinnamomum camphora* leaves. *J. Mol. Liq.* **2019**, *293*, 111571. [CrossRef]
43. Wang, Y.; O'Connor, D.; Shen, Z.; Lo, I.M.; Tsang, D.C.; Pehkonen, S.; Pu, S.; Hou, D. Green synthesis of nanoparticles for the remediation of contaminated waters and soils: Constituents, synthesizing methods, and influencing factors. *J. Clean. Prod.* **2019**, *226*, 540–549. [CrossRef]
44. Zeb, A.; Gul, M.; Nguyen, T.T.; Maeng, H.J. Controlled release and targeted drug delivery with poly (lactic-co-glycolic acid) nanoparticles: Reviewing two decades of research. *J. Pharm. Investig.* **2022**, *52*, 683–724. [CrossRef]
45. Pal, N.; Agarwal, M.; Gupta, R. Green synthesis of guar gum/Ag nanoparticles and their role in peel-off gel for enhanced antibacterial efficiency and optimization using RSM. *Int. J. Biol. Macromol.* **2022**, *221*, 665–678. [CrossRef]
46. Yildiz, S.; Gürgen, A.; Kiliç, C.; Tabbouche, S.; Kiliç, A.O.; Zehra, C.A. Antioxidant, antimicrobial and anti-quorum sensing activities of *Usnea filipendula* and *Viscum album*. *J. Anatolian Environ. Animal Sci.* **2019**, *4*, 613–620. [CrossRef]
47. Keskin, D.B.; Anandappa, A.J.; Sun, J.; Tirosh, I.; Mathewson, N.D.; Li, S.; Oliveira, G.; Giobbie-Hurder, A.; Felt, K.; Gjini, E.; et al. Neoantigen vaccine generates intratumoral T cell responses in phase Ib glioblastoma trial. *Nature* **2019**, *565*, 234–239. [CrossRef]
48. Gudkov, S.V.; Burmistrov, D.E.; Serov, D.A.; Rebezov, M.B.; Semenova, A.A.; Lisitsyn, A.B. A mini review of anti-bacterial properties of ZnO nanoparticles. *Front. Phys.* **2021**, *9*, 641481. [CrossRef]
49. Alhadrami, H.A.; Orfali, R.; Hamed, A.A.; Ghoneim, M.M.; Hassan, H.M.; Hassane, A.S.; Rateb, M.E.; Sayed, A.M.; Gamaleldin, N.M. Flavonoid-coated gold nanoparticles as efficient antibiotics against Gram-negative bacteria—Evidence from in silico-supported in vitro studies. *Antibiotics* **2021**, *10*, 968. [CrossRef] [PubMed]
50. Singh, A.; Gautam, P.K.; Verma, A.; Singh, V.; Shivapriya, P.M.; Shivalkar, S.; Sahoo, A.K.; Samanta, S.K. Green synthesis of metallic nanoparticles as effective alternatives to treat antibiotics resistant bacterial infections: A review. *Biotechnol. Rep.* **2020**, *25*, e00427. [CrossRef] [PubMed]

51. Slade, D.; Radman, M. Oxidative stress resistance in *Deinococcus radiodurans*. *Microbiol. Mol. Biol. Rev.* **2011**, *75*, 133–191. [CrossRef]
52. Rjeibi, I.; Ben Saad, A.; Ncib, S.; Souid, S.; Allagui, M.S.; Hfaiedh, N. *Brachychiton populneus* as a novel source of bioactive ingredients with therapeutic effects: Antioxidant, enzyme inhibitory, anti-inflammatory properties and LC–ESI-MS profile. *Inflammopharmacology* **2020**, *28*, 563–574. [CrossRef] [PubMed]
53. Seidavi, A.; Tavakoli, M.; Asroosh, F.; Scanes, C.G.; El-Hack, A.; Mohamed, E.; Naiel, M.A.; Taha, A.E.; Aleya, L.; El-Tarabily, K.A.; et al. Antioxidant and antimicrobial activities of phytonutrients as antibiotic substitutes in poultry feed. *Environ. Sci. Pollut. Res.* **2021**, *29*, 5006–5031. [CrossRef] [PubMed]
54. Toschi, A.; Piva, A.; Grilli, E. Phenol-rich botanicals modulate oxidative stress and epithelial integrity in intestinal epithelial cells. *Animals* **2022**, *12*, 2188. [CrossRef] [PubMed]

plants MDPI

Article

Estimation of the Controlled Release of Antioxidants from β-Cyclodextrin/Chamomile (*Matricaria chamomilla* L.) or Milk Thistle (*Silybum marianum* L.), Asteraceae, Hydrophilic Extract Complexes through the Fast and Cheap Spectrophotometric Technique

Adina Horablaga [1], Alina Şibu (Ciobanu) [2], Corina Iuliana Megyesi [3], Dina Gligor (Pane) [2], Gabriel Stelian Bujancă [4], Ariana Bianca Velciov [3], Florica Emilia Morariu [5,*], Daniel Ioan Hădărugă [2,6], Corina Dana Mişcă [4,*] and Nicoleta Gabriela Hădărugă [2,3,*]

1 Department of Sustainable Development and Environmental Engineering, University of Life Sciences "King Mihai I" from Timişoara, Calea Aradului 119, 300645 Timişoara, Romania; ahorablaga@yahoo.com or adinahorablaga@usvt.ro
2 Doctoral School "Engineering of Vegetable and Animal Resources", University of Life Sciences "King Mihai I" from Timişoara, Calea Aradului 119, 300645 Timişoara, Romania; ciobanualina1966@yahoo.com (A.Ş.); gligor_dina_bihor@yahoo.com (D.G.); daniel.hadaruga@upt.ro (D.I.H.)
3 Department of Food Science, University of Life Sciences "King Mihai I" from Timişoara, Calea Aradului 119, 300645 Timişoara, Romania; corina.megyesi@usvt.ro or cor_costescu@yahoo.com (C.I.M.); arianavelciov@usvt.ro or ariana.velciov@yahoo.com (A.B.V.)
4 Department of Food Control, University of Life Sciences "King Mihai I" from Timişoara, Calea Aradului 119, 300645 Timişoara, Romania; gabrielbujanca@yahoo.com or gabrielbujanca@usvt.ro
5 Department of Biotechnologies, University of Life Sciences "King Mihai I" from Timişoara, Calea Aradului 119, 300645 Timişoara, Romania
6 Department of Applied Chemistry, Organic and Natural Compounds Engineering, Polytechnic University of Timişoara, Carol Telbisz 6, 300001 Timişoara, Romania
* Correspondence: florica.morariu@animalsci-tm.ro (F.E.M.); miscacorina@gmail.com or corinamisca@usvt.ro (C.D.M); nicoletahadaruga@usvt.ro or nico_hadaruga@yahoo.com (N.G.H.); Tel.: +40-256-277-423 (N.G.H.)

Abstract: This is the first study on the modeling of the controlled release of the estimated antioxidants (flavonoids or flavonolignans) from β-cyclodextrin (β-CD)/hydrophilic vegetable extract complexes and the modeling of transdermal pharmaceutical formulations based on these complexes using an overall estimation by the spectrophotometric method. The Korsmeyer–Peppas model was chosen for evaluating the release mechanisms. β-CD/chamomile (*Matricaria chamomilla* L., Asteraceae) ethanolic extract and β-CD/milk thistle (*Silybum marianum* L., Asteraceae) ethanolic extract complexes were obtained by the co-crystallization method with good recovering yields of 55–76%, slightly lower than for β-CD/silibinin or silymarin complexes (~87%). According to differential scanning calorimetry (DSC) and Karl Fischer water titration (KFT), the thermal stability of complexes is similar to β-CD hydrate while the hydration water content is lower, revealing the formation of molecular inclusion complexes. In the Korsmeyer–Peppas model, β-CD/*M. chamomilla* flower extract complexes reveal Case II transport mechanisms, while the corresponding complexes with leaf extracts indicate non-Fickian diffusion for the controlled release of antioxidants in ethanol 60 and 96%. The same non-Fickian diffusion was revealed by β-CD/*S. marianum* extract and β-CD/silibinin complexes. On the contrary, almost all model transdermal pharmaceutical formulations based on β-CD/*M. chamomilla* extract complexes and all those based on β-CD/*S. marianum* extract complexes revealed non-Fickian diffusion for the antioxidant release. These results indicate that *H*-bonding is mainly involved in the diffusion of antioxidants into a β-CD based matrix, while the controlled release of antioxidants in model formulations is mainly due to hydrophobic interactions. Results obtained in this study can be further used for studying the particular antioxidants (namely rutin or silibinin, quantified, for example, by liquid chromatographic techniques) for their transdermal transport and biological effects in innovatively designed pharmaceutical formulations that can be obtained using "green" methods and materials.

Keywords: β-cyclodextrin complexes; transdermal pharmaceutical formulations; controlled release; chamomile; *Matricaria chamomilla* L.; milk thistle; *Silybum marianum* L.; ethanolic extract; spectrophotometric analysis; Korsmeyer–Peppas model

1. Introduction

Matricaria chamomilla L. (chamomile) and *Silybum marianum* L. (milk thistle) are two medicinal plant species that belong to the same botanical family, Asteraceae. They are important species of medicinal and aromatic plants (MAPs). The area cultivated with MAPs exceeding 200,000 ha in Europe [1].

Chamomile is one of the most cultivated MAPs in the world, with more than 20,000 ha [2]. Among various biological activities, chamomile flowers are used for their anti-inflammatory and antioxidant activities, for digestive, uterine muscle, menstrual, and postpartum disorders, as well as for anxiety, cancer prevention, cardiovascular diseases, diabetes, liver disorders, and neuralgic pains [3–5]. Chamomile is traditionally used as an infusion, decoction, vapor inhalation, compress, and tea [5]. On the other hand, chamomile essential oils, extracts, or cell suspension cultures are studied for their antioxidant and biological activities, which are mainly due to the presence of essential oils and/or antioxidant polyphenolic components [3,6–10]. Various extraction methods are used for obtaining essential oils or polyphenol-based extracts [4,11]. Essential oils can be extracted by classical hydrodistillation, steam distillation, solid-liquid extraction using hydrophobic solvents, or extraction using supercritical fluids [6,8,12–14]. The essential oil yields reach 1.77% for some *M. chamomilla* genotypes, but the average yield is in the range of 0.8–0.9% [12]. The main essential oil compounds are bisabolol oxides A and B (21.6–47.3% and up to 20.6%), bisabolone oxide A (up to 12.4%), as well as camazulene (<19.9%). Dicycloethers such as *cis*-tonghaosu have been determined in *M. chamomile* L. essential oils at 2.3–46% [12–15]. On the other hand, polyphenolic extracts are generally obtained by cold or hot extraction using methanol or ethanol as solvents (e.g., Soxhlet extraction), microwave-assisted extraction, ultrasound-assisted extraction, subcritical water extraction, or supercritical CO_2 extraction [3,4,6,7,11]. The polyphenolic compounds of the hydrophilic *M. chamomile* L. extracts belong to the flavonoids and phenolic acid classes. Among flavonoids, quercetin and its glycosides, rutin, chrysin, apigenin, and luteolin 7-*O*-glucoside (1.07–1.97, 1.76–4.40, 3.52–11.69, and 4.70–9.17 mg/g dry extract, respectively) were the most concentrated in the 70% ethanol extracts [3,9,16]. The main phenolic acids were chlorogenic, *p*-coumaric, cafeic, and ferulic acids, with maximum contents up to 14.8% for *p*-coumaric acid [3]. They were also quantified as total phenol content (TPC) or total flavonoid content (TFC), with values of 117.3–151.5 mg chlorogenic acid equivalents (CAE)/g and 49.7–64.3 mg rutin equivalents (RE)/g, respectively [4]. The copper excess in the nutrient solutions used for *M. chamomile* L. cultivation can enhance the accumulation of some phenolic contents [17]. The overall antioxidant activities of *M. chamomilla* L. extracts were evaluated through various methods, including the DPPH· (2,2-diphenyl-1-pycrylhydrazyl radical) assay, ferrous ion chelating ability, reducing power, hydroxyl radical scavenging activity, or $ABTS^{+\cdot}$ [2,2′-azino-bis(3-ethylbenzothiazoline-6-sulfonic acid radical cation)] assay [7,11].

Silybum marianum L. is widely used for its valuable hepatoprotective effect. This effect is mainly due to the presence of flavonolignans, which are polyphenols derived from flavonoids. A complex mixture of flavonolignans can be found in *S. marianum* extracts, including silibinins and silychristins. Silymarin is the standardized *S. marianum* fruit extract. It mainly contains the above-mentioned flavonolignans and up to 25% oil [18,19]. The extraction of bioactive flavonolignans is generally performed by classical methods using organic solvents, with some disadvantages related to selectivity and toxicity. On the other hand, supercritical fluid extraction is a green alternative for the separation of silymarin extracts [20]. The antioxidant activity of *S. marianum* extract is important. Both flavonolignans and flavonoids, such as silibinins, taxifolin, and quercetin, provide antioxidant effects. Among them, chlorogenic acids and tocopherols also have an impact [21–23]. The contents of various antioxidant compounds in European *S. marianum* genotypes were in the range of 3086–9499 mg/kg for silibinins, 126.5–395.3 mg/kg for chlorogenic acids, and 3.5–79.7 mg/kg for luteolin, with TFC values of 30–84 mg RE/100 g [24]. The total antioxidant activity of *S. marianum* extracts strongly depends on the plant's development, growing conditions, and treatments, as well as the extraction procedure. The maximum antioxidant activity was observed in 80-day-old leaves and plants (60 and 65.43% by DPPH· assay, respectively) [25]. The antioxidant activity of the achene extracts of different *S. marianum* varieties was determined by various methods. The radical scavenging activity was in the range of 1763–5303 μmol TE/100 g by the DPPH· assay, 2965–14426 μmol TE/100 g by the ORAC assay, and 692–1664 μmol TE/100 g by the FRAP assay [24].

Polyphenols are easily oxidized in proper conditions, especially if they are not appropriately stabilized. Flavonoids such as quercetin and chrysin can be transformed to free radicals and further to quinone methides by various mechanisms, including hydrogen atom transfer (HAT) or single-electron transfer (SET) mechanisms. Metal ions are also involved in the antioxidant mechanism of polyphenols [26,27]. The modulation of the antioxidant activity of polyphenols and the corresponding extracts can be achieved by nanoencapsulation [28]. Among many encapsulation matrices, cyclodextrins (CDs) are appropriate natural compounds for molecular inclusion of the most known polyphenols, including flavonoids and flavonolignans [29,30]. They can be completely or partially encapsulated into the CD cavity, depending on the structural characteristics (hydrophobicity and dimensions) [31,32]. As a consequence, the reactivity of antioxidants is apparently reduced by CD nanoencapsulation, and a controlled release supramolecular system can be obtained [33–35]. Many studies deal with the CD complexation of "pure" antioxidant flavonoids and flavonolignans [34–41], but only a few of them are related to the CD nanoencapsulation of *M. chamomilla* or *S. marianum* extracts [42–45]. Some similar complexes were also obtained and studied in our laboratory. They were especially related to the CD encapsulation of *Compositae* essential oils [13,14,46].

In this study, a comparison between the thermal stability and controlled release by spectrophotometric evaluation of β-CD/chamomile (*Matricaria chamomilla* L.) or milk thistle (*Silybum marianum* L.), Asteraceae, hydrophillic extract complexes was performed. It mainly focuses on the applicative aspects regarding the β-CD complexation of the raw ethanolic extracts and the controlled release from the complexes and transdermal pharmaceutical formulations of the main antioxidant compounds (e.g., chrysin, rutin, and silibinins/silymarin, respectively) in various media.

2. Results and Discussion

2.1. Obtaining of Chamomile (Matricaria chamomilla L.) and Milk Thistle (Silybum marianum L.) Hydrophillic Extracts

The extracts of *M. chamomilla* (flowers, leaves, roots, and stems) and *S. marianum* (seeds) were obtained at final raw volumes of 27.0–30.5 and 26.5–30.0 mL, respectively (see Table S2 in the Supplementary Material file). The extraction conditions were moderate (<60 °C for 1.5 h) in order not to promote degradation of the antioxidant compounds, namely flavonoids such as chrysin, quercetin, and rutin, or flavonolignans such as silibinins. Moreover, the estimation of the overall antioxidant contents in *M. chamomilla* and *S. marianum* extracts was performed by the spectrophotometric UV–Vis method, a fast and non-destructive technique. These antioxidant contents were expressed as chrysin or rutin in *M. chamomilla* extracts. For the *S. marianum* extracts, the flavonolignan contents were expressed as silibinin (based on either the standard silibinin diastereomer mixture of 98% or the standard silymarin mixture of 64.7%; see Section 3). Determinations were based on specific standard curves obtained at 322 nm for chrysin and rutin (a mixture that better fits the extract absorbance behavior) and 288 nm for silibinin (mean values). The wavelength for flavonoids was selected based on the specific maximum absorbance of *M. chamomilla* extract and the chrysin-rutin mixture at a mass ratio of 8.8:1 (see Equations (1) and (2) and Figure 1). On the other hand, the standard curves for silibinin (based on the silibinin diastereomer mixture and silymarin mixture) are emphasized in Equations (3) and (4), as well as Figure 2. Both chrysin-rutin and silibinin standard systems have similar spectrophotometric behavior with the *M. chamomilla* and *S. marianum* extracts, respectively. The liquid chromatographic technique was not used at this stage due to its main disadvantages regarding such a study, namely its time-consuming nature and possible degradation/hydrolysis during analysis by water/acidic based solvent mixtures.

$$\text{Conc. } (\mu g/mL) = 85.7(\pm 0.00)\cdot \text{Abs.}_{(@322)}, \text{ for rutin} \tag{1}$$

$n = 6$, $r^2 = 0.9999$, $p < 0.00001$

$$\text{Conc. } (\mu g/mL) = 29.4(\pm 0.00)\cdot \text{Abs.}_{(@322)}, \text{ for chrysin} \tag{2}$$

$n = 5$, $r^2 = 1.0000$, $p < 0.00001$

$$\text{Conc. } (\mu g/mL) = 7.136\ (\pm 0.040)\cdot \text{Abs.}_{(@288)}, \text{ for silibinin (diastereomer mixture)} \tag{3}$$

$n = 6$, $r^2 = 0.9996$, $p < 0.00001$

$$\text{Conc. } (\mu g/mL) = 13.066\ (\pm 0.038)\cdot \text{Abs.}_{(@288)}, \text{ for silymarin mixture} \tag{4}$$

$n = 9$, $r^2 = 0.9999$, $p < 0.00001$

The overall estimation of the flavonoid contents (expressed as chrysin or rutin) revealed that they significantly differed between the flowers/leaves and roots/stems extracts of fresh samples. They were determined using Equations (1) and (2) and various extract dilutions (dilution levels of 0.005 to 0.08). The variation of the estimated flavonoid content (determined as chrysin or rutin in the fresh flowers, leaves, roots, and stems) with the ethanol concentration differs for flowers and leaves in comparison with roots and stems (Figure 3a–d). The maximum flavonoid content for *M. chamomilla* flowers was obtained using ethanol 96% as extraction solvent (8.24 ± 4.69 mg chrysin/g fresh weight, FW; higher values were obtained using rutin as standard: 24.02 ± 13.66 mg/g). On the other hand, roots and stems had lower flavonoid contents, as determined from extracts obtained with ethanol at 20–96% concentrations (1.14–2.75 mg/g FW as chrysin and 4.01–8.01 mg/g as rutin). Results obtained for the estimated flavonoid content (as chrysin or rutin, mg/g FW) of *M. chamomilla* flowers, leaves, roots, and stems using ethanol of various concentrations (*v*/*v*) for extraction are presented in the Supplementary Material file (Table S3). There were only few studies regarding the quantification of polyphenolic contents in specific parts of *M. chamomilla*. They are especially focused on flowers, using liquid chromatographic techniques. Generally, flavonoid contents were determined in dry extracts, such as chamomile ligulate flower extract obtained by subcritical water extraction. Rutin was the main flavonoid glycoside compound identified in extracts at 10 bar pressure (0.23 mg/kg dry extract), while chrysin had the highest concentration in the dry extract obtained at a pressure of 30 bar (1.99 mg/kg). However, higher concentrations of flavonoid aglycones and glycosides were obtained at increased extraction pressure values (1345 mg apigenin/kg at 45 bar, and 1075 mg luteolin 7-*O*-glucoside/kg at 30 bar) [7]. These values were higher if the extraction temperature was increased from 65 to 85 °C. At higher temperatures, the flavonoid aglycones become more concentrated due to partial hydrolysis of the corresponding glycosides during extraction. It is the case of luteolin 7-*O*-glucoside (1101 mg/kg dry extract obtained at 85 °C) and its aglycone, luteolin (97.2 mg/kg at 115 °C and only 55.2 mg/kg at 85 °C) [11]. Apigenin glycosides were also quantified in dry *M. chamomilla* ligulate flowers during flowering. Apigenin 7-*O*-glucoside was determined in the range of 11.38–32.4 mg/g, but the corresponding 6″-malonyl, 6″-caffeoyl, and 4″-acetyl-6″-malonyl-glucosides were also identified [16].

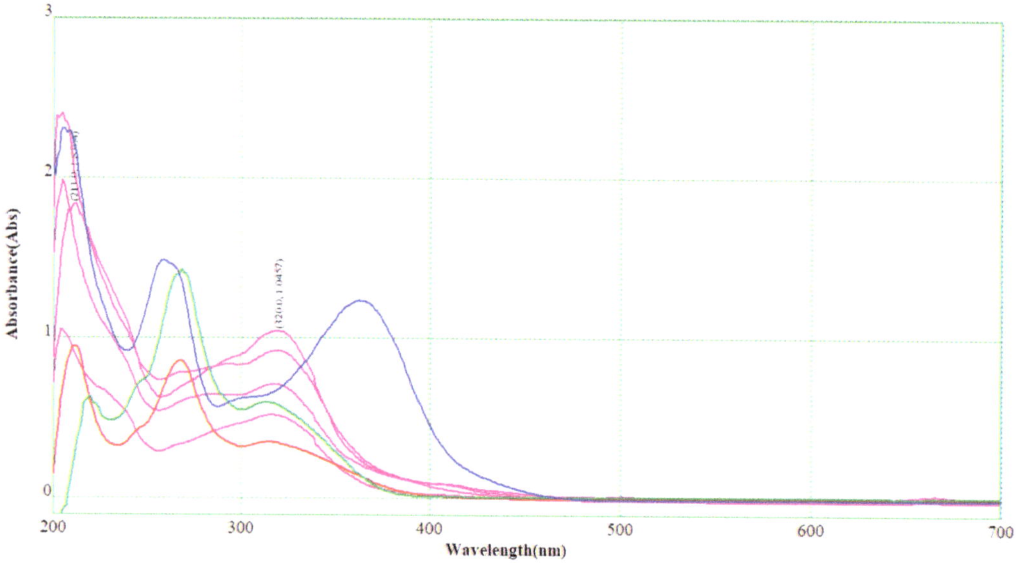

Figure 1. Superimposed UV–Vis spectra for *M. chamomilla* flower, leaf, root, and stem extracts (various dilutions, pink), standard rutin and chrysin solutions (38.0 μg/mL and 16.8 μg/mL, green and blue), respectively, and the mixture of rutin (0.76 μg/mL) and chrysin (6.72 μg/mL) (red).

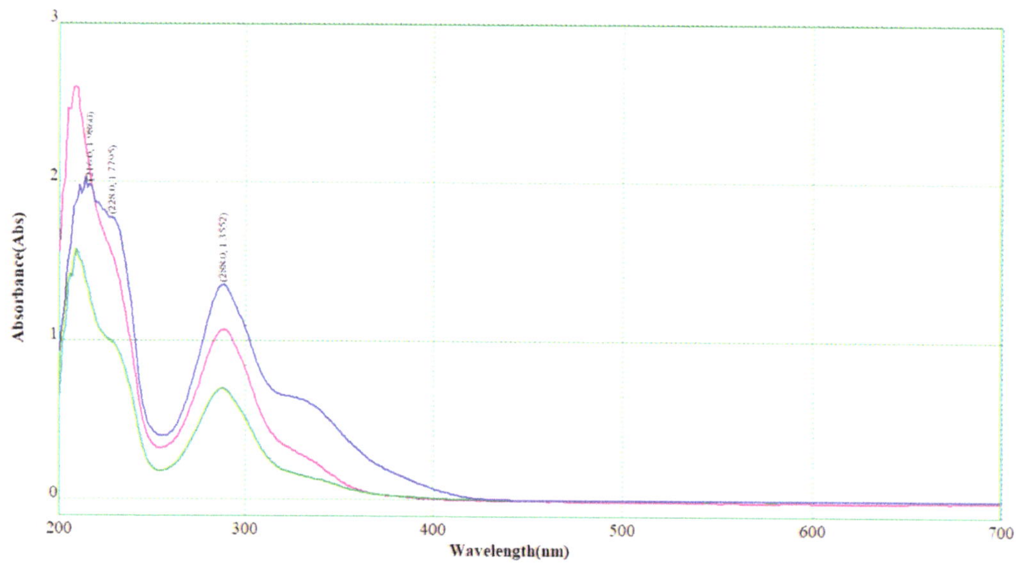

Figure 2. Superimposed UV–Vis spectra for *S. marianum* extract (example for a 33-fold diluted extract, pink), standard silibinin diastereomer mixture, and silymarin mixture solutions (7.5 μg/mL and 9.0 μg/mL, blue and green).

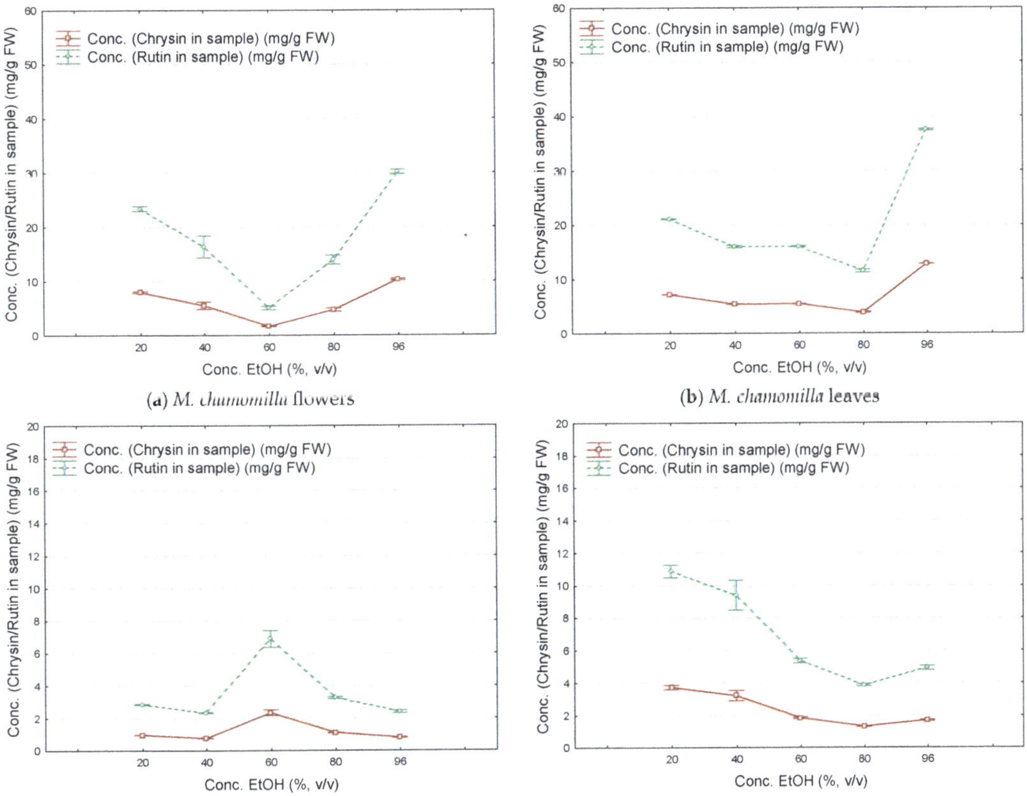

(**a**) *M. chamomilla* flowers (**b**) *M. chamomilla* leaves (**c**) *M. chamomilla* roots (**d**) *M. chamomilla* stems

Figure 3. Variation of the estimated flavonoid content in the fresh samples (as chrysin—continuous line, and as rutin—dashed line, mg/g fresh weight) with the ethanol concentration (v/v) used for obtaining the *M. chamomilla* extracts (number of replicate determinations n = 9; error bars were determined from standard errors using a coefficient of ±0.95).

In a similar manner, the estimated silibinin content of the raw *S. marianum* seeds was determined on the basis of the silibinin diastereomer mixture and silymarin mixture standard curves, using ethanol of various concentrations for extraction (Figure 4). Except for ethanol 20%, all other solvents of 40–96% provide silibinin contents in a very narrow range (mean values of 6.74 ± 0.65 mg/g FW by using silibinin diastereomer standard mixture and 7.93 ± 0.77 mg/g FW by using silymarin standard mixture). These values were approximately equal if ethanol 20% was used for extraction (3.00 ± 0.76 and 3.53 ± 0.90 mg/g FW, respectively). All results obtained for the silibinin content of *S. marianum* samples are presented in Table S4, Supplementary Material file. The results are in good agreement with the other studies related to the silibinin content of *S. marianum* seeds. The silibinin content ranges from 3086 to 9499 mg/kg, depending on genotype [24]. In another study, the silibinin content of the dry extract was in the range of 3.46–10.57 g/100 g, while the total flavonolignans were quantified in the dry extract at 20.47–34.41 mg/100 g [20].

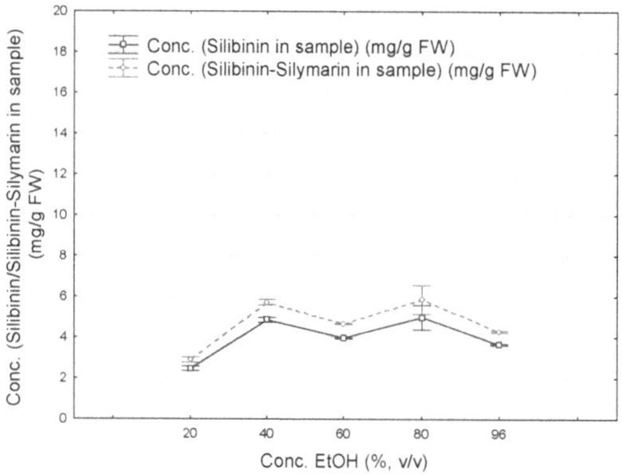

Figure 4. Variation of the estimated silibinin content in the fresh samples (as silibinin based on silibinin standard diastereomer mixture—continuous line, and as silibinin based on standard silymarin mixture—dashed line, mg/g FW) with the ethanol concentration (v/v) used for obtaining the *S. marianum* extracts (number of replicate determinations n = 9; error bars were determined from standard errors using a coefficient of ±0.95).

2.2. Obtaining of β-Cyclodextrin/M. chamomilla Extract and β-Cyclodextrin/S. marianum Extract Complexes

β-CD/vegetable extract complexes were obtained by the co-crystallization method using equal volumes of β-CD water solution and ethanolic extract. Only extracts obtained with ethanol 96% were used for β-CD complexation. Due to the complexity of antioxidant compounds in the vegetable extracts, the recovering yields of the β-CD/vegetable extract complexes were determined as the dried complex mass divided by the sum of β-CD hydrate mass and antioxidant mass (estimated as rutin for *M. chamomilla* extract and as silibinin for *S. marianum* extract) and expressed as percent. Recovering yields for β-CD/*M. chamomilla* extract complexes were in a narrow range of 55.71–61.87%. On the other hand, these yields were significantly higher for β-CD/*S. marianum* extract complexes (75.72%), which were slightly lower than for the corresponding β-CD/standard silibinin diastereomer mixture and β-CD/silymarin mixture complexes (87.03 and 86.47%, respectively). Quantities, volumes, and recovering yields for all synthesized complexes are presented in Table S4 (Supplementary Material file).

The thermal stability of complexes was determined by differential scanning calorimetry (DSC). Two main regions were identified: (1) the temperature range of 20–150 °C, where the calorimetric effect appears, which corresponds to the dissociation of water and ethanol molecules from the β-CD complex; and (2) the temperature range of ~270–320 °C, where the decomposition of β-CD and encapsulated antioxidant compounds takes place. The β-CD/*M. chamomilla* flower extract complex has a significantly lower calorimetric effect for the first region in comparison with the β-CD hydrate (792 and 1207 J/g, respectively; Figure 5). Moreover, the dissociation of the crystallization water/ethanol has the highest rate at 78.4 °C for the complex, while this value is much higher for β-CD hydrate (123.9 °C). This is especially due to the partial replacement of strongly retained water molecules into the β-CD complex in comparison with the β-CD hydrate. No significant differences were observed for the temperature corresponding to the maximum rate of decomposition (~289 °C for both cases). For the other complexes, the temperature for the first region is lower than the corresponding β-CD hydrate, but the differences are smaller, especially for β-CD complexes with leaf and root extracts. In these cases, the calorimetric effect is also closer to the β-CD hydrate. They were in the range of 119.8–123.4 °C and 1031–1155 J/g, respectively. The β-CD/*M. chamomilla* stem extract complexes have intermediate values for these parameters (103.1–85.9 °C and 804.1 J/g, respectively; see Figure 6). The temperature decomposition of these complexes was in the same narrow range of 278.7–285.2 °C. Another calorimetric peak appears at approximately 221 °C for β-CD hydrate, which is slightly modified after complexation. It is due to the transition of the anhydrous β-CD from the crystalline to the amorphous forms (see also Figures S1 and S2 in the Supplementary Material file).

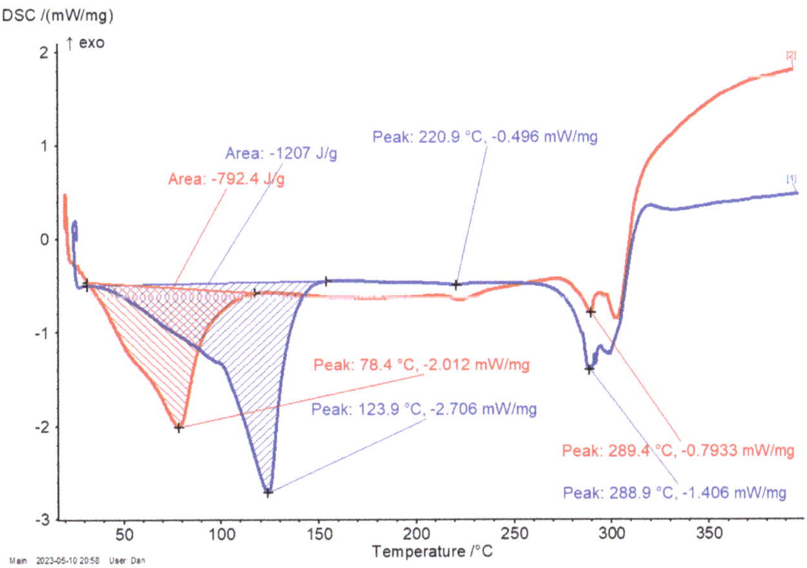

Figure 5. DSC results for the β-CD/*M. chamomilla* flower extract complex (red) and β-CD hydrate (blue); the DSC conditions were set from 20 to 400 °C, with a heating rate of 4 °C/min, under nitrogen (purge and flow).

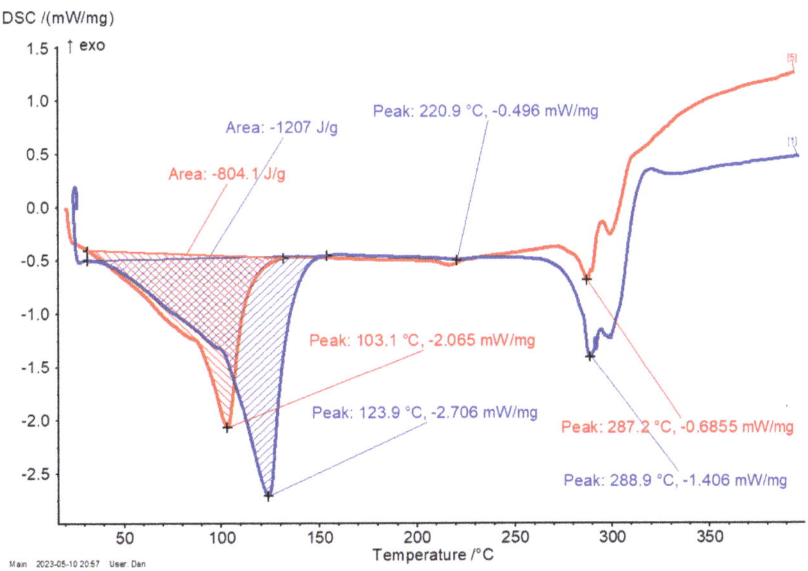

Figure 6. DSC results for the β-CD/*M. chamomilla* stem extract complex (red) and β-CD hydrate (blue); the DSC conditions were set from 20 to 400 °C, with a heating rate of 4 °C/min, under nitrogen (purge and flow).

Similar behavior was observed for β-CD/*S. marianum* seed extract complex, but with a significantly lower value for the temperature corresponding to water/ethanol molecule release. Moreover, there are two peak temperatures at 64.2 and 82.6 °C, which are due to the release of surface water and strongly retained water molecules [47,48]. The endothermal calorimetric effect is close to that corresponding to β-CD/*M. chamomilla* flower and stem extract complexes (863.8 J/g, in comparison with 792.4–804.1 J/g). The maximum decomposition rate appears at a slightly lower value of 279 °C, while the transition effects were observed at 192.4 and 220.1 °C (Figure 7).

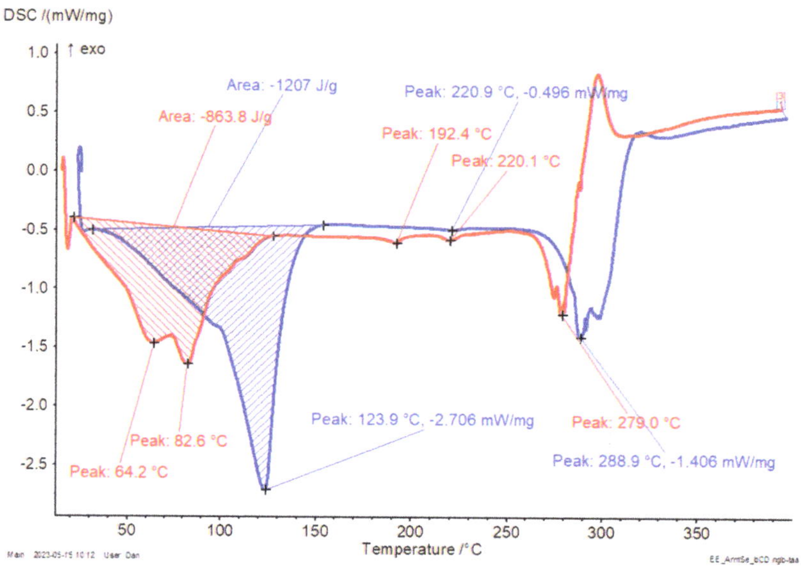

Figure 7. DSC results for the β-CD/*S. marianum* seed extract complex (red) and β-CD hydrate (blue); the DSC conditions were set from 20 to 400 °C, with a heating rate of 4 °C/min, under nitrogen (purge and flow).

This calorimetric behavior was also observed in other studies for CD/antioxidant complexes based on *Capsicum annuum* ethanolic extract, flavonoids such as chrysin, naringenin, hesperetin, apigenin, luteolin, and rutin, as well as flavonolignans, namely silibinin [32,49–51]. Moreover, CD/hydrophobic compound complexes including standard fatty acids or hydrophobic mixtures (essential oils, vegetable oils/lipid fractions, or fish oils) reveal a decrease in the calorimetric effect corresponding to water release as well as a change in the endo-exothermic calorimetric effect of the crystalline-amorphous transition processes [52–59].

The water content of β-CD complexes can be selectively determined using the volumetric Karl Fischer titration (KFT) [60]. Thus, the ethanol molecules that can remain in the complex during preparation will not be accounted by using this method. Water content values were in the range of 10.56–12.37% for complexes, significantly lower than β-CD hydrate (15.27%, Table 1). β-CD complexes with silibinin diastereomer standard mixture or silymarin standard mixture had KFT water content values close to the upper limit of the above-mentioned range (12.50 and 12.71%, respectively, Table 1). The KFT water content well correlates with the DSC endothermic calorimetric peak corresponding to the dissociation of water/ethanol molecules from the β-CD complexes, with a good determination coefficient of r^2 = 0.690 (see Equation (5)). Both KFT water content/titration behavior and KFT–DSC correlations were in accordance with other studies related to CD/flavonoid, CD/vegetable extract, CD/fish oil, or CD/essential oil complexes [32,49,55,56,59].

$$\textit{Water content}\ (\text{by KFT, \%}) = 4.529(\pm 2.567) + 0.0077(\pm 0.0026)\cdot DSC_{Area\ (20\text{-}150\ ^\circ C)} \quad (5)$$

$n = 6, r^2 = 0.690, p < 0.05, s = 1.05$

Table 1. Water content of β-CD/vegetable extract complexes and β-CD hydrate, obtained by volumetric Karl Fischer titration technique (values are expressed as mean ± standard deviation, SD, $n = 3$).

Code	Description	Water Content (%)
bCD	β-Cyclodextrin hydrate	15.27 ± 0.15
bCD_MChF	β-Cyclodextrin/*M. chamomilla* flower extract complex	11.05 ± 0.04
bCD_MChL	β-Cyclodextrin/*M. chamomilla* leaf extract complex	12.37 ± 0.68
bCD_MChR	β-Cyclodextrin/*M. chamomilla* root extract complex	11.90 ± 0.12
bCD_MChS	β-Cyclodextrin/*M. chamomilla* stem extract complex	11.31 ± 0.14
bCD_SMa	β-Cyclodextrin/*S. marianum* seed extract complex	10.56 ± 0.31
bCD_Sb	β-Cyclodextrin/silibinin complex	12.50 ± 0.28
bCD_Sm	β-Cyclodextrin/silymarin complex	12.71 ± 0.33

2.3. Controlled Release of Antioxidants from β-Cyclodextrin Complexes and Transdermal Pharmaceutical Formulations Containing β-Cyclodextrin Complexes

β-CD forms stable molecular inclusion complexes with a wide range of compounds having various structural characteristics. Flavonoid aglycones such as quercetin, naringenin, hesperetin, kaempferide, and chrysin are complexed by β-CD with stability constants of 602/3345, 235, 355, 1541, and 574 M^{-1} [36,61–63]. On the other hand, the more hydrophilic flavonoid glycoside rutin has a stability constant for the β-CD/rutin inclusion complex of 260–265 M^{-1} [39,61,64]. The last stability constant is lower due to the reduced level of hydrophobic interaction of flavonoid glycoside with the inner cavity of the β-CD in comparison with the case of less hydrophilic flavonoid aglycones. Similar behavior was observed for β-CD/silymarin (silibinin) molecular inclusion complexes, which had an apparent stability constant of 722 M^{-1} [44]. Highly hydrophobic molecules such as fatty acids, lipid sterols, and terpenes provide much higher β-CD binding constants. Thus, β-CD/lauric acid and β-CD/cholesterol complexes had a 1:1 binding constant of 28,000 and 23,190 M^{-1}, respectively [65,66]. On the other hand, monoterpenes such as β-pinene, myrcene, and limonene have β-CD formation constant values of 4646, 1267, and 2605 M^{-1}, while the sesquiterpene β-caryophyllene has a significantly higher value of 23,032 M^{-1} [67]. Consequently, more hydrophilic flavonoid aglycones and especially glycosides, as well as flavonolignans, will be easier released from the β-CD complex in proper solvents or matrices. In this study, the controlled release of flavonoids (expressed as rutin or chrysin) from β-CD/*M. chamomilla* extract complexes and flavonolignans (expressed as silibinin) from β-CD/*S. marianum* extract complexes was estimated by spectrophotometric analysis. The solvents used for the controlled release studies were ethanol at concentrations of 20, 60, and 96% for the crystalline β-CD/vegetable extract complexes, as well as saline solution (0.9% NaCl at pharmaceutical grade) and the above-mentioned ethanol solutions for the studies on transdermal pharmaceutical formulations.

The estimated flavonoids were released in higher amounts, especially in ethanol 96%, in comparison with ethanol 20 and 60%. The maximum cumulative content of flavonoids (expressed as μg rutin/mL after 15 min) obtained for the controlled release from β-CD/*M. chamomilla* flower and leaf extract complexes was 32.5 μg/mL (mean value) in ethanol 96%. On the contrary, these values were 10.4 and 0.8 μg/mL if ethanol 60 and 20% were used. By comparison, the maximum flavonoid content of only 18.0 μg/mL was obtained by controlled release from β-CD/*M. chamomilla* root and stem extract complexes (Figure 8a and Figures S3, S4, S36–S38 in the Supplementary Material file). Similar behavior was observed for the controlled release of silibinin from the β-CD/*S. marianum* extract complexes. The estimated maximum silibinin content was obtained by controlled release in ethanol 96% (0.56 μg/mL), compared with ethanol 20 and 60% (<0.24 μg/mL). By comparison, standard silibinin was easily released from the β-CD/silibinin complexes (mean value of 9.3 μg/mL in ethanol 96% after 15 min) (Figures 8b,c and S5–S7).

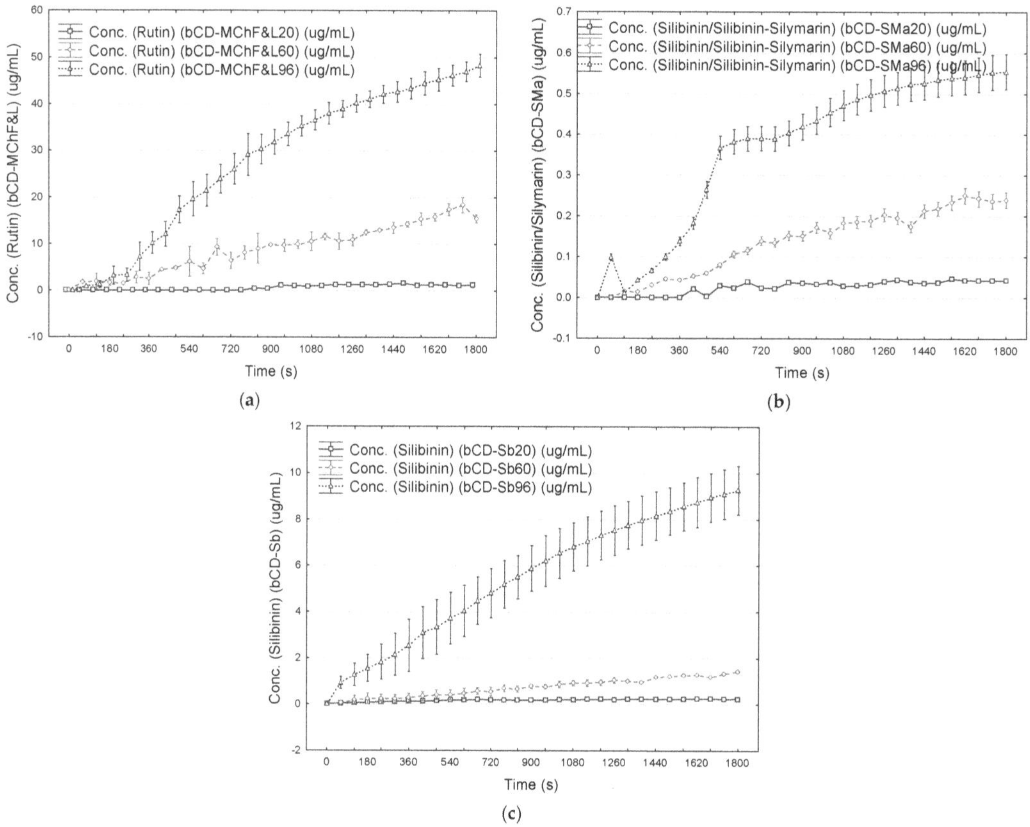

Figure 8. Controlled release of the estimated antioxidant compounds from the β-cyclodextrin/*M. chamomilla* flower and leaf extract complexes (expressed as rutin, μg/mL) (**a**), from the β-cyclodextrin/*S. marianum* seed extract complexes (expressed as silibinin, μg/mL) (**b**), and from β-cyclodextrin/silibinin or silymarin complexes (expressed as silibinin, μg/mL) (**c**) in ethanol 20% (brown), ethanol 60% (green), and ethanol 96% (bleu); number of replicate determinations $n = 2$; error bars were determined from standard errors using a coefficient of ±0.95. Codes for complexes are: *bCD_MChXY*—β-cyclodextrin/*M. chamomilla* L. extract; *bCD_SMaY*—β-cyclodextrin/*S. marianum* L. extract; X stands for the plant part, *X = F*—flowers, *L*—leaves, *R*—roots, or *S*—stems; *bCD_Sb/SmY*—β-cyclodextrin/silibinin or β-cyclodextrin/silymarin complexes; *Y* stands for the concentration of ethanol solution used for controlled release (20, 60, and 96%, *v/v*).

The mathematical modeling of the controlled release of antioxidants from β-CD complexes and transdermal pharmaceutical formulations can provide information about the releasing mechanism (Figures 8a–c, 9a–d and 10a–d). The Korsmeyer–Peppas model has been considered [68]. Values of the diffusional exponent n of 0.5 and 1 reveal Fickian diffusion and Case II transport (relaxation release), respectively. A qvasi-Fickian diffusion (partial diffusion) appears if $n < 0.5$, while the non-Fickian (anomalous) diffusion appears for $n \in (0.5, 1)$, which means that compounds are released by both diffusion and relaxation (erosion). Finally, the Super Case II transport is revealed by a diffusional exponent of $n > 1$ [68,69].

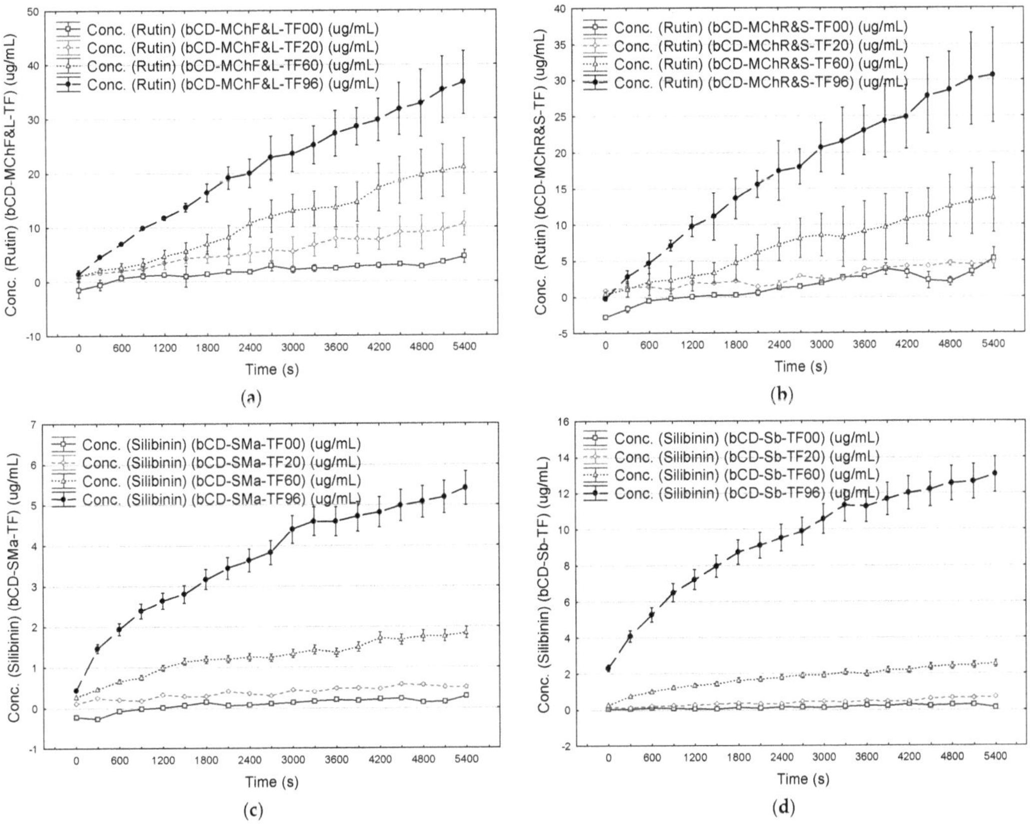

Figure 9. Controlled release of the estimated antioxidant compounds from the transdermal pharmaceutical formulations containing β-cyclodextrin/*M. chamomilla* flower and leaf extract complexes (expressed as rutin, μg/mL) (**a**), from the transdermal pharmaceutical formulations containing β-cyclodextrin/*M. chamomilla* root and stem extract complexes (expressed as rutin, μg/mL) (**b**) from the transdermal pharmaceutical formulations containing β-cyclodextrin/*S. marianum* seed extract complexes (expressed as silibinin, μg/mL) (**c**), and from the transdermal pharmaceutical formulations containing β-cyclodextrin/silibinin complexes (expressed as silibinin, μg/mL) (**d**), in saline solution (brown), ethanol 20% (green), ethanol 60% (bleu), and ethanol 96% (black); number of replicate determinations $n = 2$; error bars were determined from standard errors using a coefficient of ±0.95. Codes for the transdermal pharmaceutical formulations based on cyclodextrin complexes are: *bCD_MChX_TFY*—transdermal pharmaceutical formulation based on β-cyclodextrin/*M. chamomilla* L. extract; *bCD_SMa_TFY*—transdermal pharmaceutical formulation based on β-cyclodextrin/*S. marianum* L. extract; *X* stands for the plant part; *X* = *F*—flowers, *L*—leaves, *R*—roots, or *S*—stems; *bCD_Sb/Sm_TFY*—transdermal pharmaceutical formulation based on β-cyclodextrin/silibinin or β-cyclodextrin/silymarin complexes; *Y* stands for saline solution ("00") or the concentration of ethanol solution used for controlled release (20, 60, and 96%, *v*/*v*).

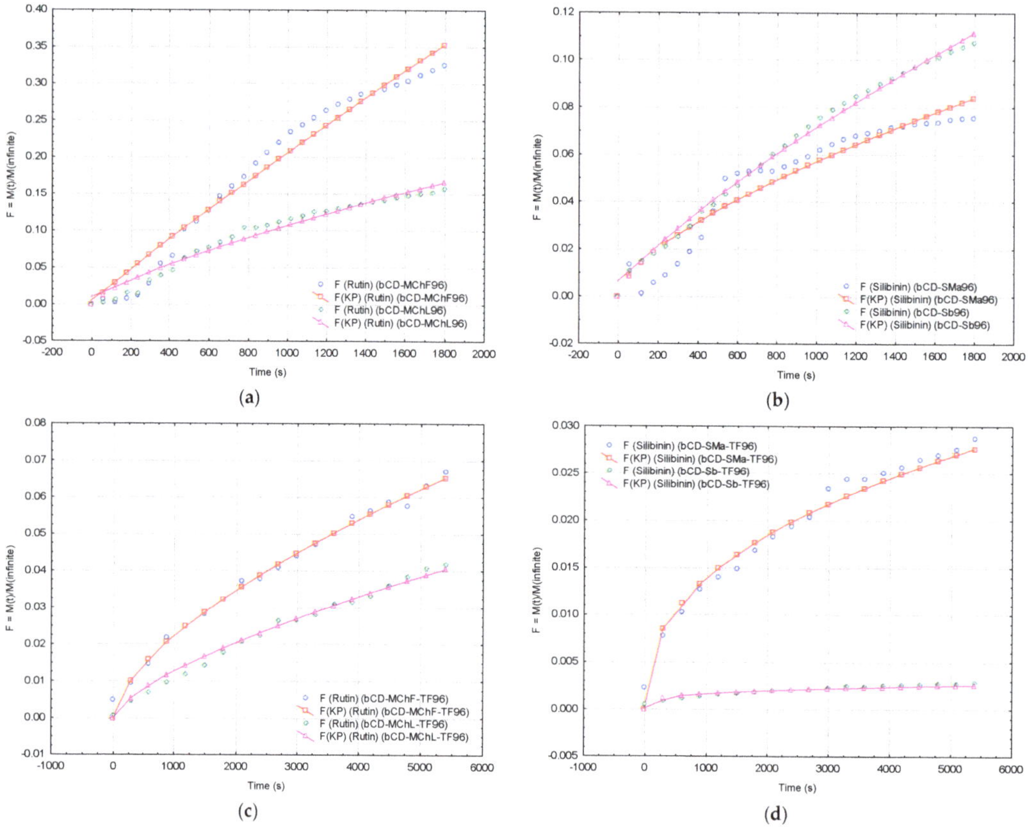

Figure 10. Korsmeyer–Peppas modeling ($M_t/M_\infty = k_{(KP)} \cdot t^n$) of the controlled release of the estimated antioxidants from β-cyclodextrin complexes and from transdermal pharmaceutical formulations based on β-cyclodextrin complexes. M_t and M_∞ stand for the amount (mg) of antioxidant released from the complex at the time t (s) and the total amount, respectively. The Korsmeyer–Peppas kinetic constant, $k_{(KP)}$, and the diffusional exponent, n, were determined by fitting the experimental data (least squares approximation; continuous lines). Codes for complexes are: *bCD_MChXY*—β-cyclodextrin/*M. chamomilla* L. extract; *bCD_SMaY*—β-cyclodextrin/*S. marianum* L. extract; codes for the transdermal pharmaceutical formulations based on cyclodextrin complexes are: *bCD_MChX_TFY*—transdermal pharmaceutical formulation based on β-cyclodextrin/*M. chamomilla* L. extract; *bCD_SMa_TFY*—transdermal pharmaceutical formulation based on β-cyclodextrin/*S. marianum* L. extract; *X* stands for the plant part, *X* = *F*—flowers, *L*—leaves, *R*—roots or *S*—stems; *bCD_Sb/Sm_TFY*—transdermal pharmaceutical formulation based on β-cyclodextrin/silibinin or β-cyclodextrin/silymarin complexes; *Y* stands for saline solution ("00") or the concentration of ethanol solution used for controlled release (20, 60, and 96%, v/v). Experimental plots and Korsmeyer–Peppas model fitting (plots and continuous line) for the controlled release of the estimated antioxidants from *bCD_MChF* and *bCD_MChL* complexes in ethanol 96% (**a**), from *bCD_SMa* and *bCD_Sb* complexes in ethanol 96% (**b**), from transdermal pharmaceutical formulations based on *bCD_MChF* and *bCD_MChL* complexes in ethanol 96% (**c**), and from transdermal pharmaceutical formulations *bCD_SMa* and *bCD_Sb* complexes in ethanol 96% (**d**).

The best results were obtained for ethanol 60 and 96% (best fitting of the experimental data; see Section 3.5). Thus, Case II transport ($n \cong 0.92$) and super-Case II transport ($n > 1$) mechanisms were observed for controlled release of estimated antioxidants from β-CD/*M.*

chamomilla flower extract complexes in ethanol 96% and ethanol 60%, respectively (r^2 of 0.985 and 0.966). On the contrary, a non-Fickian diffusion ($n \cong 0.76$) was observed for the cases of β-CD/*M. chamomilla* leaf extract complexes studied in ethanol 60 and 96% (r^2 of 0.97–0.99). The release mechanisms of the estimated antioxidants from β-CD/*M. chamomilla* root and stem extract complexes were Case II transport for releasing in ethanol 60 and 96% (root extract case, n = 1.02–1.04, $r^2 \cong 0.98$), or in ethanol 60% for the case of stem extract-based complexes (n = 0.94, r^2 = 0.97). The silibinin release mechanism from β-CD/*S. marianum* extract complexes was non-Fickian for all ethanol concentrations (n = 0.66–0.81 and r^2 = 0.90–0.98), such as in the case of β-CD/silibinin complexes in ethanol 96% (n = 0.76, r^2 = 0.998). A comparison between the Korsmeyer–Peppas model and experimental data is emphasized in Figure 10a,b for the controlled release of estimated antioxidants from β-CD/*M. chamomilla* flower and leaf extract complexes, β-CD/*S. marianum* extract, and β-CD/silibinin complexes in ethanol 96%. All values for the kinetic constant, $k_{(KP)}$, and diffusional exponent, n, as well as the values for the coefficient of determination, r^2, corresponding to the fitting of the Korsmeyer–Peppas model with the experimental data, are presented in Table S6 (Supplementary Material file).

The same Korsmeyer–Peppas model was used for the evaluation of the release mechanism of antioxidants from transdermal pharmaceutical formulations. In all cases, the diffusional exponent was $n < 1$, except in the case of β-CD/*M. chamomilla* flower extract complexes studied in ethanol 60%, where $n = 1$ (Case II transport mechanism). However, the diffusional exponent had values of 0.5–1 for all other β-CD/*M. chamomilla* extract complex-based formulations studied in ethanol 60 and 96%, revealing a non-Fickian diffusion for the antioxidant release (Figures 9a,b and 10c, as well as Table S7 and Figures S8–S23, S39 and S40 in the Supplementary Material file). On the contrary, a qvasi-Fickian diffusion of the estimated antioxidants (silibinin) in ethanol of 20–96% concentrations from all transdermal pharmaceutical formulations based on β-CD/*S. marianum* extract, β-CD/silibinin, and β-CD/silymarin complexes was observed. The highest n values and the best correlations were obtained if ethanol 96% was used (n = 0.14–0.40, r^2 = 0.88–0.99, Figures 9c,d and 10d). All results are presented in Table S7, Figures S24–S27 for the case of β-CD/*S. marianum* extract complexes, and Figures S28–S35 and S41 for the case of β-CD/silibinin or silymarin complexes (Supplementary Material file).

3. Materials and Methods

3.1. Plant Samples and Chemicals

Plant materials (*M. chamomilla* L. and *S. marianum* L.) were collected from specific locations in the west of Romania. Thus, *M. chamomilla* L. plants were harvested during autumn seasons of the last years from the Salonta wild region (46°48′0″ N, 21°39′00″ E, Bihor County, Romania). Flowers, leaves, roots, and stems were carefully separated and stored as fresh samples at −20°C until extractions. Only the seeds of *S. marianum* L. collected in the same period from the Macea region (46°23′12″ N, 21°18′39″ E, Arad County, Romania) were used. The main antioxidant compounds in extracts were chrysin, quercetin, and rutin hydrate. They were used as standard compounds from the flavonoid class and had purities of 97%, 98%, and 94%, respectively (Sigma-Aldrich, St. Louis, MO, USA). On the other hand, standard compounds from the flavonolignan class, namely silibinin and silymarin, had contents of 98% and 64.7% silibinins, respectively (silymarin composition was based on spectrophotometric analysis in this study; Sigma-Aldrich, St. Louis, MO, USA). β-Cyclodextrin hydrate was purchased from CycloLab (Budapest, Hungary) and had >98% purity and a water content of 12.4% (loss on drying, according to the manufacturer). Ethanol 96% (*v*/*v*) used for CD complexation and controlled release studies was obtained from ChemiCal® Co. (Iaşi, Romania).

3.2. Obtaining *M. chamomilla* and *S. marianum* Extracts

Fresh samples of *M. chamomilla* (flowers, leaves, roots, and stems) and *S. marianum* (seeds) were well grounded in a mortar. Approximately 5 g of every sample was immediately weighted into a 150 mL extraction flask. The samples were extracted with 70 mL of ethanol at various concentrations (20, 40, 60, 80, and 96%, *v*/*v*, a solid:solvent ratio of ~1:14) for 1.5 h by continuous stirring on a water bath at 55–60 °C [14]. The extracts were subjected to spectrophotometric analysis after filtration (see below) or were subjected to concentration at a third of the volume for CD nanoencapsulation. All data and the raw results of the extraction of extracts are presented in the Supplementary Material file (Table S1).

3.3. Obtaining of β-cyclodextrin/M. chamomilla Extract and β-cyclodextrin/S. marianum Extract Complexes

β-CD/vegetable extract complexes were obtained using the co-crystallization method. Approximately 0.5 mmoles of β-CD hydrate were dissolved in 4 mL water in a 20 mL crystallization flask, which was equipped with cooling-heating mantle, reflux condenser, and magnetic stirring system. The solution was heated to 50 °C, and then 4 mL of vegetable extract was added dropwise through the top of the condenser for 15 min. The solution became turbid during the controlled cooling to room temperature (cooling rate of ~7.5 °C/h) and the suspension of the complex was stored at 4 °C overnight. The crystallized complex was filtered through Whatman® filter paper No 1 (Merck KGaA, Darmstadt, Germany) and washed with 1 mL ethanol 96%. After drying at room temperature, the complex crystals were weighted, sealed, and stored at 4 °C until further analysis.

3.4. Obtaining Transdermal Pharmaceutical Formulations

The controlled release behavior of β-CD/vegetable extract complexes was evaluated from the standard transdermal pharmaceutical formulations (transdermal cream consisting of 10% lanolin and 90% vaseline, Herbavit, Romania). The β-CD complex was incorporated into the transdermal cream samples by mixing for 1 h at a complex:cream ratio of 1:20. All determinations were performed against a blank sample (without complexes).

3.5. Spectrophotometric (UV–Vis) Analysis and Controlled Release Measurements

Spectrophotometric analysis was performed for extracts and standard compound solutions at various concentrations, as well as for the evaluation of their release from β-CD/extract complexes and pharmaceutical formulations. This is a fast and non-destructive method for the evaluation/estimation of the overall release of biologically active compounds from nanoparticles and pharmaceutical formulations. A CamSpec 501 equipment (CamSpec Ltd., Cambridge, UK) was used. The absorbance values of the samples were monitored at specific wavelengths (322 and 288 nm for *M. chamomilla* and *S. marianum* extracts, respectively) using the Wavelength Scan Measurement and/or Time Scan Measurement modules. The monitoring time was set at 30 min for controlled release from the crystalline β-CD/vegetable extract complexes in ethanol solutions of 20, 60, and 96% concentrations (*v*/*v*). The controlled release of the estimated bioactive compounds from the transdermal pharmaceutical formulations based on β-CD/vegetable extract complexes was studied for 90 min at the same wavelengths, using even ethanol (20, 60, and 96%) or saline solution (0.9% NaCl, pharmaceutical grade) as solvent media. The acquisition and handling of the UV–Vis data were performed with the UV–Vis Analyst version 4.67 software from the same manufacturer. In all cases, the appropriate solvent mixture was used as the blank sample.

The controlled release of the estimated antioxidants from β-CD complexes or from transdermal pharmaceutical formulations based on β-CD complexes was modeled using the Korsmeyer–Peppas model (Equation 6) [68]:

$$F = M_t/M_\infty = k_{(KP)} \cdot t^n \quad (6)$$

The model uses the ratio (F) between the antioxidant (biologically active compound or drug molecule) amount that is released at the moment t (s) and the total amount of antioxidant in the sample (M_t/M_∞). This ratio is correlated with kinetic constant $k_{(KP)}$

(expressed as s^{-n}) and the time at the power of a diffusional exponent, t^n. Parameters $k_{(KP)}$ and n reveal the controlled release mechanism and were determined from the experimental data by fitting (least squares approximation). All values for these parameters, as well as those for the coefficient of determination, r^2, can be found in the Supporting Material file (Tables S6 and S7).

3.6. Differential Scanning Calorimetry (DSC)

DSC analysis was performed for all solid materials (β-CD/vegetable extract complexes and β-CD hydrate) in order to evaluate the calorimetric effects during heating. A DSC Netzsch 204 F1 Phoenix equipment (Netzsch Group, Selb, Germany) was used, with a temperature range of 20–400 °C and a heating rate of 4 °C/min. Nitrogen protection and purge flow were set at 20 mL/min. The sample masses were 4.3–8.0 mg for complexes and 9.8–10.0 mg for β-CD. DSC analysis was performed in open Al crucibles using a reference crucible of 39.4 mg. The Netzsch Proteus-Thermal Analysis ver. 6.1 software (Netzsch Group, Selb, Germany) was used for acquisition and handling of the DSC data.

3.7. Karl Fischer Water Titration (KFT)

The water content of complexes and β-CD hydrate was determined using the volumetric KFT method. A KF Titrando 701 apparatus having a Metrohm 10 dosing unit and coupled with a Ti Stand 703 mixing unit (Metrohm AG, Herisau, Switzerland) was used. The bi-component KFT technique was applied, with the component 1—Titrant 5 apura® (Merck KGaA, Darmstadt, Germany) having a titer of 3.6357 ± 0.0195 mg/g at the analysis period. The titer was determined using the water standard 1% apura® (Merck KGaA, Darmstadt, Germany). Component 2—Solvent apura®—was used as a working medium and was purchased from the same manufacturer (Merck KGaA, Darmstadt, Germany). Sample masses were in the range of 20.0–40.0 mg. The following KFT parameters were used: I(pol) 50μA, end point 250 mV, maximum titration rate 5 mL/min, stop criteria—drift with a value of 15 μL/min. The extraction time was approximately 300 s. All determinations were performed in triplicate.

3.8. Statistical Analysis

All values obtained for multiplicate determinations (analysis and/or synthesis) were provided as mean ± standard deviation (SD). The Basic Statistics&Tables and One-way ANOVA modules in Statistica 7.1 (StatSoft, Inc., Tulsa, OK, USA) software were used. The regression analysis was performed with the Multiple Linear Regression module from the same package. Pearson correlational coefficient, r, the coefficient of determination, r^2, p-values, F-test for the regression equation, and standard errors for both equations and coefficients have been considered.

4. Conclusions

A series of β-CD/chamomile (*Matricaria chamomilla* L.) and β-CD/milk thistle (*Silybum marianum* L.) extract complexes were synthesized and evaluated for their controlled release properties using a fast, cheap, and non-destructive method, namely spectrophotometric analysis. This is the first study on the controlled release of the estimated antioxidants from such complexes and pharmaceutical formulations containing the above-mentioned complexes. Extracts were obtained with ethanol solutions at various concentrations and moderate temperatures in order to reduce the level of degradation of labile antioxidant compounds. High estimates of antioxidant compounds (such as rutin or chrysin for *M. chamomilla* flowers and leaves, as well as silibinin for *S. marianum*) in extracts or raw vegetable samples were obtained if ethanol at concentrations of 80–96% was involved. These extracts were used for obtaining β-CD complexes using the co-crystallization method, which provides high recovering yields, especially for the β-CD/*S. marianum* extract complex. The thermal stability of all complexes is similar to that of β-CD hydrate, according to DSC analysis. However, the hydration water content (especially "strongly bonded", based on both DSC and KFT analyses) is much lower in complexes, which reveals the formation of molecular inclusion complexes.

The modeling of the controlled release of the estimated antioxidant compounds from β-CD complexes and transdermal pharmaceutical formulations based on these β-CD complexes was an important and innovative part of this study. Different ethanol solutions or saline solutions (releasing solvents with various hydrophobicities) were used for controlled release studies. On the other hand, both β-CD complexes (more hydrophilic) and transdermal pharmaceutical formulations based on β-CD complexes (more hydrophobic) were considered for controlled release studies. First, the release of the estimated antioxidants from β-CD complexes was performed in ethanol of 20, 60, and 96% (*v*/*v*), which have log$P$ values of −1.14, −0.74, and −0.37 (estimated from the corresponding mole fractions by using the logarithmic interpolation based on log$P$ values of pure components) [70]. The selection of these ethanol concentrations was based on the highly hydrophilic character of β-CD and its complexes (the solubility of β-CD is much higher in water than in ethanol-water mixtures). In the Korsmeyer–Peppas model, β-CD/*M. chamomilla* flower extract complexes reveal Case II transport mechanisms, while the corresponding complexes with leaf extracts indicate non-Fickian diffusion for the controlled release of the estimated antioxidants in ethanol 60 and 96%. The same non-Fickian diffusion was revealed by β-CD/*S. marianum* extract and β-CD/silibinin complexes. On the contrary, the controlled release studies for the model transdermal pharmaceutical formulations based on β-CD/vegetable extract complexes that were performed in the same ethanol (20% to 96%) as well as in saline solution (0.9% NaCl) revealed completely different releasing mechanisms. Almost all formulations based on β-CD/*M. chamomilla* extract complexes and all those based on β-CD/*S. marianum* extract complexes revealed non-Fickian diffusion for the estimated antioxidant release (expressed as rutin and silibinin, respectively). This is most likely due to the completely different hydrophilic/hydrophobic characteristics of matrices. Thus, β-CD complexes have more hydrophilic cyclic oligosaccharides and crystallization water as the main matrix components. Flavonoids and flavonolignans are mainly hydrophilic. As a consequence, *H*-bonding is especially involved in the diffusion of antioxidants into this type of matrix. For transdermal pharmaceutical formulations, fatty acid glycerides and sterol esters with long-chain fatty acids are the main components. Consequently, a competition between the hydrophobic interactions of the hydrophobic components of the matrix and the complexed antioxidants with the hydrophobic inner cavity of β-CD appears. In this case, the controlled release of antioxidants is mainly due to hydrophobic interactions. These differences have been proven in this study by the modeling of the diffusion of the estimated antioxidants into two types of matrices with different hydrophobicities. Results can be further used for studying the particular antioxidant transdermal transport and biological effects of innovatively designed pharmaceutical formulations that can be obtained using "green" methods and materials.

Supplementary Materials: The following are available online at https://www.mdpi.com/article/10.3390/plants12122352/s1. Table S1: Raw data and results for the obtaining of *M. matricaria* extracts. Table S2: Raw data and results for the obtaining of *S. marianum* extracts. Table S3: Results obtained for the flavonoid content (as chrysin or rutin, mg/g FW) of *M. chamomilla* flowers, leaves, roots, and stems using ethanol of various concentrations (*v*/*v*) for extraction. Table S4: Results obtained for the silibinin content (based on both silibinin diastereomer mixture and silymarin mixture, mg/g FW) of *S. marianum* using ethanol of various concentrations (*v*/*v*) for extraction. Table S5: Results obtained for β-cyclodextrin/vegetable extract or β-cyclodextrin/flavonolignan complexes (codes for complexes: *bCD_MChX*—β-cyclodextrin/*M. chamomilla* L. extract; *bCD_SMaX*—β-cyclodextrin/*S. marianum* L. extract; *X* stands for the plant part, *X* = *F*—flowers, *L*—leaves, *R*—roots, or *S*—stems; *bCD_Sb/Sm*—β-cyclodextrin/silibinin or β-cyclodextrin/silymarin complexes). Figure S1: DSC results for the β-CD/*M. chamomilla* leaf extract complex (red) and β-CD hydrate (blue); the DSC conditions were set from 20 to 400 °C, with a heating rate of 4 °C/min under nitrogen (purge and flow). Figure S2: DSC results for the β-CD/*M. chamomilla* root extract complex (red) and β-CD hydrate (blue); the DSC conditions were set from 20 to 400 °C, with a heating rate of 4 °C/min under nitrogen (purge and flow). Figure S3: Superimposed Absorbance versus Time (s) plots for the controlled release studies on β-CD/*M. chamomilla* flower and leaf extracts in ethanol 20% (blue), ethanol 60% (green), and ethanol 96% (turquoise); raw data obtained by monitoring the absorbance of the supernatant solution at 322 nm for 30 min. Figure S4: Superimposed Absorbance versus Time (s) plots for the controlled release studies on β-CD/*M. chamomilla* root and stem extracts in ethanol 20% (blue), ethanol 60% (green), and ethanol 96% (turquoise); raw data obtained by monitoring the absorbance of the supernatant solution at 322 nm for 30 min. Figure S5: Superimposed Absorbance versus Time (s) plots for the controlled release studies on β-CD/*S. marianum* seed extracts in ethanol 20% (blue), ethanol 60% (green), and ethanol 96% (turquoise); raw data obtained by monitoring the absorbance of the supernatant solution at 288 nm for 30 min. Figure S6: Superimposed Absorbance versus Time (s) plots for the controlled release studies on β-CD/silibinin diastereomer mixture complex in ethanol 20% (blue), ethanol 60% (green), and ethanol 96% (turquoise); raw data obtained by monitoring the absorbance of the supernatant solution at 288 nm for 30 min. Figure S7: Superimposed Absorbance versus Time (s) plots for the controlled release studies on β-CD/silymarin standard mixture complex in ethanol 20% (blue), ethanol 60% (green), and ethanol 96% (turquoise); raw data obtained by monitoring the absorbance of the supernatant solution at 288 nm for 30 min. Figure S8: Superimposed UV–Vis spectra for the controlled release studies on the transdermal pharmaceutical formulation based on β-CD/*M. chamomilla* flower extract complex in saline solution; raw data obtained by monitoring the absorbance of the supernatant solution in the range of 250–400 nm for 90 min (a number of 19 spectra were recorded every 5 min, starting from 0 min—the lowest spectra—to 90 min—the highest spectra). Figure S9: Superimposed UV–Vis spectra for the controlled release studies on the transdermal pharmaceutical formulation based on β-CD/*M. chamomilla* flower extract complex in ethanol 20% solution; raw data obtained by monitoring the absorbance of the supernatant solution in the range of 250–400 nm for 90 min (a number of 19 spectra were recorded every 5 min, starting from 0 min—the lowest spectra—to 90 min—the highest spectra). Figure S10: Superimposed UV–Vis spectra for the controlled release studies on the transdermal pharmaceutical formulation based on β-CD/*M. chamomilla* flower extract complex in ethanol 60% solution; raw data obtained by monitoring the absorbance of the supernatant solution in the range of 250–400 nm for 90 min (a number of 19 spectra were recorded every 5 min, starting from 0 min—the lowest spectra—to 90 min—the highest spectra). Figure S11: Superimposed UV–Vis spectra for the controlled release studies on the transdermal pharmaceutical formulation based on β-CD/*M. chamomilla* flower extract complex in ethanol 96% solution; raw data obtained by monitoring the absorbance of the supernatant solution in the range of 250–400 nm for 90 min (a number of 19 spectra were recorded every 5 min, starting from 0 min—the lowest spectra—to 90 min—the highest spectra). Figure S12: Superimposed UV–Vis spectra for the controlled release studies on the transdermal pharmaceutical formulation based on β-CD/*M. chamomilla* leaf extract complex in saline solution; raw data obtained by monitoring the absorbance of the supernatant solution in the range of 250–400 nm for 90 min (a number of 19 spectra were recorded every 5 min, starting from 0 min—the lowest spectra—to 90 min—the highest spectra). Figure S13: Superimposed UV–Vis spectra for the controlled release studies on the transdermal pharmaceutical formulation based on β-CD/*M. chamomilla* leaf extract complex in ethanol 20% solution; raw data obtained by monitoring the absorbance of the supernatant solution in the range of 250–400 nm for 90 min (a number of 19 spectra were recorded every 5 min, starting from 0 min—the lowest spectra—to 90 min—the highest spectra). Figure S14: Superimposed UV–Vis spectra for the controlled release studies on the transdermal pharmaceutical formulation based on β-CD/*M. chamomilla* leaf extract complex in ethanol 60% solution; raw data obtained by monitoring the absorbance of the supernatant solution in the range of 250–400 nm for 90 min (a number of 19 spectra were recorded every 5 min, starting from 0 min—the lowest spectra—to 90 min—the highest spectra). Figure S15: Superimposed UV–Vis spectra for the controlled release studies on the transdermal pharmaceutical formulation based on β-CD/*M. chamomilla* leaf extract complex in ethanol 96% solution; raw data obtained by monitoring the absorbance of the supernatant solution in the range of 250–400 nm for 90 min (a number of 19 spectra were recorded every 5 min, starting from 0 min—the lowest spectra—to 90 min—the highest spectra). Figure S16: Superimposed UV–Vis spectra for the controlled release studies on the transdermal pharmaceutical formulation based on β-CD/*M. chamomilla* root extract complex in saline solution; raw data obtained by monitoring the absorbance of the supernatant solution in the range of 250–400 nm for 90 min (a number of 19 spectra were recorded every 5 min, starting from 0 min—the lowest spectra—to 90 min—the highest spectra). Figure S17: Superimposed UV–Vis spectra for the controlled release studies on the transdermal pharmaceutical formulation based on β-CD/*M. chamomilla* root extract complex in ethanol 20% solution; raw data obtained by monitoring the absorbance of the supernatant solution in the range of 250–400 nm for 90 min (a number of 19 spectra were recorded every 5 min, starting from 0 min—the lowest spectra—to 90 min—the highest spectra). Figure S18: Superimposed UV–Vis spectra for the controlled release studies on the transdermal pharmaceutical formulation based on β-CD/*M. chamomilla* root extract complex in ethanol 60% solution; raw data obtained by monitoring the absorbance of the supernatant solution in the range of 250–400 nm for 90 min (a number of 19 spectra were recorded every 5 min, starting from 0 min—the lowest spectra—to 90 min—the highest spectra). Figure S19: Superimposed UV–Vis spectra for the controlled release studies on the transdermal pharmaceutical formulation based on β-CD/*M. chamomilla* root extract complex in ethanol 96% solution; raw data obtained by monitoring the absorbance of the supernatant solution in the range of 250–400 nm for 90 min (a number of 19 spectra were recorded every 5 min, starting from 0 min—the lowest spectra—to 90 min—the highest spectra). Figure S20: Superimposed UV–Vis spectra for the controlled release studies on the transdermal pharmaceutical formulation based on β-CD/*M. chamomilla* stem extract complex in saline solution; raw data obtained by monitoring the absorbance of the supernatant solution in the range of 250–400 nm for 90 min (a number of 19 spectra were recorded every 5 min, starting from 0 min—the lowest spectra—to 90 min—the highest spectra). Figure S21: Superimposed UV–Vis spectra for the controlled release studies on the transdermal pharmaceutical formulation based on β-CD/*M. chamomilla* stem extract complex in ethanol 20% solution; raw data obtained by monitoring the absorbance of the supernatant solution in the range of 250–400 nm for 90 min (a number of 19 spectra were recorded every 5 min, starting from 0 min—the lowest spectra—to 90 min—the highest spectra). Figure S22: Superimposed UV–Vis spectra for the controlled release studies on the transdermal pharmaceutical formulation based on β-CD/*M. chamomilla* stem extract complex in ethanol 60% solution; raw data obtained by monitoring the absorbance of the supernatant solution in the range of 250–400 nm for 90 min (a number of 19 spectra were recorded every 5 min, starting from 0 min—the lowest spectra—to 90 min—the highest spectra). Figure S23: Superimposed UV–Vis spectra for the controlled release studies on the transdermal pharmaceutical formulation based on β-CD/*M. chamomilla* stem extract complex in ethanol 96% solution; raw data obtained by monitoring the absorbance of the supernatant solution in the range of 250–400 nm for 90 min (a number of 19 spectra were recorded every 5 min, starting from 0 min—the lowest spectra—to 90 min—the highest spectra). Figure S24: Superimposed UV–Vis spectra for the controlled release studies on the transdermal pharmaceutical formulation based on β-CD/*S. marianum* seed extract complex in saline solution; raw data obtained by monitoring the absorbance of the supernatant solution in the range of 250–400 nm for 90 min (a number of 19 spectra were recorded every 5 min, starting from 0 min—the lowest spectra—to 90 min—the highest spectra). Figure S25: Superimposed UV–Vis spectra for the controlled release studies on the transdermal pharmaceutical formulation based on β-CD/*S. marianum* seed extract complex in ethanol 20% solution; raw data obtained by monitoring the absorbance of the supernatant solution in the range of 250–400 nm for 90 min (a number of 19 spectra were recorded every 5 min, starting from 0 min—the lowest spectra—to 90 min—the highest spectra). Figure S26: Superimposed UV–Vis spectra for the controlled release studies on the transdermal pharmaceutical formulation based on β-CD/*S. marianum* seed extract complex in ethanol 60% solution; raw data obtained by monitoring the absorbance of the supernatant solution in the range of 250–400 nm for 90 min (a number of 19 spectra were recorded every 5 min, starting from 0 min—the lowest spectra—to 90 min—the highest spectra). Figure S27: Superimposed UV–Vis spectra for the controlled release studies on the transdermal pharmaceutical formulation based on β-CD/*S. marianum* seed extract complex in ethanol 96% solution; raw data obtained by monitoring the absorbance of the supernatant solution in the range of 250–400 nm for 90 min (a number of 19 spectra were recorded every 5 min, starting from 0 min—the lowest spectra—to 90 min—the highest spectra). Figure S28: Superimposed UV–Vis spectra for the controlled release studies on the transdermal pharmaceutical formulation based on β-CD/silibinin diastereomer mixture complex in saline solution; raw data obtained by monitoring the absorbance of the supernatant solution in the range of 250–400 nm for 90 min (a number of 19 spectra were recorded every 5 min, starting from 0 min—the lowest spectra—to 90 min—the highest spectra). Figure S29: Superimposed UV–Vis spectra for the controlled release studies on the transdermal pharmaceutical formulation based on β-CD/silibinin diastereomer mixture complex in ethanol 20% solution; raw data obtained by monitoring the absorbance of the supernatant solution in the range of 250–400 nm for 90 min (a number of 19 spectra were recorded every 5 min, starting from 0 min—the lowest spectra—to 90 min—the highest spectra). Figure S30: Superimposed UV–Vis spectra for the controlled release studies on the transdermal pharmaceutical formulation based on β-CD/silibinin diastereomer mixture complex in ethanol 60% solution; raw data obtained by monitoring the absorbance of the supernatant solution in the range of 250–400 nm for 90 min (a number of 19 spectra were recorded every 5 min, starting from 0 min—the lowest spectra—to 90 min—the highest spectra). Figure S31: Superimposed UV–Vis spectra for the controlled release studies on the transdermal pharmaceutical formulation based on β-CD/silibinin diastereomer mixture complex in ethanol 96% solution; raw data obtained by monitoring the absorbance of the supernatant solution in the range of 250–400 nm for 90 min (a number of 19 spectra were recorded every 5 min, starting from 0 min—the lowest spectra—to 90 min—the highest spectra). Figure S32: Superimposed UV–Vis spectra for the controlled release studies on the transdermal pharmaceutical formulation based on β-CD/silymarin mixture complex in saline solution; raw data obtained by monitoring the absorbance of the supernatant solution in the range of 250–400 nm for 90 min (a number of 19 spectra were recorded every 5 min, starting from 0 min—the lowest spectra—to 90 min—the highest spectra). Figure S33: Superimposed UV–Vis spectra for the controlled release studies on the transdermal pharmaceutical formulation based on β-CD/silymarin mixture complex in ethanol 20% solution; raw data obtained by monitoring the absorbance of the supernatant solution in the range of 250–400 nm for 90 min (a number of 19 spectra were recorded every 5 min, starting from 0 min—the lowest spectra—to 90 min—the highest spectra). Figure S34: Superimposed UV–Vis spectra for the controlled release studies on the transdermal pharmaceutical formulation based on β-CD/silymarin mixture complex in ethanol 60% solution; raw data obtained by monitoring the absorbance of the supernatant solution in the range of 250–400 nm for 90 min (a number of 19 spectra were recorded every 5 min, starting from 0 min—the lowest spectra—to 90 min—the highest spectra). Figure S35: Superimposed UV–Vis spectra for the controlled release studies on the transdermal pharmaceutical formulation based on β-CD/silymarin mixture complex in ethanol 96% solution; raw data obtained by monitoring the absorbance of the supernatant solution in the range of 250–400 nm for 90 min (a number of 19 spectra were recorded every 5 min, starting from 0 min—the lowest spectra—to 90 min—the highest spectra). Figure S36: Controlled release of antioxidant compounds from the β-cyclodextrin/*M. chamomilla* flower and leaf extract complexes (expressed as chrysin, μg/mL) in ethanol 20% (brown), ethanol 60% (green), and ethanol 96% (bleu); number of replicate determinations $n = 2$; error bars were determined from standard errors using a coefficient of ±0.95. Figure S37: Controlled release of antioxidant compounds from the β-cyclodextrin/*M. chamomilla* root and stem extract complexes (expressed as rutin, μg/mL) in ethanol 20% (brown), ethanol 60% (green), and ethanol 96% (bleu); number of replicate determinations $n = 2$; error bars were determined from standard errors using a coefficient of ±0.95. Figure S38: Controlled release of antioxidant compounds from the β-cyclodextrin/*M. chamomilla* root and stem extract complexes (expressed as chrysin, μg/mL) in ethanol 20% (brown), ethanol 60% (green), and ethanol 96% (bleu); number of replicate determinations $n = 2$; error bars were determined from standard errors using a coefficient of ±0.95. Figure S39: Controlled release of antioxidant compounds (expressed as chrysin, μg/mL) from the transdermal pharmaceutical formulations containing β-cyclodextrin/*M. chamomilla* flower and leaf extract complexes in saline solution (brown), ethanol 20% (green), ethanol 60% (bleu), and ethanol 96% (black); number of replicate determinations $n = 2$; error bars were determined from standard errors using a coefficient of ±0.95. Figure S40: Controlled release of antioxidant compounds (expressed as chrysin, μg/mL) from the transdermal pharmaceutical formulations containing β-cyclodextrin/*M. chamomilla* root and stem extract complexes in saline solution (brown), ethanol 20% (green), ethanol 60% (bleu), and ethanol 96% (black); number of replicate determinations $n = 2$; error bars were determined from standard errors using a coefficient of ±0.95. Figure S41: Controlled release of antioxidant compounds (expressed as silibinin, μg/mL) from the transdermal pharmaceutical formulations containing β-cyclodextrin/silymarin complexes in saline solution (brown), ethanol 20% (green), ethanol 60% (bleu), and ethanol 96% (black); number of replicate determinations $n = 2$; error bars were determined from standard errors using a coefficient of ±0.95. Table S6: Korsmeyer–Peppas modeling ($M_t/M_\infty = k_{(KP)} \cdot t^n$) of the controlled release of antioxidants from β-cyclodextrin complexes. M_t and M_∞ stand for the amount (mg) of antioxidant released from the complex at the time t (s) and the total amount, respectively. The Korsmeyer–Peppas kinetic constant, $k_{(KP)}$ (s^{-n}), and the diffusional exponent, n, were determined by fitting the experimental data (least squares approximation). Codes for complexes are: *bCD_MChXY*—β-cyclodextrin/*M. chamomilla* L. extract; *bCD_SMaY*—β-cyclodextrin/*S. marianum* L. extract; *X* stands for the plant part, *X* = *F*—flowers,

L—leaves, *R*—roots or *S*—stems; *bCD_Sb/SmY*—β-cyclodextrin/silibinin or β-cyclodextrin/silymarin complexes; *Y* stands for the concentration of ethanol solution used for controlled release (20, 60, and 96%, *v*/*v*). Table S7: Korsmeyer–Peppas modeling ($M_t/M_\infty = k_{(KP)} \cdot t^n$) of the controlled release of antioxidants from transdermal pharmaceutical formulations based on β-cyclodextrin complexes. M_t and M_∞ stand for the amount (mg) of antioxidant released from the transdermal pharmaceutical formulation at the time *t* (s) and the total amount, respectively. The Korsmeyer–Peppas kinetic constant, $k_{(KP)}$ (s^{-n}), and the diffusional exponent, *n*, were determined by fitting the experimental data (least squares approximation). Codes for the transdermal pharmaceutical formulations based on cyclodextrin complexes are: *bCD_MChX_TFY*—transdermal pharmaceutical formulation based on β-cyclodextrin/*M. chamomilla* L. extract; *bCD_SMa_TFY*—transdermal pharmaceutical formulation based on β-cyclodextrin/*S. marianum* L. extract; *X* stands for the plant part, *X* = *F*—flowers, *L*—leaves, *R*—roots, or *S*—stems; *bCD_Sb/Sm_TFY*—transdermal pharmaceutical formulation based on β-cyclodextrin/silibinin or β-cyclodextrin/silymarin complexes; *Y* stands for saline solution ("00") or the concentration of ethanol solution used for controlled release (20, 60, and 96%, *v*/*v*).

Author Contributions: Conceptualization, D.I.H. and N.G.H.; methodology, A.H., C.I.M., D.I.H. and N.G.H.; formal analysis, D.G., F.E.M., D.I.H., C.D.M. and N.G.H.; investigation, A.Ş., C.I.M., G.S.B., A.B.V. and N.G.H.; resources, F.E.M., D.I.H., C.D.M. and N.G.H.; writing—original draft preparation, A.H., F.E.M., D.I.H., C.D.M. and N.G.H.; writing—review and editing, A.H., F.E.M., D.I.H., C.D.M. and N.G.H.; supervision, D.I.H. and N.G.H.; project administration, N.G.H. All authors have read and agreed to the published version of the manuscript.

Funding: This research received no external funding.

Data Availability Statement: Not applicable.

Acknowledgments: Authors want to thank Simona Funar-Timofei ("Coriolan Drăgulescu" Institute of Chemistry, Romanian Academy) for the help with Statistica 7.1 software and to the "Program to increase performance and innovation in doctoral and postdoctoral research excellence—PROINVENT", Contract No. 62487/03.06.2022, POCU/993/6/13, SMIS Code 153299, for support.

Conflicts of Interest: The authors declare no conflict of interest.

References

1. *Plant-Based Medicinal and Cosmetic Products*; EIP-AGRI Focus Group: Brussels, Belgium, 2020; p. 24.
2. Brabandt, H.; Ehlert, D. Chamomile harvesters: A review. *Ind. Crops Prod.* **2011**, *34*, 818–824. [CrossRef]
3. Catani, M.V.; Rinaldi, F.; Tullio, V.; Gasperi, V.; Savini, I. Comparative Analysis of Phenolic Composition of Six Commercially Available Chamomile (*Matricaria chamomilla* L.) Extracts: Potential Biological Implications. *Int. J. Mol. Sci.* **2021**, *22*, 10601. [CrossRef] [PubMed]
4. Cvetanović, A.; Švarc-Gajić, J.; Mašković, P.; Savić, S.; Nikolić, L. Antioxidant and biological activity of chamomile extracts obtained by different techniques: Perspective of using superheated water for isolation of biologically active compounds. *Ind. Crops Prod.* **2015**, *65*, 582–591. [CrossRef]
5. El-Mihyaoui, A.; da-Silva, J.C.G.E.; Charfi, S.; Castillo, M.E.C.; Lamarti, A.; Arnao, M.B. Chamomile (*Matricaria chamomilla* L.): A Review of Ethnomedicinal Use, Phytochemistry and Pharmacological Uses. *Life* **2022**, *12*, 479. [CrossRef]
6. Agatonovic-Kustrin, S.; Ortakand, D.B.; Morton, D.W.; Yusof, A.P. Rapid evaluation and comparison of natural products and antioxidantactivity in calendula, feverfew, and German chamomile extracts. *J. Chromatogr. A* **2015**, *1385*, 103–110. [CrossRef]
7. Cvetanović, A.; Švarc-Gajić, J.; Zeković, Z.; Gašić, U.; Tešić, Ž.; Zengin, G.; Mašković, P.; Mahomoodally, M.F.; Đurović, S. Subcritical water extraction as a cutting edge technology for the extraction of bioactive compounds from chamomile: Influence of pressure on chemical composition and bioactivity of extracts. *Food Chem.* **2018**, *266*, 389–396. [CrossRef]
8. Formisano, C.; Delfine, S.; Oliviero, F.; Tenore, G.C.; Rigano, D.; Senatore, F. Correlation among environmental factors, chemical composition and antioxidative properties of essential oil and extracts of chamomile (*Matricaria chamomilla* L.) collected in Molise (South-central Italy). *Ind. Crops Prod.* **2015**, *63*, 256–263. [CrossRef]
9. Nováková, L.; Vildová, A.; Mateus, J.P.; Gonçalvesd, T.; Solich, P. Development and application of UHPLC–MS/MS method for the determination of phenolic compounds in Chamomile flowers and Chamomile tea extracts. *Talanta* **2010**, *82*, 1271–1280. [CrossRef]
10. Sebai, H.; Jabri, M.-A.; Souli, A.; Hosni, K.; Rtibi, K.; Tebourbi, O.; El-Benna, J.; Sakly, M. Chemical composition, antioxidant properties and hepatoprotective effects of chamomile (*Matricaria recutita* L.) decoction extract against alcohol-induced oxidative stress in rat. *Gen. Physiol. Biophys.* **2015**, *34*, 263–275. [CrossRef]

11. Cvetanović, A.; Švarc-Gajić, J.; Zeković, Z.; Jerković, J.; Zengin, G.; Gašić, U.; Tešić, Ž.; Mašković, P.; Soares, C.; Barroso, M.F.; et al. The influence of the extraction temperature on polyphenolic profiles and bioactivity of chamomile (*Matricaria chamomilla* L.) subcritical water extracts. *Food Chem.* **2019**, *271*, 328–337. [CrossRef]
12. Ghareeb, Y.E.; Soliman, S.S.; Ismail, T.A.; Hassan, M.A.; Abdelkader, M.A.; Latef, A.A.H.A.; Al-Khayri, J.M.; ALshamrani, S.M.; Safhi, F.A.; Awad, M.F.; et al. Improvement of German Chamomile (*Matricaria recutita* L.) for Mechanical Harvesting, High Flower Yield and Essential Oil Content Using Physical and Chemical Mutagenesis. *Plants* **2022**, *11*, 2940. [CrossRef]
13. Costescu, C.I.; Hădărugă, N.G.; Hădărugă, D.I.; Lupea, A.X.; Riviş, A.; Parvu, D. Obtaining, characterization, encapsulation and the antioxidant activity evaluation of some *Matricaria chamomilla* L. extracts. In Proceedings of the 15th Symposium on Analytical and Environmental Problems, Szeged, Hungary, 22 September 2008; SZAB Kémiai Szakbizottság Analitikai és Környezetvédelmi Munkabizottsága: Szeged, Hungary, 2008; pp. 367–371.
14. Costescu, C.I.; Hădărugă, N.G.; Riviş, A.; Hădărugă, D.I.; Lupea, A.X.; Pârvu, D. Antioxidant activity evaluation of some *Matricaria chamomilla* L. extracts. *J. Agroaliment. Process. Technol.* **2008**, *14*, 417–432.
15. Hădărugă, N.G.; Hădărugă, D.I.; Tatu, C.; Gruia, A.; Costescu, C.; Lupea, A.X. Multivariate analysis (PCA) in Compositae biocompounds class. *J. Agroaliment. Process. Technol.* **2009**, *15*, 201–210.
16. Repčák, M.; Krausová, T. Phenolic glucosides in the course of ligulate flower development in diploid and tetraploid *Matricaria chamomilla*. *Food Chem.* **2009**, *116*, 19–22. [CrossRef]
17. Kováčik, J.; Grúz, J.; Bačkor, M.; Tomko, J.; Strnad, M.; Repčák, M. Phenolic compounds composition and physiological attributes of *Matricaria chamomilla* grown in copper excess. *Environ. Exp. Bot.* **2008**, *62*, 145–152. [CrossRef]
18. AbouZid, S.F.; Chen, S.-N.; McAlpine, J.B.; Friesen, J.B.; Pauli, G.F. *Silybum marianum* pericarp yields enhanced silymarin products. *Fitoterapia* **2016**, *112*, 136–143. [CrossRef]
19. Csorba, D.C.A.; Hohmann, J. Recent advances in the analysis of flavonolignans of *Silybum marianum*. *J. Pharm. Biomed. Anal.* **2016**, *130*, 301–317. [CrossRef]
20. Lukic, I.; Milovanovic, S.; Pantic, M.; Srbljak, I.; Djuric, A.; Tadic, V.; Tyśkiewicz, K. Separation of high-value extracts from *Silybum marianum* seeds: Influence of extraction technique and storage on composition and bioactivity. *LWT Food Sci. Technol.* **2022**, *160*, 113319. [CrossRef]
21. Fibigr, J.; Šatínský, D.; Solich, P. A new approach to the rapid separation of isomeric compounds in a *Silybum marianum* extract using UHPLC core-shell column with F5 stationary phase. *J. Pharm. Biomed. Anal.* **2017**, *134*, 203–213. [CrossRef]
22. Hădărugă, D.I.; Hădărugă, N.G. Antioxidant activity of hepatoprotective silymarin and *Silybum marianum* L. extract. *Chem. Bull. Polytech. Univ. Timis. Ser. Chem. Environ. Eng.* **2009**, *54*, 104–107.
23. Zholobenko, A.; Modriansky, M. Silymarin and its constituents in cardiac preconditioning. *Fitoterapia* **2014**, *97*, 122–132. [CrossRef] [PubMed]
24. Lucini, L.; Kane, D.; Pellizzoni, M.; Ferrari, A.; Trevisi, E.; Ruzickova, G.; Arslan, D. Phenolic profile and in vitro antioxidant power of different milkthistle [*Silybum marianum* (L.) Gaertn.] cultivars. *Ind. Crops Prod.* **2016**, *83*, 11–16. [CrossRef]
25. Ahmad, N.; Abbasi, B.H.; Fazal, H. Evaluation of antioxidant activity and its association with plant development in *Silybum marianum* L. *Ind. Crops Prod.* **2013**, *49*, 164–168. [CrossRef]
26. Hădărugă, D.I.; Hădărugă, N.G. Chemistry, Functionality, and Applications of Flavonols. In *Handbook of Food Bioactive Ingredients: Properties and Applications*; Jafari, S.M., Rashidinejad, A., Simal-Gandara, J., Eds.; Springer: Cham, Switzerland, 2023; pp. 1–63. [CrossRef]
27. Hădărugă, D.I.; Hădărugă, N.G. Flavanones in Plants and Humans. Chemistry, Functionality, and Applications. In *Handbook of Food Bioactive Ingredients: Properties and Applications*; Jafari, S.M., Rashidinejad, A., Simal-Gandara, J., Eds.; Springer: Cham, Switzerland, 2023; pp. 1–53. [CrossRef]
28. Hamoudi, M.C.; Bourasset, F.; Domergue-Dupont, V.; Gueutin, C.; Nicolas, V.; Fattal, E.; Bochot, A. Formulations based on alpha cyclodextrin and soybean oil: An approach to modulate the oral release of lipophilic drugs. *J. Control. Release* **2012**, *161*, 861–867. [CrossRef]
29. Kurkov, S.V.; Loftsson, T. Cyclodextrins. *Int. J. Pharm.* **2013**, *453*, 167–180. [CrossRef]
30. Pinhoa, E.; Grootveld, M.; Soares, G.; Henriques, M. Cyclodextrins as encapsulation agents for plant bioactive compounds. *Carbohydr. Polym.* **2014**, *101*, 121–135. [CrossRef]
31. Liu, J.; Zhang, S.; Zhao, X.; Lu, Y.; Song, M.; Wu, S. Molecular simulation and experimental study on the inclusion of rutin with β-cyclodextrin and its derivative. *J. Mol. Struct.* **2022**, *1254*, 132359. [CrossRef]
32. Hădărugă, D.I.; Hădărugă, N.G.; Bandur, G.N.; Isengard, H.-D. Water content of flavonoid/cyclodextrin nanoparticles: Relationship with the structural descriptors of biologically active compounds. *Food Chem.* **2012**, *132*, 1651–1659. [CrossRef]
33. Aytac, Z.; Kusku, S.I.; Durgun, E.; Uyar, T. Quercetin/β-cyclodextrin inclusion complex embedded nanofibres: Slow release and high solubility. *Food Chem.* **2016**, *197*, 864–871. [CrossRef]
34. Azzi, J.; Jraij, A.; Auezova, L.; Fourmentin, S.; Greige-Gerges, H. Novel findings for quercetin encapsulation and preservation with cyclodextrins, liposomes, and drug-in-cyclodextrin-in-liposomes. *Food Hydrocoll.* **2018**, *81*, 328–340. [CrossRef]
35. Başaran, E.; Öztürk, A.A.; Şenel, B.; Demirel, M.; Sarica, Ş. Quercetin, Rutin And Quercetin-Rutin Incorporated Hydroxypropyl-β-Cyclodextrin Inclusion Complexes. *Eur. J. Pharm. Sci.* **2022**, *172*, 106153. [CrossRef]
36. Jullian, C.; Moyano, L.; Yañez, C.; Olea-Azar, C. Complexation of quercetin with three kinds of cyclodextrins: An antioxidant study. *Spectrochim. Acta Part A* **2007**, *67*, 230–234. [CrossRef]

37. Krstić, L.; Jarho, P.; Ruponen, M.; Urtti, A.; Gonzlez-García, M.J.; Diebold, Y. Improved ocular delivery of quercetin and resveratrol: A comparative study between binary and ternary cyclodextrin complexes. *Int. J. Pharm.* **2022**, *624*, 122028. [CrossRef]
38. Liu, M.; Dong, L.; Chen, A.; Zheng, Y.; Sun, D.; Wang, X.; Wang, B. Inclusion complexes of quercetin with three β-cyclodextrins derivatives at physiological pH: Spectroscopic study and antioxidant activity. *Spectrochim. Acta Part A Mol. Biomol. Spectrosc.* **2013**, *115*, 854–860. [CrossRef]
39. Nguyen, T.A.; Liu, B.; Zhao, J.; Thomas, D.S.; Hook, J.M. An investigation into the supramolecular structure, solubility, stability and antioxidant activity of rutin/cyclodextrin inclusion complex. *Food Chem.* **2013**, *136*, 186–192. [CrossRef]
40. Kellici, T.F.; Ntountaniotis, D.; Leonis, G.; Chatziathanasiadou, M.; Chatzikonstantinou, A.V.; Becker-Baldus, J.; Glaubitz, C.; Tzakos, A.G.; Viras, K.; Chatzigeorgiou, P.; et al. Investigation of the interactions of silibinin with 2-hydroxypropyl-β-cyclodextrin through biophysical techniques and computational methods. *Mol. Pharm.* **2015**, *12*, 954–965. [CrossRef]
41. Sav, A.; Hădărugă, D.I. Silibinin-saturated fatty acid bioconjugate/natural cyclodextrin interactions: Molecular modeling, docking and QSPR experiments. *J. Agroaliment. Process. Technol.* **2017**, *23*, 5–12.
42. Onodera, R.; Hayashi, T.; Motoyama, K.; Tahara, K.; Takeuchi, H. Hydroxypropyl-β-cyclodextrin enhances oral absorption of silymarin nanoparticles prepared using PureNano™ continuous crystallizer. *Pharmaceutics* **2022**, *14*, 394. [CrossRef]
43. Spada, G.; Gavini, E.; Cossu, M.; Rassu, G.; Carta, A.; Giunchedi, P. Evaluation of the effect of hydroxypropyl-β-cyclodextrin on topical administration of milk thistle extract. *Carbohydr. Polym.* **2013**, *92*, 40–47. [CrossRef]
44. Ghosh, A.; Biswas, S.; Ghosh, T. Preparation and evaluation of silymarin β-cyclodextrin molecular inclusion complexes. *J. Young Pharm.* **2011**, *3*, 205–210. [CrossRef]
45. Kaiser, C.S.; Römpp, H.; Schmidt, P.C. Supercritical carbon dioxide extraction of chamomile flowers: Extraction efficiency, stability, and in-line inclusion of chamomile-carbon dioxide extract in β-cyclodextrin. *Phytochem. Anal.* **2004**, *15*, 249–256. [CrossRef] [PubMed]
46. Hădărugă, D.I.; Hădărugă, N.G.; Riviş, A.; Pârvu, D. Molecular modeling and docking studies on Compositae biocompounds—Cyclodextrin interactions. *J. Agroaliment. Process. Technol.* **2009**, *15*, 273–282.
47. Hădărugă, N.G.; Bandur, G.N.; David, I.; Hădărugă, D.I. A review on thermal analyses of cyclodextrins and cyclodextrin complexes. *Environ. Chem. Lett.* **2019**, *17*, 349–373. [CrossRef]
48. Hădărugă, N.G.; Hădărugă, D.I.; Isengard, H.-D. "Surface water" and "strong-bonded water" in cyclodextrins: A Karl Fischer titration approach. *J. Incl. Phenom. Macrocycl. Chem.* **2013**, *75*, 297–302. [CrossRef]
49. Hegheş, A.; Hădărugă, N.G.; Fuliaş, A.-V.; Bandur, G.N.; Hădărugă, D.I.; Dehelean, C.-A. *Capsicum annuum* extracts/β-cyclodextrin complexes. Thermal analyses—Karl Fischer water titration correlations and antioxidant activity. *J. Therm. Anal. Calorim.* **2015**, *120*, 603–615. [CrossRef]
50. Paczkowska, M.; McDonagh, A.F.; Tajber, K.B.L.; Cielecka-Piontek, J. Mechanochemical activation with cyclodextrins followed by compaction as an effective approach to improving dissolution of rutin. *Int. J. Pharm.* **2020**, *581*, 119294. [CrossRef]
51. Wu, W.; Zu, Y.; Zhao, X.; Zhang, X.; Wang, L.; Li, Y.; Wang, L.; Zhang, Y.; Lian, B. Solubility and dissolution rate improvement of the inclusion complex of apigenin with 2-hydroxypropyl-β-cyclodextrin prepared using the liquid antisolvent precipitation and solvent removal combination methods. *Drug Dev. Ind. Pharm.* **2017**, *43*, 1366–1377. [CrossRef]
52. David, I.; Orboi, M.D.; Simandi, M.D.; Chirilă, C.A.; Megyesi, C.I.; Rădulescu, L.; Lukinich-Gruia, A.T.; Muntean, C.; Hădărugă, D.I.; Hădărugă, N.G. Fatty acid profile of Romanian's common bean (*Phaseolus vulgaris* L.) lipid fractions and their complexation ability by β-cyclodextrin. *PLoS ONE* **2019**, *14*, e0225474. [CrossRef]
53. Hădărugă, D.I.; (Mitroi), C.L.B.; Gruia, A.T.; Păunescu, V.; Bandur, G.N.; Hădărugă, N.G. Moisture evaluation of β-cyclodextrin/fish oils complexes by thermal analyses: A data review on common barbel (*Barbus barbus* L.), Pontic shad (*Alosa immaculata* Bennett), European wels catfish (*Silurus glanis* L.), and common bleak (*Alburnus alburnus* L.) living in Danube river. *Food Chem.* **2017**, *236*, 49–58. [CrossRef]
54. Hădărugă, D.I.; Hădărugă, N.G.; Costescu, C.I.; David, I.; Gruia, A.T. Thermal and oxidative stability of the *Ocimum basilicum* L. essential oil/β-cyclodextrin supramolecular system. *Beilstein J. Org. Chem.* **2014**, *10*, 2809–2820. [CrossRef]
55. Hădărugă, D.I.; Ünlüsayin, M.; Gruia, A.T.; Birău-(Mitroi), C.; Rusu, G.; Hădărugă, N.G. Thermal and oxidative stability of Atlantic salmon oil (*Salmo salar* L.) and complexation with β-cyclodextrin. *Beilstein J. Org. Chem.* **2016**, *12*, 179–191. [CrossRef]
56. Hădărugă, N.G.; Hădărugă, D.I.; Isengard, H.-D. Water content of natural cyclodextrins and their essential oil complexes: A comparative study between Karl Fischer titration and thermal methods. *Food Chem.* **2012**, *132*, 1741–1748. [CrossRef]
57. Hădărugă, N.G.; Hădărugă, D.I.; Păunescu, V.; Tatu, C.; Ordodi, V.L.; Bandur, G.; Lupea, A.X. Thermal stability of the linoleic acid/alpha- and beta-cyclodextrin complexes. *Food Chem.* **2006**, *99*, 500–508. [CrossRef]
58. Hădărugă, N.G.; Szakal, R.N.; Chirilă, C.A.; Lukinich-Gruia, A.T.; Păunescu, V.; Muntean, C.; Rusu, G.; Bujancă, G.; Hădărugă, D.I. Complexation of Danube common nase (*Chondrostoma nasus* L.) oil by β-cyclodextrin and 2-hydroxypropyl-β-cyclodextrin. *Food Chem.* **2020**, *303*, 125419. [CrossRef]
59. Ünlüsayin, M.; Hădărugă, N.G.; Rusu, G.; Gruia, A.T.; Păunescu, V.; Hădărugă, D.I. Nano-encapsulation competitiveness of omega-3 fatty acids and correlations of thermal analysis and Karl Fischer water titration for European anchovy (*Engraulis encrasicolus* L.) oil/beta-cyclodextrin complexes. *LWT Food Sci. Technol.* **2016**, *68*, 135–144. [CrossRef]
60. Margolis, S.A.; Huang, P.H.; Hădărugă, N.G.; Hădărugă, D.I. Water determination. In *Encyclopedia of Analytical Science*, 3rd ed.; Elsevier: Oxford, UK, 2019; Volume 10, pp. 382–390.

61. Suvarna, V.; Bore, B.; Bhawar, C.; Mallya, R. Complexation of phytochemicals with cyclodextrins and their derivativesan update. *Biomed. Pharmacother.* **2022**, *149*, 112862. [CrossRef]
62. Tommasini, S.; Raneri, D.; Ficarra, R.; Calabrò, M.L.; Stancanelli, R.; Ficarra, P. Improvement in solubility and dissolution rate of flavonoids by complexation with β-cyclodextrin. *J. Pharm. Biomed. Anal.* **2004**, *35*, 379–387. [CrossRef]
63. You, G.-J.; Sun, L.-L.; Cao, X.-X.; Li, H.-H.; Wang, M.; Liu, Y.-N.; Ren, X.-L. Comprehensive evaluation of solubilization of flavonoids by various cyclodextrins using high performance liquid chromatography and chemometry. *LWT Food Sci. Technol.* **2018**, *94*, 172–177. [CrossRef]
64. Haiyun, D.; Jianbin, C.; Guomei, Z.; Shaomin, S.; Jinhao, P. Preparation and spectral investigation on inclusion complex of β-cyclodextrin with rutin. *Spectrochim. Acta Part A* **2003**, *59*, 3421–3429. [CrossRef]
65. Parker, K.M.; Stalcup, A.M. Affinity capillary electrophoresis and isothermal titration calorimetry for the determination of fatty acid binding with beta-cyclodextrin. *J. Chromatogr. A* **2008**, *1204*, 171–182. [CrossRef]
66. Kashino, E.; Fujii, T. Interaction of Cyclodextrins with Oily Substances. *J. Home Econ. Jpn.* **2005**, *56*, 533–539. [CrossRef]
67. Kfoury, M.; Auezova, L.; Greige-Gerges, H.; Fourmentin, S. Promising applications of cyclodextrins in food: Improvement ofessential oils retention, controlled release and antiradical activity. *Carbohydr. Polym.* **2015**, *131*, 264–272. [CrossRef] [PubMed]
68. Korsmeyer, R.W.; Gurny, R.; Doelker, E.; Buri, P.; Peppas, N.A. Mechanisms of solute release from porous hydrophilic polymers. *Int. J. Pharm.* **1983**, *15*, 25–35. [CrossRef]
69. Machín, R.; Isasi, J.R.; Vélaz, I. Hydrogel matrices containing single and mixed natural cyclodextrins. Mechanisms of drug release. *Eur. Polym. J.* **2013**, *49*, 3912–3920. [CrossRef]
70. Leo, A.; Hansch, C.; Elkins, D. Partition coefficients and their uses. *Chem. Rev.* **1971**, *71*, 525–616. [CrossRef]

plants

MDPI

Article

Influence of the Growing Region on the Phytochemical Composition and Antioxidant Properties of North American Cranberry Fruit (*Vaccinium macrocarpon* Aiton)

Liang Xue [1], Maureen Otieno [1], Kimberly Colson [2] and Catherine Neto [1,*]

[1] Department of Chemistry and Biochemistry, University of Massachusetts Dartmouth, North Dartmouth, MA 02747, USA; lxue1@umassd.edu (L.X.); motieno1@umassd.edu (M.O.)
[2] Bruker BioSpin Corp., Billerica, MA 01821, USA; kimcolson.sebago@gmail.com
* Correspondence: cneto@umassd.edu

Abstract: The impact of the growth environment on the production of health-promoting phytochemicals in cranberry fruit (*Vaccinium macrocarpon* Aiton) is not well established despite increased production worldwide. We investigated the secondary metabolite composition among the cranberry fruit of nine cultivars produced in two major coastal North American growing regions that differ in climate. Using ^{1}H NOESY NMR to generate metabolic fingerprints, principal component analysis revealed variation between the two regions and identified likely contributing metabolites. Triterpenoids ursolic and oleanolic acid, as well as citric and malic acids, were quantified using ^{1}H qNMR, and anthocyanins and flavonols were determined by HPLC-DAD. Total proanthocyanidins (PACs), total soluble phenolics, and DPPH free-radical scavenging antioxidant activity were also evaluated. Across all cultivars, anthocyanins, flavonols, and total phenolic content were significantly higher in West Coast fruit than East Coast fruit, correlating with a regional trend of higher antioxidant activity in fruit grown on the West Coast. The opposite trend was observed for triterpenoids and organic acids, which were significantly higher across cultivars in East Coast fruit. These trends persisted over two growing seasons. The study demonstrates that climate plays an important role in the production of antioxidant and anti-inflammatory phytochemicals in cranberry plants.

Keywords: cranberry; *Vaccinium*; flavonoids; antioxidant; triterpenoids; anti-inflammatory; growing region; climate

1. Introduction

Cranberry (*Vaccinium macrocarpon*) is a native fruit of North America belonging to the Ericaceae family, with increasing worldwide production and consumption due to its unique flavor and potential health benefits. The United States ranks as the top producer worldwide of cranberries (*Vaccinium macrocarpon*), followed by Canada, Chile, Turkiye, and Azerbaijan [1], and within the U.S., the top four producers by state are Wisconsin, Massachusetts, New Jersey, and Oregon [2]. Potential health benefits associated with cranberry consumption include promotion of urinary, oral, and gut health, as well as antioxidant and anti-inflammatory properties [3]. Clinical research studies indicate that consumption of cranberry, like other polyphenol-rich fruit, can positively impact serum lipid levels, blood pressure, oxidative damage to neurons and the vascular endothelium, and reduce the risk of cardiovascular disease [4,5]. Preclinical studies also suggest that cranberry has the potential to limit some cancers, including esophageal and inflammation-induced colon cancers [6–9].

Cranberry bioactivities are linked to the presence of multiple phytonutrients in cranberry fruits, notably flavonoids and triterpenoids. Anthocyanins, flavonols, and proanthocyanidins (PACs) are the major flavonoids found in cranberries [10]. Anthocyanins from various berries, including cranberries, were reported to possess powerful antioxidant and anti-inflammatory properties, for example, reducing inflammatory and oxidative stresses in BV-2 microglia [11]. Quercetin and its glycosides, abundant flavonols in cranberry fruit, are known for their antioxidant and anti-inflammatory properties and are likely to contribute to their health benefits [10]. Cranberry's most abundant polyphenols are A-type PACs, of considerable interest due to their ability to interfere with adhesion and colonization of the bladder and urinary tract by extraintestinal pathogenic *Escherichia coli* and promote urinary and gut health [3]. We have recently demonstrated the ability of both polyphenol-rich and non-polyphenol-rich cranberry extracts to reduce inflammation and tumorigenesis in a mouse model of colitis [12]. Non-polyphenols of interest from a health perspective include ursolic acid and its isomer, oleanolic acid, two major triterpenoids present in the outer layer of the cranberry fruit that have recognized anti-inflammatory properties [13,14], and other health benefits related to mitigating oxidative stress [15]. Among *Vaccinium* fruit, cranberries are a rich source of ursolic acid and its esters, for which in vitro anti-tumor activities were previously reported [16]. These likely contribute to cranberries' observed colon-protective properties.

Over 100 cultivars of cranberries are grown in North America [17], with a wide variety cultivated in Wisconsin and on the Eastern and Western coasts of the United States, primarily in Massachusetts, New Jersey, and Oregon. Massachusetts, the second largest producer of cranberries, grew approximately 23% of the cranberry crop in 2016 and 2017; Oregon contributed about 4–6% [18]. Phytochemical content in plants can be expected to vary due to genetic factors. Several previously published studies have compared the phytochemical profiles of different cranberry cultivars, but little data exists on regional influences. Previous studies have demonstrated that North American cranberry cultivars can vary significantly in the content of PACs, anthocyanins, flavonols, and organic acids [19–21]. For example, Canadian researchers characterized the phytochemical content diversity among five cranberry fruit cultivars grown in British Columbia, particularly anthocyanins, using HPLC and UPLC-MS-TOF profiling with chemometrics [19]. Three newer cranberry hybrids developed at Rutgers, Crimson Queen (Stevens × Ben Lear), Demoranville (Franklin (Early Black × Howes) × Ben Lear), and Mullica Queen (No. 35 (Howes × Searles) × Lemunyon), and other cultivars were examined for flavonoid, PAC, and organic acid content [20]. Triterpenoid content may also vary, but limited data is available [16]. In addition to genetic differences between fruit cultivars, variation in seasonal growing conditions between regions is also expected to affect the content of secondary metabolites in the fruit, since differences in climate may impact metabolic pathways in the plant. However, few studies have examined how regional factors impact cranberry phytochemical content. A recent study of eight different berries harvested in both Romania and Russia found significant differences in polyphenol content between samples from the two regions, which impacted their antioxidant and antibacterial properties [22].

The goal of this study was to determine how secondary metabolite composition varies between cranberry fruit produced in two major North American growing regions. Massachusetts on the East Coast and Oregon on the West Coast are approximately 3000 miles (4800 km) apart and vary substantially in climate, including rainfall, temperature, and humidity, providing an opportunity to study the resulting differences in fruit phytochemistry of common cultivars. To characterize variations in metabolite composition between regions, ^{1}H NOESY NMR spectra of extracts were acquired, and principle component analysis (PCA) was used to identify key metabolites contributing to variation between sample sets. These metabolites were quantitated using qNMR and other established methods, and regional data were compared. The antioxidant activities of the extracts were evaluated using the DPPH free-radical scavenging assay and correlated with metabolite content. The data illustrates regional influences on phytochemicals in cranberry fruit associated with potential health benefits.

2. Results and Discussion

2.1. Non-Targeted PCA of Cranberry ^{1}H NMR Reveals Regional Differences

The present study examined variations in both polyphenol and non-polyphenol constituents of cranberry between multiple cultivars in two major growing regions. The genetic origins of the cultivars are shown in Figure 1. We employed ^{1}H NMR in a non-targeted approach to reveal some differences between sample sets, then followed up with quantification of key metabolites using qNMR and other quantitative methods. Non-polyphenol metabolites of interest in cranberries include triterpenoids, due to their anti-inflammatory properties and potential impacts on carcinogenesis [16,23]. For PCA, samples were grouped in sets either by region or by cultivar.

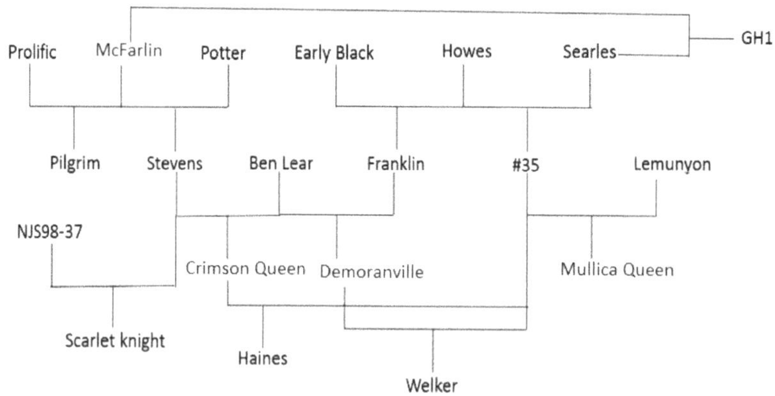

Figure 1. Breeding map showing the origins of Vaccinium macrocarpon cultivars included in the study and other common cultivars.

NMR can provide a profile of all metabolites present in plant extracts in a single experiment without requiring chromatographic separations. Chemometric analysis of ^{1}H NMR spectra has been used successfully to detect metabolic variations among plant sources, for example, ginsenoside and sucrose content across Ontario ginseng landraces [24], and for fingerprinting of Vaccinium species [25]. We recently reported the application of ^{1}H NMR to compare commercial cranberry supplements to authentic cranberry standards and to identify constituents contributing to sample variation [26]. Principal component analysis considers all included data points within the spectrum and determines individual sources of variation (principal components, or PCs), which contribute to overall differences between populations. ^{1}H NOESY NMR data were collected for all fruit extracts (n = 5 per treatment) and analyzed using Bruker AssureNMR™ 2.0 and AMIX™ 3.9.15, using methods we developed previously [26].

PCA of all cranberry fruit samples (Figure 2) revealed variations associated with the two different growing regions, for which PC1 accounts for 25.88% and PC2 for 21.44% of the sample variance. The separation of the two groups is evident in the score plot. Cranberry metabolites contributing to these differences were identified by comparing chemical shift buckets on the loadings plot with data in the SBASE. Buckets appearing prominently in the loadings plot and contributing significantly to variations between MA and OR cranberry fruit include chemical shifts associated with ursolic acid, oleanolic acid, malic acid, and citric acid. Their contributions were subsequently confirmed by quantification of the analytes using ^{1}H NMR, discussed below. Chemical shifts between 3.3 and 3.8 ppm are associated with sugars, which also appear to play a role in the variations between the two regions; however, the individual sugars were not identified. A 2018 study of ten cranberry cultivars harvested in New Jersey used nontargeted metabolomic analysis of UPLC-HRMS data to characterize variations in phenolics among the cultivars [27]. In the present study, when data on samples from both regions were combined, no significant separation of samples by cultivar was observed.

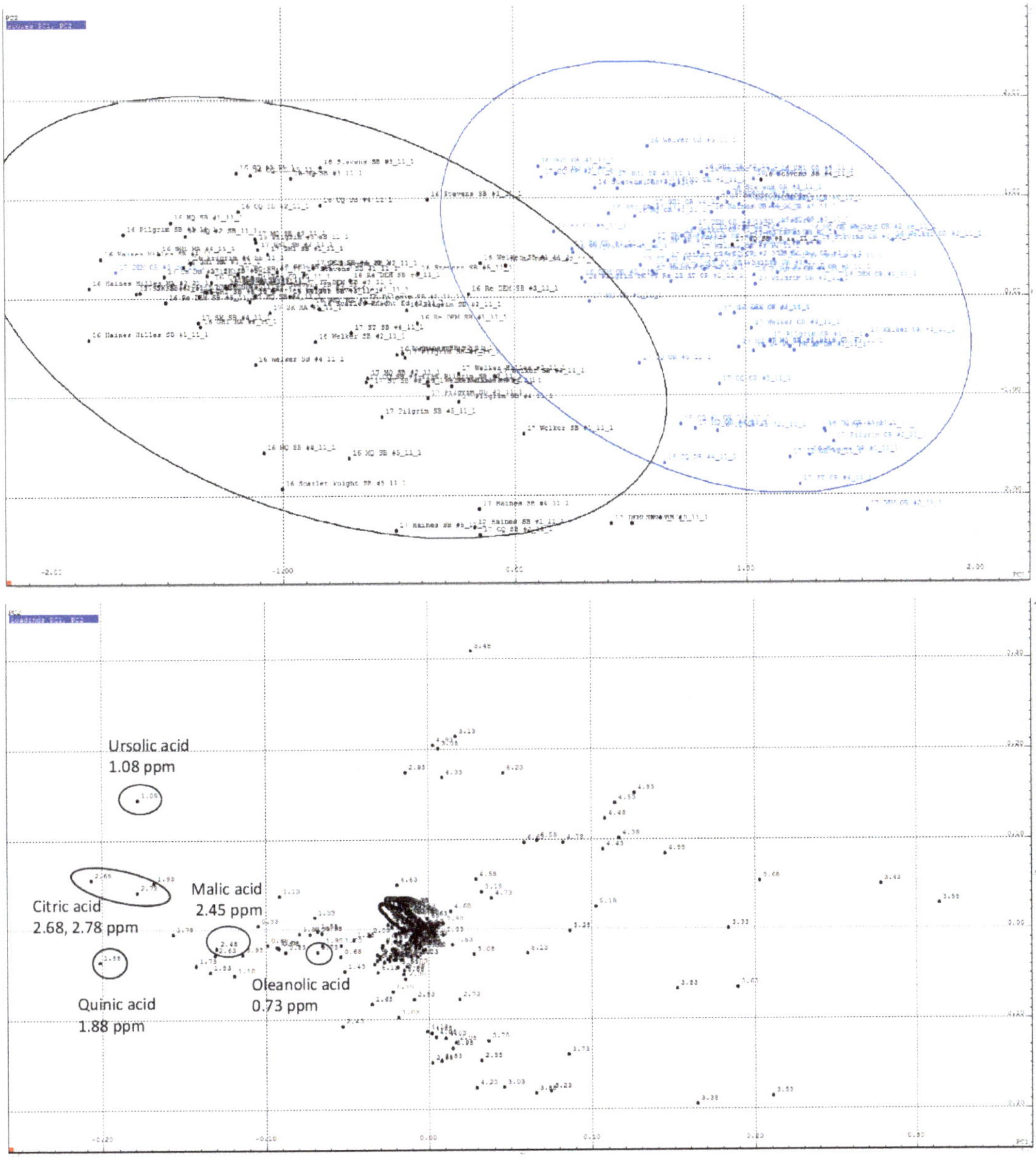

Figure 2. Principal component analysis (PCA) of ^{1}H NOESY NMR spectra in the chemical shift range of 0.6–12.5 ppm. (**Top**) Scores plot showing a comparison of all samples by growing region (PC1 vs. PC2). Black dots represent MA samples; blue dots represent OR samples. (**Bottom**) Loadings plot showing the contributions of each chemical shift bucket (black dots) to the regional variance and identification of selected metabolites. Chemical shift buckets corresponding to the labeled metabolites are circled.

PCA of ^{1}H NMR data in the chemical shift range of 5.5 ppm to 13 ppm (Figure 3) indicates that some sample variation between regions can also be attributed to polyphenols. Some clustering of the MA and OR samples was evident along PC3 (6.77%) and PC4 (3.40% of variance), with OR samples tending towards higher PC4 scores and MA samples skewing lower. From the loadings plot, quercetin, cyanidin, peonidin, and coumaric acid could be identified as metabolites associated with chemical shift buckets contributing to the variance in NMR spectra between MA and OR samples. The contribution of the flavonoids to regional variation was subsequently confirmed by quantitative analysis of anthocyanins and flavonols using HPLC-DAD, discussed below.

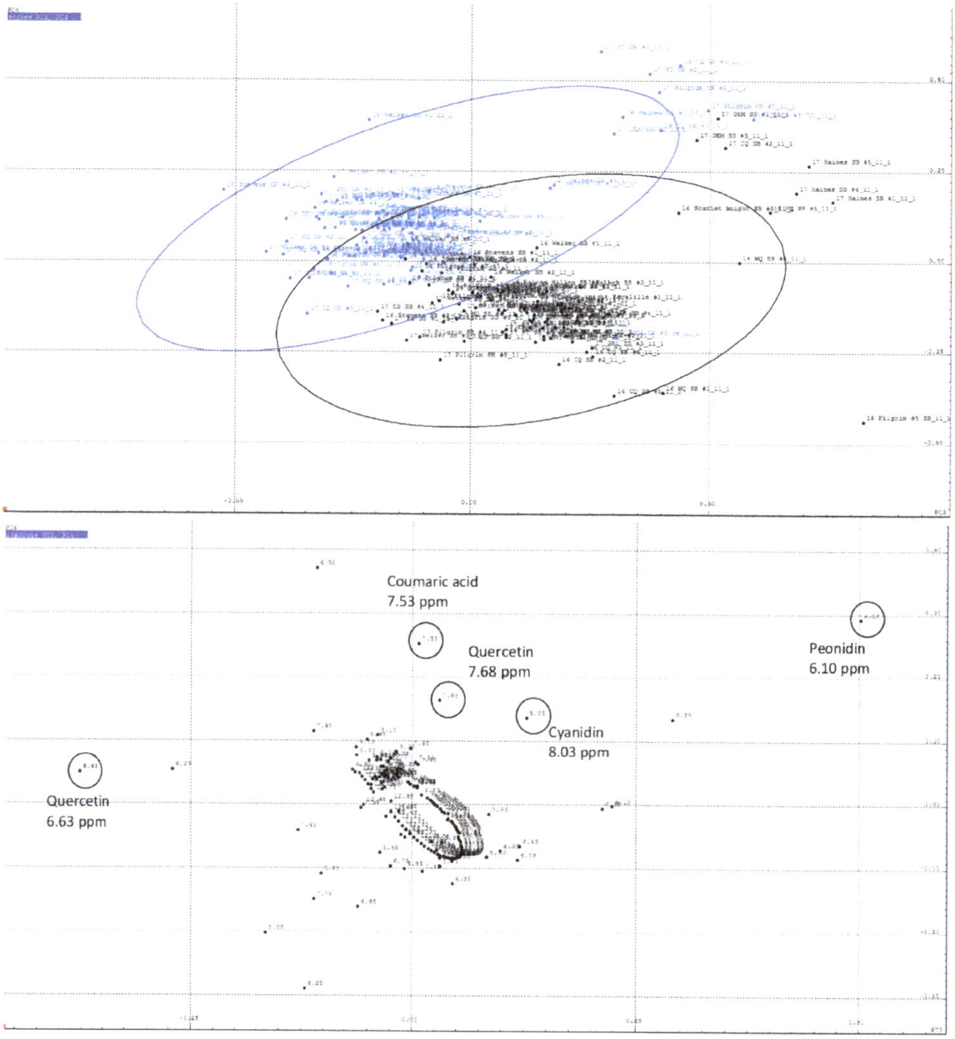

Figure 3. Principal component analysis (PCA) of ^{1}H NOESY NMR spectra in the aromatic chemical shift range of 5.5–13.5 ppm. (**Top**) Scores plot showing a comparison of all samples by growing region (PC3 vs. PC4). Black dots represent MA samples; blue dots represent OR samples. (**Bottom**) Loadings plot showing contributions of each chemical shift bucket (black dots) to the regional variance and identification of selected metabolites. Chemical shift buckets corresponding to the labeled metabolites are circled.

2.2. Triterpenoids and Organic Acids Content

Quantitative proton ^{1}H NMR (qNMR) methods were developed as a means to rapidly quantify triterpenoids and compare their content in cranberry fruit samples and supplements [26]. Reverse phase HPLC with UV or diode array detection is commonly used to detect phenolic acids and other flavonoids, including (+)-catechin, p-coumaric acid, myricetin, and quercetin, in cranberry samples through carefully designed separations [28,29]. However, it can be difficult to detect compounds lacking a chromophore (such as organic acids and triterpenoids) or to separate and identify isomers. Ursolic acid and oleanolic acid are isomeric triterpenes, differing only in the position of a single methyl group, making them harder to detect and distinguish through absorbance or LC-based quantification methods.

Quantitative proton ^{1}H NMR (qNMR) has the advantages of universal detection of compounds without the need for chromatographic separation, simple, rapid analysis, and the ability to differentiate nearly identical compounds [30]. Hicks et al. quantified chlorogenic acid and hyperoside from crude blueberry extract with a LOQ of 0.01 mM using qNMR analysis [31]. QNMR has also been applied to the quantification of triterpenoids in Ganoderma resinaceum mushrooms used medicinally in China and Nigeria [32]. For our samples, qNMR worked best for analytes present at concentrations over 0.03 mM.

PCA identified ursolic and oleanolic acids as contributors to the regional variation in the secondary metabolite composition of the fruit. As these triterpenoids are also of interest from a health perspective, quantification was performed. Peak signals of selected metabolites were determined by comparison with previously reported NMR data by Turbitt and coworkers [26]. Table 1 shows the content of ursolic acid and oleanolic acid found in cranberry samples (average of five replicates) from each cultivar and region. Figure 1 shows relationships between common cultivars [27,33]. Ursolic acid was the predominant triterpenoid in all cranberry cultivars, with content ranging from 2.72 to 16.80 mg/g DW, followed by oleanolic acid, with content ranging from 1.07 to 7.89 mg/g DW. The cultivars Welker (MA) and Demoranville (MA) had the highest ursolic acid and oleanolic acid, respectively. Pilgrim fruit (OR) contained the lowest oleanolic acid. Triterpenoid content depended significantly on growing regions (Figure 4).

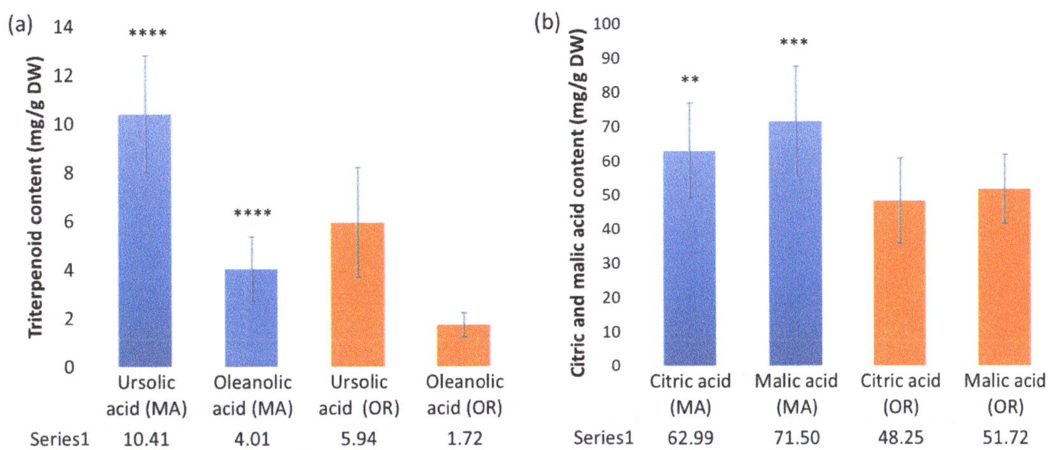

Figure 4. Regional comparison of (**a**) triterpenoids ursolic acid and oleanolic acid and (**b**) citric and malic acid in cranberry fruit, average of all cultivars over two growing seasons for MA vs. OR fruit (** = $p \leq 0.01$, *** = $p \leq 0.001$, **** = $p \leq 0.0001$).

Table 1. Content of triterpenoids and organic acids in cranberry fruit samples (mg analyte/g dry weight ± standard deviation of mean, n = 5).

Cultivar/Region	Growing Season	Ursolic Acid	Oleanolic Acid	Citric Acid	Malic Acid
ST/MA	2016	9.63 ± 2.39	4.23 ± 0.55	57.61 ± 9.00	43.78 ± 10.25
	2017	10.44 ± 3.28	3.83 ± 1.54	81.67 ± 10.91	54.98 ± 7.17
ST/OR	2016	7.09 ± 1.05	2.11 ± 0.68	38.69 ± 3.69	40.36 ± 8.91
	2017	7.01 ± 1.53	2.23 ± 1.18	47.86 ± 9.20	48.54 ± 7.73
MQ/MA	2016	9.02 ± 2.50	4.24 ± 1.18	50.89 ± 4.96	78.37 ± 12.07
	2017	9.61 ± 2.85	3.01 ± 0.90	67.31 ± 10.91	78.63 ± 9.16
MQ/OR	2016	6.25 ± 1.55	1.54 ± 0.28	47.81 ± 15.67	60.56 ± 7.72
	2017	12.32 ± 5.28	2.76 ± 1.58	34.97 ± 8.95	54.59 ± 8.29
DM/MA	2016	9.90 ± 2.57	3.59 ± 2.59	44.11 ± 9.49	75.68 ± 10.75
	2017	8.62 ± 2.66	4.27 ± 1.09	59.27 ± 12.83	79.92 ± 8.63
DM/OR	2016	3.18 ± 1.25	1.27 ± 0.06	45.86 ± 12.79	52.21 ± 6.24
	2017	6.68 ± 2.39	2.05 ± 0.36	53.85 ± 13.96	59.36 ± 5.81
PI/MA	2016	12.83 ± 1.78	5.80 ± 0.68	52.40 ± 9.48	51.43 ± 14.32
	2017	8.94 ± 2.35	2.49 ± 0.93	76.49 ± 18.91	66.00 ± 12.48
PI/OR	2016	4.31 ± 1.39	1.32 ± 0.99	52.41 ± 6.64	52.08 ± 8.70
	2017	3.09 ± 1.72	1.09 ± 0.34	63.30 ± 13.73	51.46 ± 7.87
CQ/MA	2016	9.18 ± 2.34	3.84 ± 1.22	64.30 ± 4.60	86.43 ± 10.69
	2017	9.50 ± 3.92	3.72 ± 1.70	84.59 ± 21.09	81.98 ± 10.72
CQ/OR	2016	2.72 ± 1.18	2.25 ± 0.65	25.95 ± 6.22	39.72 ± 4.46
	2017	5.60 ± 0.67	1.75 ± 0.60	58.71 ± 23.22	73.67 ± 12.36
HA/MA	2016	15.39 ± 4.31	7.60 ± 2.48	57.59 ± 8.39	88.27 ± 29.78
	2017	7.34 ± 1.20	2.60 ± 0.88	40.42 ± 15.04	104.78 ± 15.76
HA/OR	2016	5.92 ± 1.84	2.30 ± 0.81	64.37 ± 17.62	56.59 ± 12.12
	2017	6.10 ± 2.61	1.66 ± 0.89	30.97 ± 13.32	63.23 ± 8.55
SK/MA	2016	10.00 ± 2.27	3.39 ± 1.38	75.26 ± 16.41	58.19 ± 12.07
	2017	9.46 ± 3.41	3.25 ± 1.38	80.65 ± 14.47	57.94 ± 6.52
SK/OR	2016	6.01 ± 2.38	1.32 ± 0.60	61.52 ± 13.71	38.94 ± 6.82
	2017	8.20 ± 3.73	1.66 ± 0.57	51.80 ± 7.94	40.84 ± 6.99
WK/MA	2016	16.80 ± 9.95	5.99 ± 2.82	44.02 ± 10.32	76.37 ± 11.90
	2017	10.76 ± 2.87	3.05 ± 0.98	53.35 ± 2.87	88.29 ± 0.98
WK/OR	2016	4.40 ± 3.31	1.61 ± 1.11	34.46 ± 9.18	51.17 ± 9.33
	2017	4.72 ± 0.89	1.10 ± 0.27	33.50 ± 5.02	65.45 ± 11.28
GH1/MA	2016	11.46 ± 4.75	4.04 ± 1.38	68.13 ± 9.44	58.19 ± 1.38
	2017	8.45 ± 1.90	2.46 ± 1.16	75.84 ± 19.68	57.83 ± 9.26
GH1/OR	2016	5.16 ± 3.26	1.07 ± 0.32	65.68 ± 8.16	41.27 ± 3.94
	2017	8.20 ± 3.97	1.86 ± 0.82	56.71 ± 8.73	40.97 ± 5.21

The average content of triterpenoids over both the 2016 and 2017 seasons across all cultivars from MA was significantly higher than the average across OR cultivars. Ursolic acid averaged 10.4 mg + 2.4 per g dry weight of fruit in MA samples and 5.94 + 2.27 mg/g dry weight in OR fruit. Likewise, oleanolic acid averaged 4.01 + 1.3 mg/g dry weight in MA fruit and 1.72 + 0.49 mg/g dry weight in OR fruit. Our previous study reporting on Early Black, Howes, and mixed cultivar fruit grown in MA found between 0.460 and 1.09 g of ursolic acid per kg of fresh weight of cranberry fruit, roughly equivalent to 5.0–11 mg/g dry weight [16]. A 2022 study of Red Star cultivar fruit grown in Lithuania reported 5.74 mg UA/g dry weight and 1.60 mg OA/g dry weight, comparable to the average quantity in our Oregon samples [34]. Somewhat lower quantities of UA and OA were reported for several cultivars grown at a cranberry plantation in Latvia [35]. A report on the 2016 cranberry fruit of several cultivars grown in Poland found ursolic acid at 1.0–1.76 mg/g dry matter and oleanolic acid from 0.89 to 1.14 mg/g of dry matter [36]. These triterpenoid levels were several times lower than what we detected in North American fruit, a difference which could be attributed to climate or other environmental factors. There is precedent for regional variation in the triterpenoid content of other fruit; a 2019 study of two apple cultivars grown across four locations in northeastern Europe, from Poland to Estonia, reported a significant regional influence on the content of ursolic and oleanolic acid in the fruits, with concentrations increasing from south to north [37]. The present study on East and West Coast cranberry fruit is the first report suggesting regional variation in content of two major triterpenoids associated with the anti-inflammatory and anti-cancer properties of cranberry fruit. This variation may be associated with climate, pest pressure, or other differences among the regions; further study is needed to elucidate specific contributing factors.

Using PCA, citric and malic acid were found to contribute to regional variation in phytochemical content, and this was confirmed by qNMR using methods we developed for cranberry supplements. Content ranged from 38.94 to 104.78 mg/g DW and 25.95 to 84.59 mg/g DW, respectively (Table 1). SK and GH1 cultivars averaged the highest citric acid; the HA cultivar was highest in malic acid, followed by CQ and WK. Citric acid was significantly higher in the MA samples overall (Figure 4), averaging 63 + 14 mg/g DW for Massachusetts (all cultivars) vs. 48.3 + 12.6 mg/g DW for Oregon cranberry fruit. Malic acid was also higher in MA fruit (71.5 + 16 mg/g DW) vs. OR fruit (51.7 + 10 mg/g DW).

2.3. Content of Anthocyanins and Flavonols

Cranberries are a rich source of anthocyanins and flavonol glycosides. PCA of ^{1}H NMR data in the chemical shift range of 5.5 ppm to 13 ppm indicated that some of the region-to-region sample variation can be attributed to quercetin glycosides, cyanidins, and peonidins. We have found ^{1}H NMR less suitable for quantitative determination of cranberry flavonoids due to lower sensitivity and overlapping signals preventing reliable quantification [26]. HPLC-DAD analysis was instead employed to determine the content of individual anthocyanins and flavonols in the cranberry samples. Five predominant anthocyanin peaks were identified by HPLC-DAD (monitoring at 520 nm) eluting between 16 and 23 min; these were identified by matching retention time and UV absorbance patterns with those of authentic commercial standards and previously published data [38]: cyanidin-3-galactoside, cyanidin-3-glucoside, cyanidin-3-arabinoside, peonidin-3-galactoside, and peonidin-3-arabinoside. A representative HPLC chromatogram is shown (Figure 5). Cyanidins were quantified based on a cyanidin-3-galactoside standard, and peonidins were quantified based on a peonidin-3-galactoside standard (Table 2). Peonidin-3-galactoside was the most abundant anthocyanin. Peonidins in all fruit samples were higher than cyanidins. Total anthocyanins varied between 0.05 and 14.53 mg/g DW. Total monomeric anthocyanins in the fruit of multiple varieties harvested in New Jersey reportedly ranged from 12 to 86 mg/100 g of fresh fruit; when approximated for dry weight, this range is consistent with most of the cultivars analyzed in this study [20].

Seven flavonol glycosides were identified as eluting between 23 and 32 min (monitoring at 355 nm) by matching retention time and absorbance patterns with those of authentic commercial standards and previously published data [28,39]: myricetin-3-galactoside, myricetin-3-arabinoside, quercetin-3-galactoside, quercetin-3-xyloside, quercetin-3-arabinopyranoside, quercetin-3-arabinofuranoside, and quercetin-3-rhamnopyranoside. Total flavonol content was determined based on a quercetin-3-galactoside standard (Table 2). The most abundant flavonol was hyperoside (1.10–7.62 mg/g DW). Total flavonol content across all treatments ranged from 4.29–23.30 mg/g DW. Differences in anthocyanin and flavonol content were evident between cultivars, but there was no clear trend across all treatments. The Scarlet Knight cultivar was consistently the highest in total anthocyanins and flavonols for 2016 in both regions and among the highest in 2017. A UPLC-HRMS metabolomic study on fruit grown in New Jersey during 2017 found anthocyanins to be a strong contributor to cross-cultivar variations [27]. In the present study, metabolomic analysis of NMR spectra did not reveal significant differences in flavonoid content or clear groupings among the cultivars studied, but it did indicate significant variation in anthocyanin and flavonol content between growing regions.

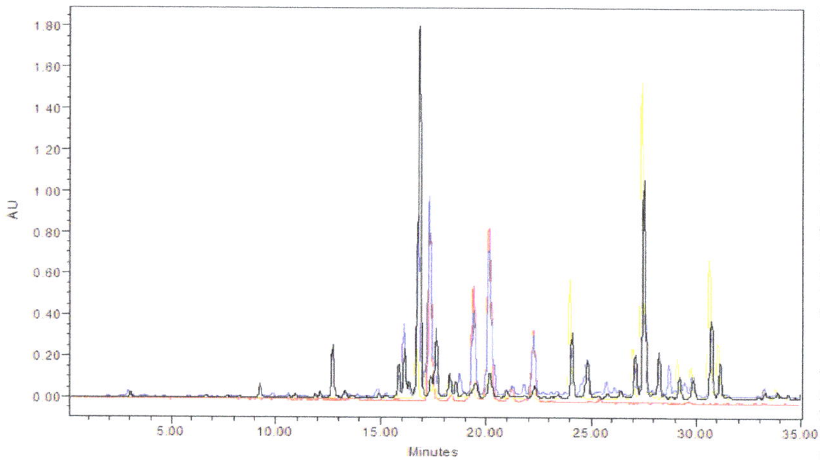

Figure 5. Representative HPLC-DAD chromatogram of ethanolic extract (ST MA 2016) showing the presence of anthocyanins (red, 520 nm), flavonols (yellow, 355 nm), and other phenolic constituents (blue, 280 nm, and black, 310 nm) detected at characteristic absorbance wavelengths.

Table 2. Content of flavonoids, PACs, and total phenolics in cranberry fruit samples (mg analyte/g dry weight ± standard deviation, n = 5).

Cultivar/Region	Year	Cyanidin Glycosides	Peonidin Glycosides	Total Antho-cyanin	Quercetin-3-Galactoside	Total Flavonols	Total Phenolics [a]	Total PACs
ST/MA	2016	0.87 ± 0.16	2.01 ± 0.38	2.88 ± 0.54	2.78 ± 0.67	6.88 ± 1.46	13.46 ± 0.59	41.71 ± 4.20
	2017	0.20 ± 0.13	0.42 ± 0.25	0.62 ± 0.24	2.58 ± 0.35	8.17 ± 1.07	12.32 ± 0.84	53.20 ± 5.15
ST/OR	2016	1.22 ± 0.28	4.29 ± 1.20	5.50 ± 1.48	3.77 ± 0.76	10.44 ± 2.94	16.25 ± 0.33	47.71 ± 7.67
	2017	0.55 ± 0.22	1.65 ± 0.67	2.20 ± 1.02	4.07 ± 0.90	13.37 ± 1.89	18.22 ± 0.35	70.17 ± 9.11
MQ/MA	2016	0.68 ± 0.11	1.57 ± 0.29	2.25 ± 0.39	3.75 ± 0.64	8.92 ± 1.65	15.10 ± 0.49	89.94 ± 11.68
	2017	0.09 ± 0.02	0.20 ± 0.03	0.29 ± 0.05	1.64 ± 0.37	5.42 ± 1.28	14.29 ± 0.74	81.77 ± 12.68
MQ/OR	2016	2.15 ± 0.59	7.13 ± 1.33	9.28 ± 1.90	5.00 ± 2.30	15.99 ± 5.71	17.83 ± 1.65	108.22 ± 22.1
	2017	1.05 ± 0.18	3.31 ± 0.30	4.36 ± 0.36	4.03 ± 0.77	13.01 ± 2.78	19.23 ± 0.24	106.58 ± 6.85
DM/MA	2016	0.79 ± 0.22	1.76 ± 0.50	2.55 ± 0.71	2.77 ± 0.65	7.86 ± 2.72	13.22 ± 0.69	51.69 ± 3.29
	2017	0.79 ± 0.42	1.59 ± 1.22	2.68 ± 1.61	1.95 ± 0.25	7.20 ± 0.79	12.48 ± 0.33	49.32 ± 9.54
DM/OR	2016	1.88 ± 0.89	7.33 ± 3.02	9.41 ± 3.90	4.11 ± 1.99	14.76 ± 7.38	15.72 ± 1.36	55.69 ± 4.37
	2017	1.39 ± 0.37	4.97 ± 1.44	6.36 ± 1.79	1.69 ± 0.29	7.12 ± 0.44	15.55 ± 0.53	56.64 ± 4.6
PI/MA	2016	0.95 ± 0.11	1.81 ± 0.17	2.75 ± 0.26	2.59 ± 0.51	7.04 ± 0.91	14.57 ± 1.39	82.39 ± 19.19
	2017	0.10 ± 0.03	0.16 ± 0.05	0.25 ± 0.06	2.09 ± 0.31	7.04 ± 0.66	14.64 ± 0.87	84.38 ± 13.16
PI/OR	2016	1.76 ± 0.52	4.20 ± 1.03	5.96 ± 1.56	4.54 ± 1.83	14.13 ± 5.44	15.07 ± 0.57	48.07 ± 5.59
	2017	0.94 ± 0.16	2.34 ± 0.30	3.28 ± 0.45	3.48 ± 0.39	12.15 ± 1.59	16.12 ± 0.96	71.40 ± 4.26
CQ/MA	2016	0.69 ± 0.14	2.03 ± 0.32	2.72 ± 0.46	3.16 ± 0.72	7.55 ± 1.78	12.95 ± 0.24	46.21 ± 4.93
	2017	0.52 ± 0.15	1.44 ± 0.59	1.97 ± 0.73	2.09 ± 0.58	7.63 ± 1.54	13.23 ± 0.62	50.49 ± 8.28
CQ/OR	2016	1.72 ± 0.36	5.76 ± 0.81	7.48 ± 0.86	6.62 ± 2.94	18.63 ± 5.84	18.58 ± 1.75	62.46 ± 13.68
	2017	0.85 ± 0.31	3.59 ± 1.82	4.44 ± 1.65	2.92 ± 1.25	10.11 ± 2.22	15.86 ± 0.42	54.84 ± 6.64
HA/MA	2016	1.15 ± 0.25	2.87 ± 0.40	4.02 ± 0.65	3.56 ± 0.82	10.19 ± 3.25	12.69 ± 0.29	52.63 ± 4.82
	2017	0.51 ± 0.09	1.33 ± 0.35	1.84 ± 0.40	1.10 ± 0.26	4.29 ± 0.79	12.89 ± 0.07	51.11 ± 2.97
HA/OR	2016	2.31 ± 0.10	6.76 ± 2.27	9.45 ± 3.11	6.00 ± 1.25	17.03 ± 3.33	17.15 ± 0.81	53.76 ± 3.78
	2017	0.55 ± 0.33	1.51 ± 0.95	2.05 ± 1.27	2.97 ± 0.65	10.55 ± 1.63	17.69 ± 0.47	81.77 ± 5.97
SK/MA	2016	1.81 ± 0.46	4.28 ± 1.08	6.10 ± 1.51	4.50 ± 0.96	11.37 ± 3.86	14.33 ± 0.67	65.86 ± 4.59
	2017	0.71 ± 0.39	1.53 ± 0.90	2.24 ± 1.28	1.77 ± 0.50	5.69 ± 1.78	15.58 ± 1.27	62.32 ± 7.82
SK/OR	2016	3.60 ± 0.72	10.93 ± 2.07	14.53 ± 2.72	7.62 ± 2.22	23.30 ± 8.25	20.94 ± 0.19	78.30 ± 3.03
	2017	2.03 ± 0.36	5.42 ± 0.98	7.45 ± 1.33	4.09 ± 0.74	11.72 ± 1.98	21.95 ± 1.01	87.71 ± 7.27
WK/MA	2016	0.88 ± 0.28	1.68 ± 0.42	2.56 ± 0.66	1.90 ± 0.52	5.65 ± 1.56	12.98 ± 0.31	51.48 ± 5.73
	2017	0.85 ± 0.11	2.71 ± 0.61	3.57 ± 0.67	1.68 ± 0.32	5.25 ± 1.63	13.51 ± 0.56	49.11 ± 3.83
WK/OR	2016	1.19 ± 0.38	4.37 ± 1.53	4.98 ± 1.69	2.13 ± 0.07	9.36 ± 1.57	15.22 ± 0.66	41.83 ± 5.14
	2017	0.85 ± 0.19	2.25 ± 0.62	3.11 ± 0.81	2.64 ± 0.91	10.18 ± 2.77	18.12 ± 1.05	73.02 ± 5.42
GH1/MA	2016	0.47 ± 0.13	0.95 ± 0.41	1.37 ± 0.57	2.31 ± 0.50	6.45 ± 1.25	14.27 ± 0.41	58.00 ± 5.07
	2017	0.24 ± 0.06	0.49 ± 0.15	0.73 ± 0.22	1.30 ± 0.52	4.41 ± 1.41	13.44 ± 1.24	61.66 ± 7.07
GH1/OR	2016	1.17 ± 0.20	4.26 ± 1.90	5.42 ± 2.05	3.26 ± 0.93	11.25 ± 1.90	17.29 ± 0.11	68.61 ± 5.92
	2017	1.33 ± 0.24	4.29 ± 0.92	5.62 ± 1.15	3.59 ± 0.82	10.35 ± 2.05	18.40 ± 0.34	66.10 ± 7.45

[a] Total phenolics are expressed as gallic acid equivalents (GAE).

In contrast with the triterpenoids and organic acids, the content of anthocyanins and flavonols, on average, was significantly higher in OR fruit than MA fruit (Figure 6). Total anthocyanin content across all OR cultivars (6.2 + 3.1 mg/g DW) was more than double that of the fruit harvested in MA (2.3 + 1.4 mg/g DW). Likewise, total flavonol content was much higher in OR fruit (13.0 + 3.9 mg/g DW) than MA fruit (7.1 + 1.9 mg/g DW). The trend in anthocyanin content can likely be explained by the influence of the environment on horticultural practices. Due to differences in climate and pest pressure, cranberries can remain in the bog longer after ripening without significant rot in coastal OR than in MA, allowing fruit to be harvested in OR into November while fruit in MA is typically harvested by early October. Season-to-season differences in the weather patterns may also impact phytochemical content; thus, we collected data on fruit from two growing seasons. MA experienced above-average temperatures and rainfall both years [40]. OR had average temperatures and rainfall during most of 2016 and 2017, but also experienced some of its highest recorded temperatures and very low humidity during August 2017 [41]. The effects of short-term weather extremes on cranberry phytochemical development are unknown at this time.

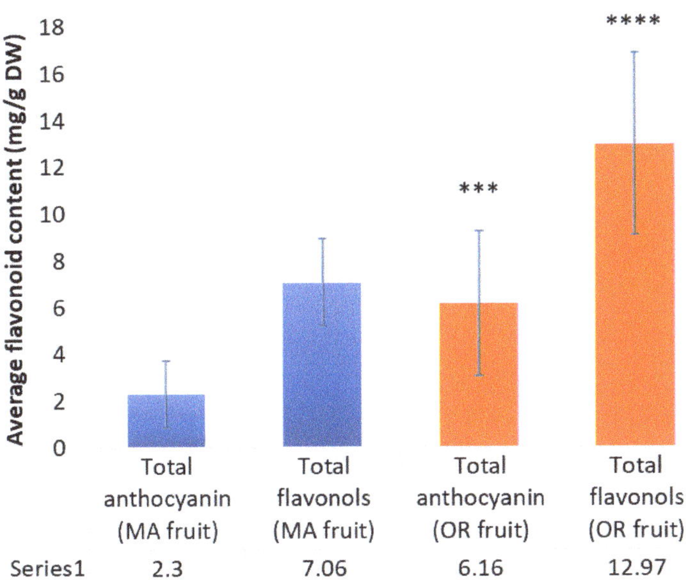

Figure 6. Regional comparison of the average total anthocyanins and flavonols for all cultivars over two growing seasons, OR vs. MA (*** = $p \leq 0.001$, **** = $p \leq 0.0001$).

2.4. Proanthocyanidin Content

The total PAC content of cranberry fruit was determined by the BL-DMAC assay using a cranberry PAC standard. Samples ranged between 41.7 and 108 mg PAC/g cranberry dry weight. Cultivar-based differences in PACs were observed (Figure 7), consistent with previous reports [20], but the averages across all cultivars did not differ significantly between the two regions. MQ exhibited the highest average PAC content (97 + 17 mg/g DW) across regions among the nine cultivars, significantly higher than the average, followed by SK (74 + 12 mg/g DW) and PI (72 + 21 mg/g DW). ST, CQ, and DM cultivars had the lowest average PAC content, around 53 mg/g DW. Across all cultivars, the average total PACs were nearly equal for 2016 OR (62.7 + 20.5 mg/g DW) and MA (60.0 + 16.4 mg/g DW) fruit. A higher average PAC content for 2017 OR (74.3 + 16.1 mg/g DW) vs. MA (60.4 + 13.8 mg/g DW) fruit was not statistically significant. The genetic background of various cultivars has been observed to influence PAC content in cranberries. Previous work by Rutgers University researchers determined the concentration of total PACs in eight cranberry cultivars grown in New Jersey using LC-MS [20]. The study found that Howes and its offspring #35 (Figure 1) exhibited the highest level of PACs. MQ, an offspring of #35, was also high in PACs. A 2014 study by Carpenter and coworkers examining proanthocyanidin content and composition in cranberry fruit of various cultivars grown in MA, NJ, WI, and British Columbia also reported cultivar-related variation in PACs. Some regional variation in PAC content was noted for MQ fruit, whereas CQ fruit had similar PAC content across three regions [42]. In the present study, MQ was highest in PACs among the nine cultivars, consistent with its genetic lineage (derived from Howes and #35) and with previous study data. We found CQ and DM among the lowest for PAC content, which is also consistent with Rutgers data. ST is the parent of CQ, which could explain its low PAC concentration. Another Rutgers study correlated fruit size with PAC concentration for ST and Ben Lear cranberries, showing an inverse correlation between PAC content and fruit size [39]. CQ and DM berries were the largest among MA fruit samples, and their PAC content was among the lowest. Cultivar appears to play a more significant role in determining PAC content, as demonstrated by the present study and others [20,27].

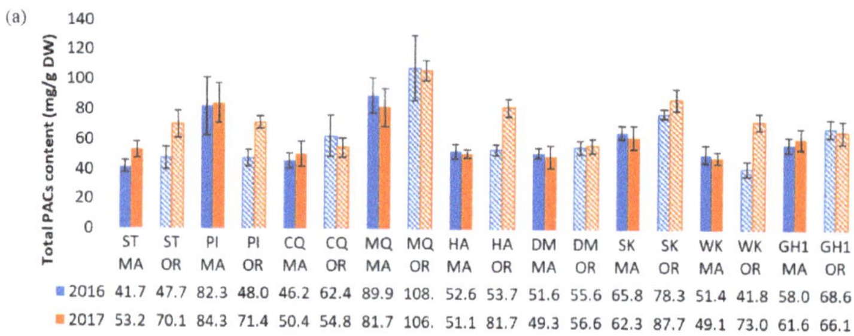

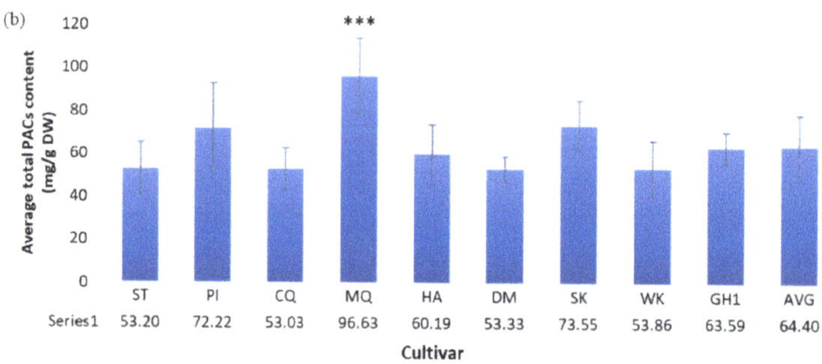

Figure 7. (**a**) Total proanthocyanidin (PAC) content by cultivar, growing region, and season, n = 5; (**b**) Average PAC content for each cultivar across all treatments, n = 20 (mg/g cranberry dry weight), *** = $p \leq 0.001$.

2.5. Trends in Total Phenolic Content

The Folin–Ciocalteu assay was used to estimate the total content of soluble phenolic compounds, or TPC (Figure 8a). This was significantly higher on average for OR fruit (17.51 + 1.91 mg GAE/g DW) than for MA fruit (13.66 + 0.94 mg GAE/g DW). OR fruit ranged from 15.07 + 0.57 mg GAE/g DW for PI cultivar to 20.94 + 0.19 mg GAE/g DW for SK cultivar in 2016, and from 15.55 + 0.53 mg GAE/g DW for DM to 21.95 + 1.01 mg GAE/g DW for SK in 2017. By comparison, 2016 MA fruit ranged from 12.69 + 0.29 mg GAE/g DW for HA to 15.09 + 0.49 mg GAE/g DW for MQ, and 2017 fruit ranged from 12.32 + 0.84 mg GAE/g DW for the ST cultivar to 15.58 + 1.27 mg GAE/g DW for SK. The averages across cultivars for OR vs. MA fruit are shown in Figure 9a. Scarlet Knight fruit from OR exhibited the highest TPC for both growing seasons, whereas Haines and Stevens fruit from MA had the lowest TPC in 2016 and 2017, respectively. The regional trend in TPC is consistent with the regional trend in the content of anthocyanins and flavonols. Cultivar-related variations in TPC have been previously reported for cranberries grown in Poland [43,44] or New Jersey [45]. Wang and Stretch [45] also reported an increase in TPC over three months in storage at temperatures above zero °C. Taken together, these studies demonstrate that phenolic content can depend on many factors, including genetic background, regional environment, cultivation conditions, maturation, and storage. The effect of pedigree on TPC was demonstrated in our study by DM and its offspring WK, both having similar TPC in 2016 MA fruit, and by CQ and its offspring HA for both growing seasons in MA (Table 2). SK was consistently among the highest cultivars for TPC. We also analyzed samples of Early Black fruit from 2017, a cultivar traditionally grown in MA. Extensive data on Early Black is not included in this study as the cultivar was not available at our OR site. Notably, Early Black fruit had the highest TPC (16.22 + 0.34 mg GAE/g DW) among all MA cultivars in our study, consistent with the results of Wang and Stretch [45].

2.6. Correlation between Antioxidant Activity and Metabolite Content

Cranberries are rich in flavonoids and other phenolics, which are likely to contribute to their antioxidant activity. The antioxidant capacities of cranberry crude extracts were determined by the DPPH free radical scavenging method. All samples were compared at the same extract concentration (250 μg/mL). In general, for both seasons and for most cultivars, extracts of the OR fruit samples inhibited free radical scavenging more effectively (53.32–85.93%) than their MA counterparts (26.05–60.56%) (Figure 8b). The average percent inhibition of DPPH radical by OR fruit extracts of all cultivars (70.8% + 9.9%) was significantly higher than the MA fruit (43% + 11%), as shown in Figure 9b. This was consistent with the regional trend in the content of anthocyanins, flavonols, and total phenolics and in contrast with the content of triterpenoids in cranberry fruit.

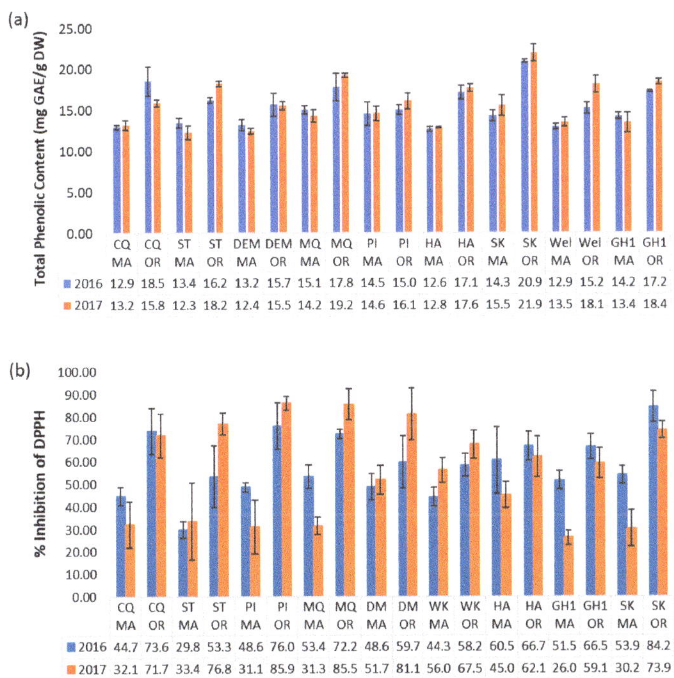

Figure 8. (**a**) Total phenolic content and (**b**) percent inhibition of DPPH radical by cranberry fruit extracts by cultivar, growing region, and season (n = 5).

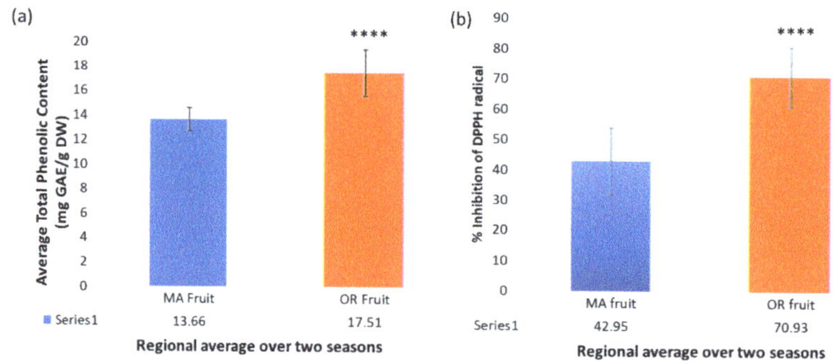

Figure 9. Regional comparison across all cultivars for (**a**) total phenolic content and (**b**) % inhibition of DPPH radical scavenging (**** = $p \leq 0.0001$).

Linear correlation analyses were performed for antioxidant activity vs. key secondary metabolite groups (Figures S1–S6). Significant positive correlation was observed between antioxidant activity and the sum of total anthocyanins and flavonols content in all cranberry fruit samples during both the 2016 and 2017 growing seasons ($R^2 = 0.7189$, $p < 0.0001$ and $R^2 = 0.7631$, $p < 0.0001$, respectively) and over all treatments ($R^2 = 0.5043$, $p < 0.00001$). This finding is consistent with positive correlations reported in a study of cranberry fruit grown in Poland [36]. In our study, SK and DM fruit samples from OR were among the highest in both antioxidant activity and anthocyanin content. Positive correlations were also observed between antioxidant activity and the total soluble phenolic content for the 2016 and 2017 growing seasons ($R^2 = 0.6351$, $p < 0.001$; and $R^2 = 0.4364$, $p < 0.01$, respectively) and over all treatments ($R^2 = 0.4808$, $p < 0.00001$). Fruits with low total phenolic content demonstrated lower antioxidant activity, a trend also reported by Wang and Stretch [44]. However, no correlation was found between antioxidant activity and total PACs content either during the growing season ($R^2 = 0.1449$ for 2016 and 0.0824 for 2017, respectively) or overall ($R^2 = 0.0868$). This was surprising, as PACs are widely regarded as antioxidants; their contributions to radical scavenging may be secondary to those of other phenolics in the fruit.

In a related study of anti-inflammatory properties, we reported that concentrated extracts of the 2016 and 2017 SK and ST cranberry fruit from MA and OR significantly inhibited the LPS-induced expression of IL-6 in human THP-1 monocytes at 100 µg/mL [46]. Several of the extracts were inhibitory at 0.1–10 µg/mL. The data suggested that the observed anti-inflammatory activities were influenced by the content of flavonoids, total polyphenols, and triterpenoids. The data from the present study indicates a strong positive correlation between free-radical scavenging antioxidant activity and total anthocyanins and flavonols. In contrast, we observed that free-radical scavenging antioxidant activity in our samples demonstrated a slight negative correlation with the content of triterpenoids ursolic and oleanolic acid ($R^2 = 0.3513$, $p < 0.05$ and $R^2 = 0.2707$, $p < 0.05$ for 2016 and 2017 fruit, respectively; $R^2 = 0.224$, $p < 0.01$ over all treatments). Reported antioxidant mechanisms for UA, OA, and related triterpenoids include upregulation of endogenous antioxidants such as catalase, superoxide dismutase, and glutathione, activation of the Nrf-2 pathway, and reduction of intracellular ROS production [15,47]. Further investigation of the antioxidant and anti-inflammatory mechanisms of cranberry triterpenoids is underway.

3. Materials and Methods

3.1. Plant Materials

Cranberry fruit from multiple cultivars was harvested from east coast Massachusetts (MA) and west coast Oregon (OR) cranberry bogs during the 2016 and 2017 growing seasons. Nine cranberry cultivars have been selected for this study, including Crimson Queen (CQ), Demoranville (DM), Mullica Queen (MQ), Pilgrim (PI), Welker (WK), Haines

(HA), Grygleski Hybrid #1 (GH1), Scarlet Knight (SK), and Stevens (ST). Cranberry fruit from MA was harvested at the UMASS Cranberry Experiment Station in East Wareham, Massachusetts (41.76° N, 70.67° W) in late September and authenticated by Krystal DeMoranville. Cranberry fruit from OR was collected in late October from Bandon, OR (Coos County, 43.02° N, 124.43° W) and authenticated by Cassie Bouska (Oregon State University Extension Service, OR). Cranberry samples were selected randomly within each cultivar to yield five field replicates. All cranberry fruits were flash frozen with liquid nitrogen and stored at −20 °C until use.

3.2. Chemicals

4,4-dimethyl-4-silapentane-1-sulfonic acid (DSS) and deuterated dimethylsulfoxide (DMSO-d6) were obtained from Cambridge Isotope Laboratories, Inc. (Andover, MA, USA). Reagent-grade methanol and ethanol were purchased from Fisher Scientific (Hampton, NH, USA). Reagent-grade glacial acetic acid, hydrochloric acid, and acetone were purchased from Pharmco-AAPER (Brookfield, CT, USA), along with UV/HPLC-grade acetonitrile and methanol. HPLC-grade water was purchased from Honeywell (Morristown, NJ). Quercetin-3-galactoside (hyperoside) standard was purchased from ChromaDex, Inc. (Irvine, CA, USA). Standards of cyanidin-3-galactoside (assay ≥ 97%), and peonidin-3-galactoside (assay ≥ 95%) for HPLC analysis were obtained from Extrasynthese (Genay, France). Standards of ursolic acid (assay ≥ 90%), oleanolic acid (assay ≥ 97%) and 3,5-dinitrobenzoic acid were purchased from Sigma-Aldrich (St. Louis, MO, USA), along with N,N-dimethylaminocinnamaldehyde (DMAC) powder, 2, 2-diphenyl-1-picrylhydrazyl (DPPH), reagent-grade formic acid, gallic acid, and Folin–Ciocalteu reagent. Standards of benzoic acid and citric acid were obtained from J.T. Baker Chemical Co. (Phillipsburg, NJ, USA). 6-hydroxy-2,5,7,8-tetramethylchroman-2-carboxylic acid (Trolox) for the DPPH assay was purchased from Aldrich Chemical Company (Milwaukee, WI, USA). Cranberry proanthocyanidins (PAC) standard for the DMAC analysis was previously prepared from an isolated cranberry PAC fraction containing A-type flavan-3-ol oligomers [48], which was characterized by matrix-assisted laser desorption/ionization time-of-flight (MALDI-TOF) mass spectrometry.

3.3. Sample Extraction and Preparation

Whole cranberry fruits were partitioned into units of about 50 g (n = 5), freeze-dried via lyophilizer (LABCONCO FreeZone 4.5 L −84 °C), and ground to a fine powder (5~9 g). For quantitative analysis, cranberry powder from each replicate was split into sub-reps of 2.0 g and extracted in 100 mL of reagent-grade ethanol. Ethanol was chosen as a reliable universal solvent for most cranberry phytochemicals, including different classes of polyphenols and terpenoids. Solutions were stirred for an hour at room temperature and stored overnight at 0 °C. After vacuum filtration, the solid residue was re-extracted with 25 mL of ethanol, stirred for an hour at room temperature, then combined with the other extracts and rotary evaporated (Büchi Rotovapor R-200). Fruit extracts were freeze-dried to remove residual water and stored at −20 °C until use. These ethanol extracts were used for NMR, HPLC analysis, and DPPH assays.

3.4. Metabolomic Fingerprinting and Determination of Triterpenoids and Organic Acids Using Proton (^1H) qNMR

A metabolomic fingerprint of each cranberry extract was acquired using ^{1}H NOESY NMR, and a qNMR method was established to determine triterpenoids and organic acids in the whole cranberry extracts using the methods of Turbitt et al., 2020 [26]. Briefly, samples of each extract were prepared at 75 mg/mL in DMSO-d_6 solvent. spiked with 0.1 mg/mL DSS as the internal chemical shift calibration standard, vortexed, and sonicated using a Fisher Scientific FS20H Sonicator (Pittsburgh, PA, USA) until fully dissolved. Samples were centrifuged, transferred to NMR tubes, and analyzed using a one-dimensional proton nuclear Overhauser effect spectroscopy experiment (1D ^{1}H NOESY). NMR spectra were collected (ns = 8) at 298.0 K with gas flow at 400 lph on a Bruker AVANCE III 400 MHz NMR using Bruker TopSpin™ 3.5. Metabolite identification, quantification, and chemometrics were performed with Bruker AssureNMR ™ 2.0 and AMIX ™ 3.9.15, using an internal NMR spectral database (SBASE, Bruker BioSpin, Billerica, MA, USA) as a reference. 3,5-dinitrobenzoic acid was chosen as an external calibration standard according to the PULCON approach [48]. Triterpenoids, including ursolic and oleanolic acids, were identified and quantified by methods described by Turbitt and coworkers [26], with automated matching of sample signals to well-characterized signals of purchased authentic standards in the SBASE. The calibration curves of these metabolites, including ursolic acid, and oleanolic acid produced suitable linearity, R^2 = 0.9995 and 0.9998, respectively.

3.5. Determination of Flavonols and Anthocyanins by HPLC-DAD

High-Performance Liquid Chromatography with Diode-Array Detection (HPLC-DAD) was used to identify and quantify anthocyanins and flavonols within cranberry crude samples, according to previously published methods, with minor modifications [26]. Analysis was performed on a Waters Millennium binary HPLC system with two Waters 515 pumps at a flow rate of 0.9 mL/min, Millennium 32 version 32.0 software, coupled with a 996 photodiode array (PDA) detector. An Atlantis T3 column (4.6 × 150 mm, 3 μm; Waters) was used. Flavonols and anthocyanins were detected at 355 nm and 520 nm, respectively. The mobile phase consisted of 0.5% phosphoric acid/water as solvent A and water: acetonitrile: glacial acetic acid: phosphoric acid (50.0:48.5:1.0:0.5, *v/v/v/v*) as solvent B. Samples were dissolved in 25% methanol in solvent A and analyzed at room temperature with an injection volume of 20 μL. Anthocyanins and flavonols were identified by matching their HPLC peak retention time patterns and UV absorbance with those of authentic commercial standards and previously published data [26,28,38]. Quantification of cyanidins and peonidins was performed in comparison to external standard curves for cyanidin-3-galactoside and peonidin-3-galactoside, respectively; quantification of flavonol glycosides was based on an external standard curve for quercetin-3-galactoside (hyperoside).

3.6. Determination of Total Proanthocyanidin Content by DMAC Method

Total proanthocyanidins (PACs) content was determined using the BL-DMAC method described by Prior et al. with the cranberry PAC standard as previously described [49,50]. All samples of each replicate were tested in triplicate (5 replicates × 3, n = 15). Results are expressed as mg PAC per g DW.

3.7. Determination of Total Phenolic Content by Folin–Ciocalteu Assay

The total soluble phenolic content of freeze-dried cranberry samples was determined in triplicate using a microplate Folin–Ciocalteu method as described previously [51]. Briefly, 20 mg of each sample was sonicated in 95% ice-cold methanol (2 mL) for 30 min, then incubated for 24 h in the dark at room temperature. Samples were centrifuged at 13,000 g for five minutes. The supernatant (100 μL) was mixed with Folin–Ciocalteu reagent (10% *v/v*, 200 μL), briefly vortexed, then mixed with 700 mM aqueous Na_2CO_3 (800 μL) and incubated at room temperature for two hours. Samples (200 μL) were transferred to a 96-well microtiter plate, and the absorbance was read at 765 nm vs. a 95% methanol blank using a SpectraMax 190 microplate reader (Molecular Devices, San Jose, CA, USA). Total phenolic content was determined as gallic acid equivalents by comparison to a gallic acid standard curve.

3.8. Determination of Antioxidant Activity by DPPH Free-Radical Scavenging Assay

The free radical scavenging capacities of cranberry samples were determined by the 2,2-diphenyl-1-picrylhydrazyl (DPPH) free radical scavenging assay using a modified microplate method [52]. Cranberry ethanol extracts were dissolved in methanol to yield a concentration of 250 μg/mL, and then 200 μL aliquots were transferred to a 96-well plate

(Costar, Corning, NY, USA). An amount of 50 μL of DPPH methanol solution (100 μg/mL) was added to each well, followed by a 30 min incubation in the dark. The final test sample concentration was 200 μg/mL, and the final DPPH concentration was 20 μg/mL. Absorbance was measured at 517 nm using a SpectraMax 190 microplate reader (Molecular Devices, San Jose, CA, USA). The blank was 250 μL of methanol, and the control contained 50 μL of DPPH methanol solution in 200 μL of methanol. Each field replicate was tested in triplicate (5 replicates × 3, n = 15). The free-radical scavenging antioxidant activity was expressed as percent inhibition using the following equation:

$$\% \text{ inhibition} = ((A_0 - A_b) - (A_1 - A_b))/(A_0 - A_b) \times 100\%$$

A_0 is the absorbance of the control solution, A_b is the absorbance of the blank, and A_1 is the absorbance of the remaining DPPH• in the presence of a cranberry sample.

3.9. Chemometric and Statistical Analyses

Non-targeted principal component analysis (PCA) of ^{1}H NOESY NMR data was performed using AMIX ™ 3.9.15 software, as previously described [26]. Briefly, spectra were partitioned into 0.05 ppm buckets, solvent regions were excluded, and variances were optimized via Pareto scaling. The principal contributors to variance between sample sets appearing in the Scores plots were identified by selecting the signal intensities that differ significantly between spectra and appear as outliers in the loadings' plots. The confidence ellipsoids and minimum explained variance were set at 95% [25]. Metabolites contributing to the variance were identified by matching these signals to standards in the SBASE. Linear regression analysis was performed as conducted previously [46] using XLMiner to determine significant correlations between antioxidant activity and metabolite content. A one-way analysis of variance (ANOVA) was used to determine statistically significant differences between sample sets.

4. Conclusions

The present study illustrates a significant effect of the growing region in North America on the secondary metabolite composition and antioxidant properties of cranberries. PCA of ^{1}H NOESY NMR data revealed specific metabolites (triterpenoids, organic acids, and flavonoids) that contributed to the variance between Oregon and Massachusetts fruit. Quantitation of these metabolites confirmed significant regional variation. Fruit from Oregon, a major growing region on the West Coast, exhibited significantly higher total flavonol and anthocyanin content over two harvest seasons than fruit from Massachusetts, a major growing region on the East Coast. By contrast, we found that MA fruit was consistently richer in triterpenoids and organic acids than fruit from OR. The influence of growing region on PAC content among fruit in our study was less evident, but PACs varied substantially among the cultivars. A strong positive correlation was observed for the antioxidant activity of cranberry fruits with the content of flavonols and anthocyanins, as well as with total soluble phenolic content, with OR fruit possessing greater overall anti-radical scavenging potential than fruit from MA. We report for the first time a regional influence (MA > OR) on the content of the triterpenoids ursolic and oleanolic acid, compounds that are likely to contribute to the anti-inflammatory benefits of cranberry fruit. Understanding patterns in cranberry metabolite production will contribute to improving harvest strategies and product development relating to the health-promoting properties of cranberry fruit. Cranberry flavonoids and triterpenoids have demonstrated potential for human health by protecting against oxidative stress and inflammation. The development of horticultural practices that maximize the content of these important compounds can provide greater dietary health benefits.

Supplementary Materials: The following supporting information can be downloaded at: https://www.mdpi.com/article/10.3390/plants12203595/s1, Figures S1–S6: Linear correlation data for antioxidant activity with total phenolic content, anthocyanins and flavonols, proanthocyanidins and triterpenoids.

Author Contributions: Conceptualization, L.X. and C.N.; Methodology, L.X., M.O. and K.C.; Software, K.C.; Formal Analysis, L.X. and M.O.; Resources, C.N.; Data Curation, L.X.; Writing—Original Draft Preparation, L.X. and C.N.; Writing—Review and Editing, C.N.; Supervision and Project Administration, C.N.; Funding Acquisition, C.N. and K.C. All authors have read and agreed to the published version of the manuscript.

Funding: This work was funded by the Massachusetts Department of Health Award to the UMass Cranberry Health Research Center (INTF4005HH2W18082101), Bruker Biospin Inc., and the Leo and Anne Albert Charitable Trust.

Data Availability Statement: Data are available in the article, Supplementary Materials, or by request to the corresponding authors.

Acknowledgments: The authors wish to acknowledge the assistance of Krystal Demoranville (UMass Cranberry Experiment Station) and Cassie Bouska (Oregon State University Extension) in obtaining fruit samples and Andrew Milstead for guidance in using the Bruker AssureNMR TM 2.0 and AMIX TM 3.9.15 NMR software.

Conflicts of Interest: The authors declare no conflict of interest. The funders had no role in the design of the study; in the collection, analyses, or interpretation of data; in the writing of the manuscript; or in the decision to publish the results.

References

1. Fresh Cranberry Production and Top Producing Countries. Available online: https://www.tridge.com/intelligences/cranberry/production (accessed on 15 September 2023).
2. Cranberry Production in Top-Producing States to Increase Modestly in 2021. Available online: http://www.ers.usda.gov/data-products/chart-gallery/gallery/chart-detail/?chartId=102649 (accessed on 28 June 2022).
3. Blumberg, J.B.; Camesano, T.A.; Cassidy, A.; Kris-Etherton, P.; Howell, A.; Manach, C.; Ostertag, L.M.; Sies, H.; Skulas-Ray, A.; Vita, J.A. Cranberries and Their Bioactive Constituents in Human Health. *Adv. Nutr.* **2013**, *4*, 618–632. [CrossRef] [PubMed]
4. McKay, D.L.; Blumberg, J.B. Cranberries (*Vaccinium macrocarpon*) and Cardiovascular Disease Risk Factors. *Nutr. Rev.* **2007**, *65*, 490–502. [CrossRef] [PubMed]
5. Cai, X.; Han, Y.; Gu, M.; Song, M.; Wu, X.; Li, Z.; Li, F.; Goulette, T.; Xiao, H. Dietary Cranberry Suppressed Colonic Inflammation and Alleviated Gut Microbiota Dysbiosis in Dextran Sodium Sulfate-Treated Mice. *Food Funct.* **2019**, *10*, 6331–6341. [CrossRef]
6. Seeram, N.P.; Adams, L.S.; Hardy, M.L.; Heber, D. Total Cranberry Extract versus Its Phytochemical Constituents: Antiproliferative and Synergistic Effects against Human Tumor Cell Lines. *J. Agric. Food Chem.* **2004**, *52*, 2512–2517. [CrossRef] [PubMed]
7. Kresty, L.A.; Clarke, J.; Ezell, K.; Exum, A.; Howell, A.B.; Guettouche, T. MicroRNA Alterations in Barrett's Esophagus, Esophageal Adenocarcinoma, and Esophageal Adenocarcinoma Cell Lines Following Cranberry Extract Treatment: Insights for Chemoprevention. *J. Carcinog.* **2011**, *10*, 34. [CrossRef]
8. Weh, K.M.; Clarke, J.; Kresty, L.A. Cranberries and Cancer: An Update of Preclinical Studies Evaluating the Cancer Inhibitory Potential of Cranberry and Cranberry Derived Constituents. *Antioxidants* **2016**, *5*, 27. [CrossRef]
9. Narayansingh, R.; Hurta, R.A. Cranberry Extract and Quercetin Modulate the Expression of Cyclooxygenase-2 (COX-2) and IκBα in Human Colon Cancer Cells. *J. Sci. Food Agric.* **2009**, *89*, 542–547. [CrossRef]
10. Côté, J.; Caillet, S.; Doyon, G.; Sylvain, J.-F.; Lacroix, M. Bioactive Compounds in Cranberries and Their Biological Properties. *Crit. Rev. Food Sci. Nutr.* **2010**, *50*, 666–679. [CrossRef]
11. Ma, H.; Johnson, S.L.; Liu, W.; DaSilva, N.A.; Meschwitz, S.; Dain, J.A.; Seeram, N.P. Evaluation of Polyphenol Anthocyanin-Enriched Extracts of Blackberry, Black Raspberry, Blueberry, Cranberry, Red Raspberry, and Strawberry for Free Radical Scavenging, Reactive Carbonyl Species Trapping, Anti-Glycation, Anti-β-Amyloid Aggregation, and Microglial Neuroprotective Effects. *Int. J. Mol. Sci.* **2018**, *19*, 461. [CrossRef]
12. Wu, X.; Xue, L.; Tata, A.; Song, M.; Neto, C.C.; Xiao, H. Bioactive Components of Polyphenol-Rich and Non-Polyphenol-Rich Cranberry Fruit Extracts and Their Chemopreventive Effects on Colitis-Associated Colon Cancer. *J. Agric. Food Chem.* **2020**, *68*, 6845–6853. [CrossRef]
13. Wang, C.; Gao, Y.; Zhang, Z.; Chen, C.; Chi, Q.; Xu, K.; Yang, L. Ursolic Acid Protects Chondrocytes, Exhibits Anti-Inflammatory Properties via Regulation of the NF-KB/NLRP3 Inflammasome Pathway and Ameliorates Osteoarthritis. *Biomed. Pharmacother.* **2020**, *130*, 110568. [CrossRef] [PubMed]
14. Mlala, S.; Oyedeji, A.O.; Gondwe, M.; Oyedeji, O.O. Ursolic Acid and Its Derivatives as Bioactive Agents. *Molecules* **2019**, *24*, 2751. [CrossRef]
15. Wang, X.-T.; Gong, Y.; Zhou, B.; Yang, J.-J.; Cheng, Y.; Zhao, J.-G.; Qi, M.-Y. Ursolic Acid Ameliorates Oxidative Stress, Inflammation and Fibrosis in Diabetic Cardiomyopathy Rats. *Biomed. Pharmacother.* **2018**, *97*, 1461–1467. [CrossRef] [PubMed]

16. Kondo, M.; MacKinnon, S.L.; Craft, C.C.; Matchett, M.D.; Hurta, R.A.R.; Neto, C.C. Ursolic Acid and Its Esters: Occurrence in Cranberries and Other *Vaccinium* Fruit and Effects on Matrix Metalloproteinase Activity in DU145 Prostate Tumor Cells: Anti-Tumor Activity and Content of Ursolic Acid from *Vaccinium* Fruit. *J. Sci. Food Agric.* **2011**, *91*, 789–796. [CrossRef] [PubMed]
17. Cranberry Marketing Committee. Cranberry Varieties. Available online: https://reports.uscranberries.com/TheCranberryStory/Varieties/ (accessed on 19 January 2023).
18. USDA/NASS QuickStats Ad-Hoc Query Tool. Available online: https://quickstats.nass.usda.gov/results/4CD92CD5-FC42-347C-ABFD-3487DC6D149D (accessed on 28 June 2022).
19. Brown, P.N.; Murch, S.J.; Shipley, P. Phytochemical Diversity of Cranberry (*Vaccinium macrocarpon* Aiton) Cultivars by Anthocyanin Determination and Metabolomic Profiling with Chemometric Analysis. *J. Agric. Food Chem.* **2012**, *60*, 261–271. [CrossRef]
20. Wang, Y.; Johnson-Cicalese, J.; Singh, A.P.; Vorsa, N. Characterization and Quantification of Flavonoids and Organic Acids over Fruit Development in American Cranberry (*Vaccinium macrocarpon*) Cultivars Using HPLC and APCI-MS/MS. *Plant Sci. Int. J. Exp. Plant Biol.* **2017**, *262*, 91–102. [CrossRef]
21. Lu, Y.; Pekerti, B.N.; Toh, Z.S.; Broom, F.; Savage, G.; Liu, S.Q.; Huang, D. Physico-Chemical Parameters and Proanthocyanidin Profiles of Cranberries Cultivated in New Zealand. *J. Food Compos. Anal.* **2017**, *63*, 1–7. [CrossRef]
22. Georgescu, C.; Frum, A.; Virchea, L.-I.; Sumacheva, A.; Shamtsyan, M.; Gligor, F.-G.; Olah, N.K.; Mathe, E.; Mironescu, M. Geographic Variability of Berry Phytochemicals with Antioxidant and Antimicrobial Properties. *Molecules* **2022**, *27*, 4986. [CrossRef]
23. Murphy, B.T.; MacKinnon, S.L.; Yan, X.; Hammond, G.B.; Vaisberg, A.J.; Neto, C.C. Identification of Triterpene Hydroxycinnamates with in Vitro Antitumor Activity from Whole Cranberry Fruit (*Vaccinium macrocarpon*). *J. Agric. Food Chem.* **2003**, *51*, 3541–3545. [CrossRef]
24. Yuk, J.; McIntyre, K.L.; Fischer, C.; Hicks, J.; Colson, K.L.; Lui, E.; Brown, D.; Arnason, J.T. Distinguishing Ontario Ginseng Landraces and Ginseng Species Using NMR-Based Metabolomics. *Anal. Bioanal. Chem.* **2013**, *405*, 4499–4509. [CrossRef]
25. Markus, M.A.; Ferrier, J.; Luchsinger, S.M.; Yuk, J.; Cuerrier, A.; Balick, M.J.; Hicks, J.M.; Killday, K.B.; Kirby, C.W.; Berrue, F.; et al. Distinguishing *Vaccinium* species by Chemical Fingerprinting Based on NMR Spectra, Validated with Spectra Collected in Different Laboratories. *Planta Med.* **2014**, *80*, 732–739. [CrossRef] [PubMed]
26. Turbitt, J.R.; Colson, K.L.; Killday, K.B.; Milstead, A.; Neto, C.C. Application of 1 H-NMR-Based Metabolomics to the Analysis of Cranberry (*Vaccinium macrocarpon*) Supplements. *Phytochem. Anal. PCA* **2020**, *31*, 68–80. [CrossRef] [PubMed]
27. Wang, Y.; Vorsa, N.; Harrington, P.d.B.; Chen, P. Nontargeted Metabolomic Study on Variation of Phenolics in Different Cranberry Cultivars Using UPLC-IM—HRMS. *J. Agric. Food Chem.* **2018**, *66*, 12206–12216. [CrossRef] [PubMed]
28. Vvedenskaya, I.O.; Rosen, R.T.; Guido, J.E.; Russell, D.J.; Mills, K.A.; Vorsa, N. Characterization of Flavonols in Cranberry (*Vaccinium macrocarpon*) Powder. *J. Agric. Food Chem.* **2004**, *52*, 188–195. [CrossRef]
29. Chen, H.; Zuo, Y.; Deng, Y. Separation and Determination of Flavonoids and Other Phenolic Compounds in Cranberry Juice by High-Performance Liquid Chromatography. *J. Chromatogr. A* **2001**, *913*, 387–395. [CrossRef]
30. Simmler, C.; Napolitano, J.G.; McAlpine, J.B.; Chen, S.-N.; Pauli, G.F. Universal Quantitative NMR Analysis of Complex Natural Samples. *Curr. Opin. Biotechnol.* **2014**, *25*, 51–59. [CrossRef]
31. Hicks, J.M.; Muhammad, A.; Ferrier, J.; Saleem, A.; Cuerrier, A.; Arnason, J.T.; Colson, K.L. Quantification of Chlorogenic Acid and Hyperoside Directly from Crude Blueberry (*Vaccinium angustifolium*) Leaf Extract by NMR Spectroscopy Analysis: Single-Laboratory Validation. *J. AOAC Int.* **2012**, *95*, 1406–1411. [CrossRef]
32. Huang, Y.; Li, X.; Peng, X.; Adegoke, A.T.; Chen, J.; Su, H.; Hu, G.; Wei, G.; Qiu, M. NMR-Based Structural Classification, Identification, and Quantification of Triterpenoids from Edible Mushroom Ganoderma Resinaceum. *J. Agric. Food Chem.* **2020**, *68*, 2816–2825. [CrossRef]
33. Fajardo, D.; Morales, J.; Zhu, H.; Steffan, S.; Harbut, R.; Bassil, N.; Hummer, K.; Polashock, J.; Vorsa, N.; Zalapa, J. Discrimination of American Cranberry Cultivars and Assessment of Clonal Heterogeneity Using Microsatellite Markers. *Plant Mol. Biol. Report.* **2012**, *31*, 264–271. [CrossRef]
34. Sedbare, R.; Raudone, L.; Zvikas, V.; Viskelis, J.; Liaudanskas, M.; Janulis, V. Development and Validation of the UPLC-DAD Methodology for the Detection of Triterpenoids and Phytosterols in Fruit Samples of *Vaccinium macrocarpon* Aiton and *Vaccinium oxycoccos* L. *Molecules* **2022**, *27*, 4403. [CrossRef]
35. Šedbarė, R.; Siliņa, D.; Janulis, V. Evaluation of the Phytochemical Composition of Phenolic and Triterpene Compounds in Fruit of Large Cranberries (*Vaccinium macrocarpon* Aiton) Grown in Latvia. *Plants* **2022**, *11*, 2725. [CrossRef]
36. Oszmiański, J.; Kolniak-Ostek, J.; Lachowicz, S.; Gorzelany, J.; Matłok, N. Phytochemical Compounds and Antioxidant Activity in Different Cultivars of Cranberry (*Vaccinium macrocarpon* L.). *J. Food Sci.* **2017**, *82*, 2569–2575. [CrossRef] [PubMed]
37. Viškelis, J.; Uselis, N.; Liaudanskas, M.; Lanauskas, J.; Bielicki, P.; Univer, T.; Lepsis, J.; Kviklys, D. Location Effects across Northeastern Europe on Bioactive Compounds in Apple Fruit. *Agric. Food Sci.* **2019**, *28*, 93–100. [CrossRef]
38. Brown, P.N.; Shipley, P.R. Determination of Anthocyanins in Cranberry Fruit and Cranberry Fruit Products by High-Performance Liquid Chromatography with Ultraviolet Detection: Single-Laboratory Validation. *J. AOAC Int.* **2011**, *94*, 459–466. [CrossRef] [PubMed]
39. Vvedenskaya, I.; Vorsa, N. Flavonoid Composition over Fruit Development and Maturation in American Cranberry, *Vaccinium macrocarpon* Ait. *Plant Sci.* **2004**, *167*, 1043–1054. [CrossRef]

40. Cranberry Weather Summaries Archive. Available online: https://ag.umass.edu/cranberry/weather-summaries-archive (accessed on 5 July 2022).
41. Bureau of Reclamation. AgriMet Columbia-Pacific Northwest Region. Available online: https://www.usbr.gov/pn/agrimet/webarcread.html (accessed on 5 July 2022).
42. Carpenter, J.L.; Caruso, F.L.; Tata, A.; Vorsa, N.; Neto, C.C. Variation in Proanthocyanidin Content and Composition among Commonly Grown North American Cranberry Cultivars (*Vaccinium macrocarpon*): Proanthocyanidin Content of Cranberries. *J. Sci. Food Agric.* **2014**, *94*, 2738–2745. [CrossRef]
43. Borowska, E.; Mazur, B.; Gadzal, R.; Kopciuch, V.; Buszewski, B. Polyphenol, Anthocyanin and Resveratrol Mass Fractions and Antioxidant Properties of Cranberry Cultivars. *Food Technol. Biotechnol.* **2009**, *47*, 56–61.
44. Narwojsz, A.; Tańska, M.; Mazur, B.; Borowska, E.J. Fruit Physical Features, Phenolic Compounds Profile and Inhibition Activities of Cranberry Cultivars (*Vaccinium macrocarpon*) Compared to Wild-Grown Cranberry (*Vaccinium oxycoccus*). *Plant Foods Hum. Nutr.* **2019**, *74*, 300–306. [CrossRef]
45. Wang, S.Y.; Stretch, A.W. Antioxidant Capacity in Cranberry Is Influenced by Cultivar and Storage Temperature. *J. Agric. Food Chem.* **2001**, *49*, 969–974. [CrossRef]
46. Xue, L.; Liu, C.; Ma, H.; Seeram, N.P.; Neto, C.C. Anti-Inflammatory Activities of Cranberry Fruit Extracts in Human THP-1 Monocytes Are Influenced by Their Phytochemical Composition. *ACS Food Sci. Technol.* **2022**, *2*, 75–83. [CrossRef]
47. Gudoityte, E.; Arandarcikaite, O.; Mazeikiene, I.; Bendokas, V.; Liobikas, J. Ursolic and Oleanolic Acids: Plant Metabolites with Neuroprotective Potential. *Int. J. Mol. Sci.* **2021**, *22*, 4599. [CrossRef]
48. Dreier, L.; Wider, G. Concentration Measurements by PULCON Using X-Filtered or 2D NMR Spectra. *Magn. Reson. Chem. MRC* **2006**, *44* (Suppl. S1), S206–S212. [CrossRef]
49. Feliciano, R.P.; Shea, M.P.; Shanmuganayagam, D.; Krueger, C.G.; Howell, A.B.; Reed, J.D. Comparison of Isolated Cranberry (*Vaccinium macrocarpon* Ait.) Proanthocyanidins to Catechin and Procyanidins A2 and B2 for Use as Standards in the 4-(Dimethylamino)Cinnamaldehyde Assay. *J. Agric. Food Chem.* **2012**, *60*, 4578–4585. [CrossRef]
50. Prior, R.L.; Fan, E.; Ji, H.; Howell, A.; Nio, C.; Payne, M.J.; Reed, J. Multi-Laboratory Validation of a Standard Method for Quantifying Proanthocyanidins in Cranberry Powders. *J. Sci. Food Agric.* **2010**, *90*, 1473–1478. [CrossRef]
51. Ainsworth, E.A.; Gillespie, K.M. Estimation of Total Phenolic Content and Other Oxidation Substrates in Plant Tissues Using Folin-Ciocalteu Reagent. *Nat. Protoc.* **2007**, *2*, 875–877. [CrossRef]
52. Liu, J.; Zhao, J.; Dai, Z.; Lin, R.; Wang, G.; Ma, S. A Pair of New Antioxidant Phenolic Acid Stereoisomers Isolated from Danshen Injection (Lyophilized Powder). *Molecules* **2014**, *19*, 1786–1794. [CrossRef]

plants

MDPI

Article

The Chemical Characterisation of the Maritime Pine Bark Cultivated in Northern Portugal

Diana Barros [1,2,3,*], Élia Fernandes [2,3], Meirielly Jesus [2,3], Lillian Barros [4,5], José Ignacio Alonso-Esteban [4,6,7], Preciosa Pires [2,3] and Manuela Vaz Velho [2,3]

1 EDIUS—International PhD School of the USC, University of Santiago de Compostela, 15705 Santiago de Compostela, Spain
2 CISAS—Center for Research and Development in Agrifood Systems and Sustainability, Instituto Politécnico de Viana do Castelo, Rua Escola Industrial e Comercial de Nun'Álvares, 4900-347 Viana do Castelo, Portugal; eliaf@estg.ipvc.pt (É.F.); meiriellyjesus@ipvc.pt (M.J.); ppires@estg.ipvc.pt (P.P.); mvazvelho@estg.ipvc.pt (M.V.V.)
3 ESTG IPVC—Escola Superior de Tecnologia e Gestão, Instituto Politécnico de Viana do Castelo, Avenida Atlântico, 644, 4900-348 Viana do Castelo, Portugal
4 Centro de Investigação de Montanha (CIMO), Instituto Politécnico de Bragança, Campus de Santa Apolónia, 5300-253 Bragança, Portugal; lillian@ipb.pt (L.B.); jignacio.alonso@uah.es (J.I.A.-E.)
5 Laboratório Associado para a Sustentabilidade e Tecnologia em Regiões de Montanha (SusTEC), Instituto Politécnico de Bragança, Campus de Santa Apolónia, 5300-253 Bragança, Portugal
6 Departamento de Nutrición y Ciencia de los Alimentos, Facultad de Farmacia, Universidad Complutense de Madrid, Plaza Ramón y Cajal, s/n, 28040 Madrid, Spain
7 Departamento de Ciencias Biomédicas, Facultad de Farmacia, Universidad de Alcalá, Carretera Madrid-Barcelona, Km 33.600, 28805 Madrid, Spain
* Correspondence: dib@ipvc.pt

Abstract: Maritime pine, scientifically known as *Pinus pinaster*, holds a vital role in Iberian Peninsula forests, primarily as a source of wood for panels, paper, and cellulose production. Recently, there has been a growing interest in utilising agroforestry by-products to yield valuable chemicals for applications in various sectors, including the food, pharmaceutical, and bioenergy industries. This study aimed to assess the value of the primary by-product of *Pinus pinaster* from the Minho region of northwestern Portugal, i.e., the bark. The research extensively examined the bark's chemical and thermal characteristics, including ash content, extractives, lignin, cellulose, hemicellulose, fatty acids, and mineral composition. Additionally, various analytical techniques like FTIR, SEM, DSC, DTG, and XRD were used to observe chemical structure differences. The results reveal that the *Pinus pinaster* bark primarily consists of lignin (51.15%) and holocellulose (46.09%), with extractives mainly soluble in toluene–ethanol, followed by water, and a small amount of them are soluble in ethanol. The bark contained around 0.44% ash, and heavy metals such as Cd and Pb were not found. During degradation, *Pinus pinaster* experienced a 10% mass loss at 140 °C. In terms of crystallinity, holocellulose and cellulose showed similar percentages at approximately 25.5%, while α-cellulose displayed the highest crystallinity index at 41%.

Keywords: bark; *Pinus pinaster* Aiton subsp. *atlantica*; extractives; holocellulose; cellulose; FTIR

1. Introduction

The forest industry is an important productive sector based on renewable natural resources. Wood and tree bark contain lignocellulosic materials—cellulose, hemicelluloses, and lignin—as well as extractives [1,2]. Wood, as the primary forest resource, finds extensive application in various sectors such as furniture, construction, and papermaking, and is also used as a renewable energy source [3]. The sawmill industry generates substantial quantities of the bark as a by-product during wood production [4,5]. One of the most available by-products or residues derived in wood-based industries is the bark because it is removed from the tree before processing it (de-barking). Considering the fact that the bark comprises about 9–15% of the total volume of the tree, the amount of waste generated is high [5,6]. Over time, the epidermis gradually accumulates a layer of dead cells, which is commonly known as "bark". Depending on the species, this layer can weigh between 5% and 28% of the tree's total weight and acts as a protective structure for trees regarding the environment they are living in, with hazards such as fire, insect pests, and fungi [7,8]. However, due to its woody and outer nature, the bark is typically only utilised for chemical extraction and low-value applications (such as compost amendment) or altogether discarded [7,9]. The defence mechanisms of pine trees (*Pinus* spp.) consist of both mechanical and chemical mechanisms that can be present either inherently or triggered upon challenge. These mechanisms involve a variety of specialised chemicals, including terpenoids found in oleoresin and phenolics, as well as structural features such as thick bark, large resin ducts, and specialised phloem parenchyma cells [10]. The bark contains unique compounds that are not present or vary from those present in the wood [1,11]. The biomass recalcitrance is composed of these components, forming a structure analogous to "reinforced concrete" at various scales [12]. In this analogy, cellulose fibres serve as reinforcement bars, hemicelluloses function as the wire mesh or cables, and lignin acts as the concrete component [13]. The utilisation of lignocellulosic biomass and its primary constituents, namely cellulose, hemicellulose, and lignin, presents a highly attractive opportunity for a multitude of applications. These applications can be found in various areas, including biofuels, cosmetics, detergents, adhesives, biomedical and pharmaceutical devices, flocculants, and food additives [2,14]. These recycled products have low economic value, despite several bioactive compounds that have been identified in bark [15]. Therefore, there exists a great potential for the forest industry to convert and take advantage of these by-products, transforming them into high-value products.

It is clear that the valorisation of tree bark from the major European forests will lead to a more efficient use of resources, with environmental and economic benefits. The potential availability of pine bark is significant; in Europe alone, it is estimated that ca. 38 Mm3 of bark is available from harvested trees [16].

The chemical characterisation of bark, which is essential in assessing its potential uses, has not received as much attention as that of wood. Bark differs from wood in that it has a high concentration of extractives, such as compounds soluble in organic solvents and water, in addition to a significant amount of polyphenolics and inorganic constituents known as ash [17,18]. Analysing the chemical composition of the bark is more complex than that of wood, and the standard methods used for wood analysis may not be suitable for bark due to cross-interference between its components, such as polyphenols and suberin, in the determination of lignin and holocellulose content. Additionally, the structural and anatomical complexity of bark compared to wood makes sampling, characterisation, and processing more difficult.

Pinus spp. is an extensive group of conifers from the *Pinaceae* family, mainly distributed around the Northern Hemisphere. Within the more than 100 species of the pine family, *Pinus pinaster*, or maritime pine, is the most commonly used for nutritional and medicinal purposes. Maritime pine (*Pinus pinaster* subsp. *atlantica*) is a conifer native to Mediterranean countries such as France, Spain, and Portugal, as well as some North African countries [19]. Pine is an evergreen, coniferous, resinous, extended-living, and mostly monoecious tree [20]. The subspecies of *P. pinaster* differ from each other in cold and salt tolerance, in addition to their content of terpenes and proteins [21].

The highest area of the Portuguese *P. pinaster* forest was recorded in 1995 (978×10^3 ha), and it has been reduced since that time, with an area of 714×10^3 ha recorded in 2010. Despite this, this species represents an important part of the total forest (about 23%), considered the third most important species after eucalyptus (812×10^3 ha, 26%) and cork oak (737×10^3 ha, 23%) [22,23].

Considering the importance of this pine species for forestry and forest sustainability in Portugal, as well as in the rest of the Iberian Peninsula and other parts of Europe, it is essential to explore all potential uses beyond the wood industry. This involves moving up the value chain to utilise all parts of the tree. Such an approach is closely tied to the principles of the circular economy, which is an economic system that prioritises sustainability by maximising resource usage and minimising waste generation [24].

Considering the interest in the valorisation of agroforest by-products, the present work aims to deepen the knowledge regarding the bark's chemical composition.

2. Results and Discussion

2.1. Chemical Composition

There is an abundance of information available on wood, whereas knowledge regarding pine bark is comparatively limited, specifically the chemical composition of maritime pine grown in Portugal. The chemical makeup of pine bark differs from that of wood, with extractives considered the predominant group of compounds [25]. Additionally, the composition of the bark is influenced by various factors such as the tree's age, location, and growing conditions, as noted by Vázquez et al. (1987) [26]. Other authors, namely Fradinho et al. (2002) [27], Santo et al. (2021) [28], and Elmas and Yilgor (2020) [29], studied dry maritime pines and obtained low values for ash, namely 0.80, 0.40, and 0.38% *w*/*w*, respectively, in line with Alonso-Esteban et al. (2022) [30], who reported an ash content of less than 1% in pine bark. In this study (Table 1), the average ash content was 0.44%, in agreement with the values presented in the literature, namely by Santos et al. (2021) [28]. The lipid fraction, or fat, and the protein content represented 0.98% and 1.94% of the total pine bark composition, respectively, with the value obtained for the fat of the same order of magnitude as that presented by Santos et al. (2021) [28].

Table 1. Chemical composition of dry pine bark from *P. pinaster* Aiton subsp. *atlantica*.

	Composition (% *w*/*w*)						
Parameter	**This Study**	**Nunes et al. (1996) [31]**	**Simões et al. (2021) [10]**	**Ferreira-Santos et al. (2020) [32]**	**Fradinho et al. (2002) [27]**	**Santos et al. (2021) [28]**	**Vieito et al. (2019) [33]**
Moisture	6.67 ± 0.01			8.15 ± 0.02			12.31 ± 0.20
Protein	1.94 ± 0.22			1.64 ± 0.03			12.82 ± 0.01
Total fat	0.98 ± 0.00			2.54 ± 0.26			1.60 ± 0.04
Ash	0.44 ± 0.01			0.87 ± 0.00		0.40 ± 0.02	1.75 ± 0.03
Holocellulose	46.09 ± 0.48				48.4 [d]		
Cellulose	23.21 ± 0.57			17.39 ± 0.37 [c]			
α-Cellulose	22.88 ± 0.02					23.2 ± 2.0	
Hemicellulose	23.21 ± 0.02 [a]			12.31 ± 0.20		16.1 ± 3.8	
Lignins	51.15 ± 0.35	43.7 ± 2.4		41.65 ± 0.24			
Acid-soluble	2.85 ± 0.35			0.60 ± 0.00			
Klason	48.31 ± 0.31			41.05 ± 0.24	33.2 [e]	46.2 ± 3.4	
Total extractives	9.75 ± 0.13	11.4 ± 2.2			16.6	11.97	
Toluene–ethanol (T:E)	5.61 ± 0.05						
Lipophilic (L)		2.3 ± 1.3	11.3 ± 1.7	2.54 ± 0.26	3.1	1.17 ± 0.09	
Ethanol (E)	1.54 ± 0.06	6.5 ± 1.2	19.5 ± 2.0	13.20 ± 0.31	10.3	8.71 ± 0.16	
(T:E) + (E) or (L) + (E)	7.15 [b]	8.8	30.8	15.74	13.4	9.88	
Hydrophilic (H)	2.59 ± 0.10	2.6 ± 0.2	6.9 ± 1.7		3.2	2.09 ± 0.02	
%/[(L) + (E)] in total extractives	76.4	77.2	81.7		81.7	82.5	
[(L) + (E)]/(H)	2.9	3.4	4.7		4.5	4.2	

[a] Hemicellulose (α-cellulose–holocellulose); [b] extractives: toluene–ethanol + ethanol; [c] estimated from the glucan content; [d] holocellulose determined after solvent extraction and alkaline extraction with 2% NaOH (0.5 h, 100 °C, 1 g/10 mL); [e] Klason lignin determined after solvent extraction and alkaline extraction with 2% NaOH (0.5 h, 100 °C, 1 g/10 mL).

Softwood barks, like the bark of maritime pine, were found to have a higher lignin content than the majority of hardwood barks. Additionally, they demonstrated a lower content of acid-soluble lignin, which aligns with the findings observed in wood [34,35].

According to the results (Table 1), the primary component of the bark's overall composition is lignin, which accounts for 51.15% (comprising 48.31% Klason lignin and 2.85% lignin soluble in acid). The high amount of Klason lignin obtained for the maritime pine bark is similar to that reported by Santos et al. (2021) [28], with a Klason lignin percentage of 46%, as well as that presented by Ferreira-Santos et al. (2020) [32], who found lignin to be the predominant component in the total composition of the pine bark, with 41.65% (41.05% Klason lignin and 0.60% acid-soluble lignin). The cellulose content was 23.21%, a higher value than the 17.39% obtained by Ferreira-Santos et al. (2020) [32], although the α-cellulose content of 22.88% was similar to the 23.2% reported by Santos et al. (2021) [28]. The amount of hemicellulose (23.21%) was higher than the levels obtained by other authors, namely Ferreira-Santos et al. (2020) [32] and Santos et al. (2021) [28], with 12.31% and 16.1%, respectively.

Total extractives represented 9.75% of the sample (w/w), most of which were soluble in the toluene–ethanol (T:E) mixture (5.61%). Despite being lower, the value of the total extractives obtained was of the same order of magnitude as those presented by Nunes et al. (1996) [31] and Santos et al. (2021) [28], who reported 11.4% and 11.97%, respectively. As in this study, ethanol (E) was used in the first extraction, and the comparison of the extractives was carried out considering the sum of the lipophilic and ethanolic ((L) + (E)) extractives and the hydrophilic (H) ones. This extractive combination is acceptable, as the toluene–ethanol mixture and ethanol are used to remove more hydrophobic extractives (containing lipophilic (L) extractives), such as waxes, fats, resins, and wood gums [36]. When the value found in the combination of extractives of toluene–ethanol and ethanol (7.45%) was compared with the combination of lipophilic and ethanol extractives in the literature, the same tendency was found for total extractives. However, the obtained value for this study was the lowest, with the closest value (8.8%) reported by Nunes et al. (1996) [31]. Despite the differences found for the values in the literature, when the percentage of this extractive fraction in the total extractives was calculated, it was verified that the values were relatively close to the value of this study (76.4%), that is, within an interval of 77.2% [31] to 8.5% [28]. These extractives possibly contained resin acids, such as pimaric and abietic acids, fatty acids, and some other phenolic compounds [37]. Vangeel et al. (2023) [38] also reported that lipophilic extracts contained predominantly resinous acids in the softwood bark, with triterpenoids constituting an important class of lipophilic extracts in the bark. On the other hand, extraction with water allows for the removal of hydrophilic compounds such as tannins, gums, sugars, starches, and pigments. The value found for hydrophilic extractives (2.54%) was very similar to that reported by Nunes et al. (1996) [31]. In the composition of hydrophilic water extractives, there was the possibility of the presence of carbohydrates (that is, arabinose and ribose derivatives) and levulinic acid, as reported by Fernandes (2020) [37]. Thus, the ethanolic extracts obtained from the maritime pine bark probably have phenolic compounds, and the aqueous extracts likely have some short-chain sugars. The ratio of lipophilic and ethanol extracts to hydrophilic extracts was 2.9, in agreement with other authors who obtained values in the range of 3.4 to 4.7. These results seem to indicate that the amount of hydrophilic extractives is much lower than the other extractives.

The complete extraction used a combination of different solvent polarities; thus, toluene–ethanol removed nonpolar groups more significantly, while water removed polar structures and ethanol groups with similar chemical solubility [39]. In other words, the combination of sequential extractions using solvents with a wide range of polarities seems to be advantageous for the extraction of both lipophilic and hydrophilic fractions.

The lignocellulosic material cellulose is normally obtained through bleaching and alkaline treatment processes. In this study, two processes were carried out to obtain cellulose, an acid hydrolysis using a CH_3COOH-HNO_3 mixture to obtain crude cellulose and an alkaline treatment using NaHO to obtain α-cellulose from holocellulose (Table 2).

The colour observed for the crude cellulose was brownish-yellow. In the alkaline hydrolysis process, the brownish-yellow colour of holocellulose changed to the yellowish-white colour of α-cellulose, with this colour change possibly resulting from the removal of noncellulosic materials such as residual lignin and hemicellulose. The results showed variations in the brightness and whiteness index (WI) with the type of hydrolysis performed. The results pointed to lower whiteness index values for holocellulose, followed by crude cellulose, with α-cellulose having the highest values for that parameter. The use of different hydrolysis agents showed different whiteness indexes for the cellulose obtained, with the alkaline hydrolysis process resulting in a higher degree of whiteness (higher L* value; brightness). It should be noted, however, that this alkaline hydrolysis was performed on a holocellulose sample that had already undergone a lignin removal process. Thus, the degree of whiteness seems to have been influenced by the content of noncellulosic materials such as lignin and hemicellulose [40], that is, the lower the lignin content in the cellulose, the greater the degree of whiteness [41].

Table 2. The CIELab system value of celluloses from *P. pinaster* Aiton subsp. *atlantica*.

Sample	Process.	L*	a*	b*	WI (%) [1]
Crude cellulose	Hydrolysis with a mixture of acetic acid (CH_3COOH) and nitric acid (HNO_3)	81.21 ± 0.10	3.93 ± 0.12	19.27 ± 0.06	72.8
Holocellulose	Digestion with a mixture of acetic acid ($C1H_3COOH$) and sodium chlorite ($NaClO_2$)	79.92 ± 0.19	6.23 ± 0.20	20.11 ± 0.03	70.9
α-Cellulose	Hydrolysis with sodium hydroxide (NaHO)	82.71 ± 0.19	3.31 ± 0.20	11.13 ± 0.03	79.2

[1] The whiteness index (WI%) was calculated using the equation $100 - \sqrt{(100 - L^*)^2 + a^{*2} + b^{*2}}$; L* represents perceptual lightness; a* axis represents the red–green colour spectrum, with positive values indicating red and negative values indicating green; b* axis represents the yellow–blue colour spectrum, with positive values indicating yellow and negative values indicating blue.

2.2. Fatty Acid Composition

Thirteen fatty acids were identified in the pine bark lipid extract (Table 3), and among the identified fatty acids, eleven were saturated, one was monounsaturated (MUFA—oleic acid (18:1*n*-9)), and one was polyunsaturated (PUFA—α-linoleic acid (18:2*n*-6)). The chain length of the fatty acids ranged from 12 to 24 carbon atoms. The saturated fatty acid content was much higher than the unsaturated fatty acid content, ca. 11.9-fold. Thus, the fatty acid composition showed mainly saturated fatty acids, representing approximately 92% of the total fatty acid profile, in line with the 90% registered in the study by Sousa et al. (2018) [42]. Indeed, lignoceric acid (24:0) and behenic acid (22:0) were the major fatty acids, both classified as very long-chain saturated fatty acids, with 34.48% and 32.47% of the total fatty acids, respectively. These were followed by palmitic acid (16:0) and arachidic acid (20:0), with 9.84% and 5.91%, respectively. In other words, only four fatty acids contributed more than 80% (82.7%) of the total fatty acids found. The saturated fatty acids found in higher amounts coincided with those observed in the studies by Kurtça and Tumen (2020) [43], Simões et al. (2021) [10], and Sousa et al. (2018) [42], however, their relative abundance was different, being only similar to those found by Sousa et al. (2018) [42]. Oleic acid (18:1*n*-9) was the unsaturated fatty acid with the highest concentration, representing 5.43% of the total fatty acids. In the studies by Sousa et al. (2018) [42], Kurtça and Tumen (2020) [43], and Simões et al. (2021) [10], oleic acid was also the most representative of the unsaturated fatty acids, although in a higher percentage than that shown in this study. The fatty acids detected in this study were similar to those detected in other studies; however, the percentage composition of the fatty acid profile differed from study to study, with the study by Sousa et al. (2018) [42] presenting more similar values. The results of the comparison showed that, although the main components were similar in general, the fatty acid content was different. The differences shown in the fatty acid profile of this study and those reported by other authors may be due to the time of year in which the pine bark was collected, as verified by Kurtca and Tumen (2020) [43].

Table 3. Fatty acid composition of maritime pine bark.

	This Study	Kurtça & Tumen (2020) [43]	Simões et al. (2021) [10]	Sousa et al. (2018) [42]
	(Relative Abundance %)			
Caproic acid (6:0)	-	-	-	0.403
Caprylic acid (8:0)	-	-	-	0.134
Pelargonic acid (9:0)	-	-	-	0.134
Lauric acid (12:0)	0.51 ± 0.003	-	-	-
Myristic acid (14:0)	1.10 ± 0.08	0.20 ± 0.20	-	0.134
Pentadecylic acid (15:0)	0.74 ± 0.02	-	-	0.134
Palmitic acid (16:0)	9.84 ± 0.38	6.50 ± 3.15	2.2 ± 1.7	7.38
Margaric acid (17:0)	0.54 ± 0.02	-	-	0.537
Stearic acid (18:0)	3.47 ± 0.24	0.45 ± 0.37	0.7 ± 0.3	3.09
Oleic acid (18:1*n*-9)	5.43 ± 0.24	13.91 ± 7.33	2.9 ± 1.0	6.44
Elaidic acid (18:1*n*-9)	-	-	-	0.671
Linoleic acid (18:2*n*-6)	2.35 ± 0.08	8.84 ± 4.72	-	2.95
α-Linolenic acid (18:3*n*-3)	-	0.37 ± 0.20	1.2 ± 0.0	-
Arachidic acid (20:0)	5.91 ± 0.22	1.92 ± 1.22	6.9 ± 1.8	8.99
Heneicosanoic acid (21:0)	0.71 ± 0.03	-	-	-
Behenic acid (22:0)	32.47 ± 0.21	3.10 ± 4.18	1.6 ± 0.3	33.7
Tricosylic acid (23:0)	2.45 ± 0.17	0.31 ± 0.06	0.8 ± 0.2	4.03
Lignoceric acid (24:0)	34.48 ± 0.28	3.19 ± 4.75	1.7 ± 0.5	28.6
Cerotic acid (26:0)	-	0.59 ± 0.76	-	2.68
Saturated fatty acid	92.22	20.41 ± 11.90	13.7 ± 1.1	89.9
Unsaturated fatty acid	7.78	23.31 ± 12.38	5.1 ± 1.0	10.1

2.3. Mineral Analysis

As the SEM-EDS (scanning electron microscopy–energy-dispersive spectroscopy) method is a semiquantitative method to determine elemental concentrations, to verify the acceptability of its quantification, the P concentration was also determined using the MAE-Vis (molecular absorption spectrophotometry) method, and the values were compared. The values obtained were 57.9 and 61.8 mg/kg dry matter using the MAE-Vis and SEM-EDS methods, respectively (Table 4). Considering the value obtained for P using MAE-Vis as the reference value, the relative error of the SEM-EDS determination of that element was 6.7%. Since the relative error of the SEM-EDS determination was <10%, the values obtained with that method were considered acceptable.

In the inorganic fraction of pine bark (ash), 18 elements were determined, and 16 were quantified, with their concentrations in descending order as follows: O, Ca, Al, Mg, C (inorganic), K, Si, Na, S, Fe, P, Mn, Zn, Cu, Cr, and Ni (Table 4). The elements with the highest concentration were O, Ca, Al, Mg, and C (inorganic), in the range between 2088 and 199 mg/kg dry matter. Those elements were followed by a set of elements consisting of K, Si, and Na, with concentrations of 134, 125, and 93 mg/kg dry matter, respectively. Each of the aforementioned elements presented a global contribution of 4167 mg/kg of dry matter (0.42 g/100 dry matter), that is, 0.42% of the total dry matter, which represents 95% of the inorganic fraction (ash) of pine bark.

The elements S, Fe, P, and Mn showed concentrations in the range of 67 to 34 mg/kg dry matter. Fe and Mn are important elements, as plant photosynthesis, redox processes, nitrogen metabolism, and nucleic acid metabolism in conifers are influenced by these elements. The excessive presence of manganese hinders the absorption and movement of iron in plant tissues. That antagonistic relationship can also be observed, where a high iron concentration decreases the absorption and effectiveness of other metals [44,45]. It should be noted that sulphur is an important element both for the structure of proteins [46] and for the functioning of enzymes [47], in addition to playing roles in maintaining the homeostasis of essential micronutrients such as iron (Fe), copper (Cu), zinc (Zn), and manganese (Mn) [48]. Furthermore, this element plays an important role in plant defence against stress and pests [47].

Table 4. Mineral composition of maritime pine bark.

Element	Method	(mg/Kg Dry Matter)
Carbon (inorganic) (C)	SEM-EDS	199 ± 5.7
Oxygen (O)	SEM-EDS	2088 ± 16
Sodium (Na)	SEM-EDS	92.9 ± 1.4
Magnesium (Mg)	SEM-EDS	249 ± 1.2
Aluminium (Al)	SEM-EDS	456 ± 2.9
Silicon (Si)	SEM-EDS	125 ± 1.3
Phosphorus (P) [1]	MAS-Vis	57.9 ± 0.63
Sulphur (S)	SEM-EDS	66.8 ± 0.51
Potassium (K)	SEM-EDS	134.2 ± 0.40
Calcium (Ca)	SEM-EDS	823 ± 13
Manganese (Mn)	SEM-EDS	34.1 ± 0.47
Iron (Fe)	SEM-EDS	65.4 ± 1.5
Cadmium (Cd)	AAS-Flame	<0.060
Lead (Pb)	AAS-Flame	<0.75
Copper (Cu)	AAS-Flame	1.76 ± 0.037
Chromium (Cr)	AAS-Flame	0.429 ± 0.0077
Nickel (Ni)	AAS-Flame	0.332 ± 0.017
Zinc (Zn)	AAS-Flame	3.11 ± 0.050
Total mineral content [2]		4397 (0.44%)

[1] SEM-EDS result: 61.8 ± 1.7 mg P/kg dry matter; [2] $\sum$ mineral concentrations; minerals: C, O, Na, Mg, Al, Si, P, S, K, Ca, Mn, Fe, Cu, Cr, Ni, and Zn.

The elements Cu, Cr, Ni, and Zn showed concentration values lower than 3.2 mg/kg dry matter, and Cd and Pb were below the detection limit. Heavy metals, such as Cd, Pb, Cu, Zn, Cr, Ni, and Hg, are normally present in the environment. Although some of these metals serve as essential micronutrients for plant growth, Cd, Pb, and Hg are exceptions to this rule [45,49]. In the analysed material, the Cu content found was 1.76 ± 0.037 mg/kg; however, previous studies reported higher Cu levels in pine bark, ranging from 2.54 to 4.03 mg/kg [50]. Conifers demonstrate resilience to elevated copper levels in the environment and have the capacity to accumulate substantial quantities of this metal [51]. The Zn content was 3.11 ± 0.050 mg/kg. Compared to other heavy metals, zinc is typically found in higher proportions in the environment [45]. In this study, the Ni content was 0.33 mg/kg. However, Poikolainen (1997) [50] reported a nickel content ranging from 0.79 to 2.43 mg/kg in pine bark. Nickel is classified as a micronutrient, particularly for urease and hydrogenases [45]. Coniferous trees demonstrate resilience to high environmental nickel concentrations, with pine trees belonging to the group of Ni hyperaccumulators [51]. The Cr concentration was 0.429 ± 0.0077, in agreement with the values obtained by Kirchner et al. (2008) [51], that is, chromium contents below 1 mg/kg in pine wood (*Pinus jeffreyi*). Likewise, Liu et al. (2018) [52] and Poikolainen (1997) [50] reported chromium of 0.5 mg/kg and 0.12–7.31 mg/kg contents in the pine bark, respectively. The elements Cu, Zn, Ni, and Cr are classified as micronutrients and play a crucial role as constituents of numerous plant enzymes [53] and plant growth regulators [45]. In the tested samples, the concentrations of Cd and Pb were below the detection limit (<0.060 mg/kg and 0.75 mg/kg, respectively), which seems to indicate that the tested maritime pine bark was not contaminated with these metals. For comparison, Kirchner et al. (2008) [51] found cadmium levels of 0.02–0.17 mg/kg in pine wood, and Poikolainen (1997) [50] found levels of 0.10–0.23 mg/kg in the bark of pine trees growing in areas with different degrees of contamination. The results obtained for this set of heavy metals generally showed lower values than those reported in the literature, which seems to indicate that the pine bark tested in this study did not suffer major contamination from them. The total mass of elements determined was 4397 mg/kg of dry matter, representing 99.94% of the ash content, which seems to indicate that the quantification of elements in the sample inorganic fraction was almost complete.

The heavy metal concentration is influenced by the age and specific anatomical part of the plant, as well as by environmental pollution, mainly dust levels [49]. Thus, heavy metals in wood waste may originate from natural uptake by plant roots or from dust settling on tree surfaces such as bark, needles, and cones [49,54–57]. Consequently, the surfaces of coniferous trees tend to exhibit higher levels of atmospheric dust-derived metals [55,58]. It is important to consider the presence of soil particles and other unwanted fine materials in the bark, especially when it comes to pine bark, as it has a macroscopic structure that is more likely to capture foreign particles, which can be released during the milling process.

2.4. Fourier Transform Infrared (FTIR) Analysis

The FTIR spectra of *Pinus pinaster* and its fractions ranging from 4000 to 600 cm^{-1} are illustrated in Figure 1. The spectra depict the intricate nature of this biomass, resulting in a diverse array of characteristic peaks corresponding to its constituents. As stated by Wang et al. (2020) [59], the peaks at 1425, 1169, and 890 cm^{-1} observed in all samples except for lignin are attributed to typical cellulose structures. The transmittance between 3316 and 3339 cm^{-1} in the samples corresponds to the hydroxyl bond of the water molecule [60,61]. Additionally, it is associated with OH stretching, revealing hydrogen bonding and intramolecular interactions within the cellulose component of the wood [62]. The peaks in the range of 2945–2878 cm^{-1} could be ascribed to symmetric and asymmetric CH-stretching vibrations originating from the methyl ($-CH_3$) and methylene ($>CH_2$) functional groups [63]. The peaks within the 738–1614 cm^{-1} range were assigned to the stretching vibrations of C=O in carboxyl, and certain segments of the acetyl group were found in hemicellulose, as well as vibrations related to the aromatic skeleton. The symmetric stretching bands of the carboxyl group within the 1436–1319 cm^{-1} range were associated with CH deformation (methyl and methylene), CH_2 bending vibrations, and CH_3 stretching. Additionally, this transmittance could be linked to the crystalline structure of cellulose and its derivatives [64,65]. The peaks within the range of 1261–1204 cm^{-1} were ascribed to COC vibrations in both cellulose and hemicellulose. For the pine bark biomass, peaks at 1257 cm^{-1} were detected, which can be attributed to C–O stretching in lignin and the C–O linkage in the aromatic methoxyl groups, particularly guaiacil, and this may have a direct association with softwoods [29]. These characteristic peaks are indicative of the presence of cellulose, hemicellulose, and lignin. The prominent bands falling within the range of 1003–1029 cm^{-1} originated from the CO-stretching vibration in cellulose, hemicelluloses, and lignin. Slight variations in the peak heights were noted between holocellulose and pine bark biomass samples, a finding that was consistent with observations by Elmas and Yilgor (2020) [29].

For the lignin sample, the spectra exhibited peaks at 1601, 1511, and 1452 cm^{-1} corresponding to vibrations of the aromatic ring in the phenylpropane skeleton. Furthermore, the broadband at 3337 cm^{-1} was directly linked to both aromatic and aliphatic OH groups, while the peaks observed at 2945 and 1457 cm^{-1} corresponded to CH vibration in the CH_2 groups. The bands at 1262, 875, and 798 cm^{-1} were associated with guaiacil (G) type units, which are abundant in softwoods like *Pinus* [29,66]. A lesser-intensity peak at 1212 cm^{-1} was likely a result of the OH-bending vibration. These findings align with those reported in the literature for lignin derived from pine and other similar softwoods [59,67–69].

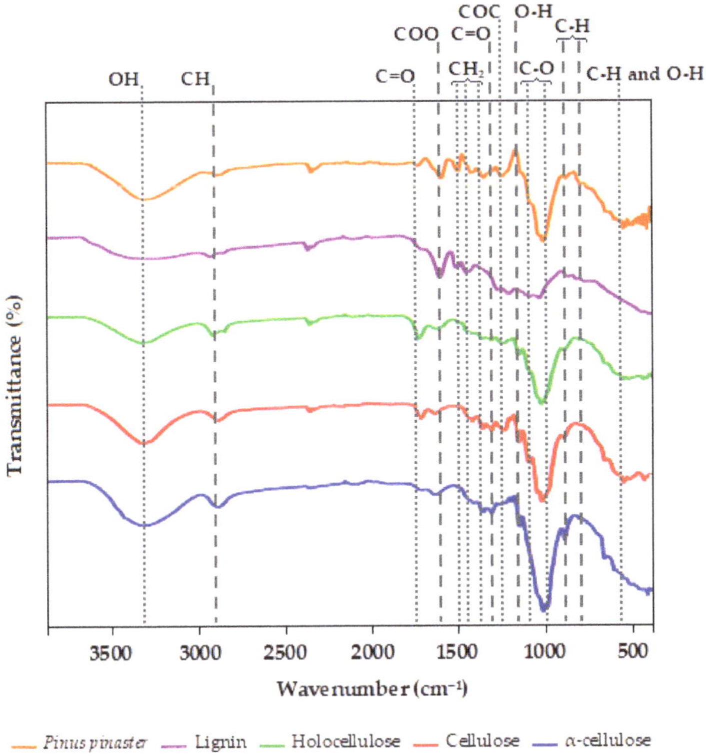

Figure 1. Characterisation of *Pinus pinaster* and their fractions: lignin, holocellulose, cellulose, and α-cellulose.

2.5. Scanning Electron Microscopy (SEM) Analysis

The morphology of pine bark biomass (A) and its fractions (holocellulose (B), cellulose (C), α-cellulose (D), and lignin (E)) were examined through scanning electron microscopy (SEM), as illustrated in Figure 2. The pine bark sample (Figure 2A) exhibits a compact, smooth, and rigid surface structure attributed to the presence of lignin, which envelops the integral fibrils of hemicellulose and cellulose [63,70]. In Figure 2B, the micrograph of holocellulose reveals a uniform and distinctly defined structure. These characteristics are likely attributed to its composition, primarily consisting of cellulose and all hemicelluloses. Holocellulose was obtained by removing extractives and lignin from the original natural material, resulting in the complete polysaccharide fraction of lignocellulosic biomass [71].

Figure 2C seems to indicate that the isolation procedure may have had a significant impact on the morphology of the cellulose, resulting in a rough and irregular surface. This may have occurred due to the fragmentation of the lignocellulosic structure and the removal of the hemicellulose and lignin components. The removal of these elements from the cell wall may have directly affected the organisation that interconnected the cellulose fibrils [72]. These findings appear to be in line with FTIR analysis, which points to the removal of lignin and hemicellulose. Notably, α-cellulose, shown in Figure 2D, exhibits aggregated and elongated fibres. These results are consistent with the literature, as some authors described cellulose and its derivatives as having a fibrous morphology with a rough and irregular appearance [73]. The micrograph of lignin isolated from pine bark and presented in Figure 2E shows a heterogeneous morphology with significant ruggedness, in agreement with findings in the literature, which characterised lignin as having an irregular surface with variations and undulations [74].

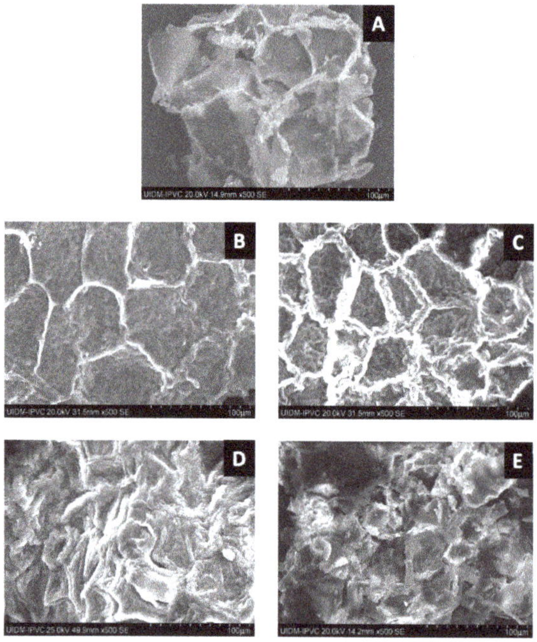

Figure 2. Displays micrographs at 500× magnification of maritime pine bark biomass (**A**) along with its isolated components, including holocellulose (**B**), cellulose (**C**), α-cellulose (**D**), and lignin (**E**).

2.6. *Thermal Analysis of Initial Material (Pinus pinaster Cortex)*

Thermogravimetric analysis is a cost-effective method used to evaluate the thermal behaviour of lignocellulosic biomass under specific atmospheres and either isothermal or nonisothermal conditions [75]. Gaining insight into the thermal properties of lignocellulosic biomass is essential for grasping the physicochemical alterations that occur during pretreatment [76]. Thermogravimetric analysis is usually depicted through differential thermogravimetry (DTG) curves, which depict mass loss concerning temperature and time. On the other hand, differential scanning calorimetry (DSC) serves as a nonspecific diagnostic instrument employed to capture the entire spectrum of phase transitions occurring during a temperature scan in a single step [77,78]. The thermogravimetric analysis of the Pinus pinaster bark is illustrated in Figure 3.

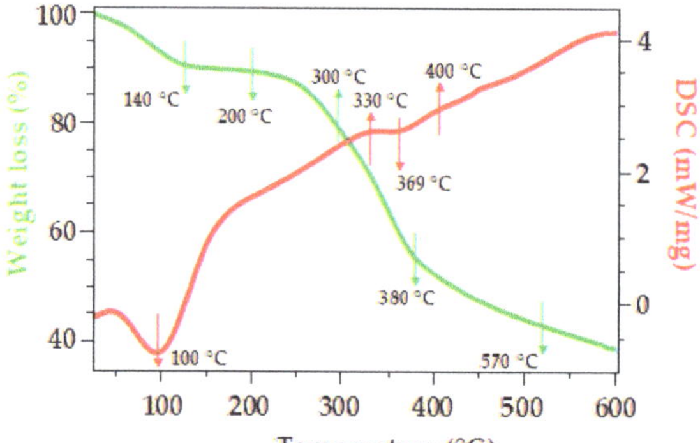

Figure 3. DSC and DTG curves of Pinus pinaster bark.

In the Pinus pinaster bark, the DSC temperature ramp revealed the initial reaction at approximately 100 °C, primarily associated with the elimination of moisture within the sample. Further analysis indicated two exothermic reactions occurring at 330 °C and 400 °C, corresponding to hemicellulose and lignin, along with an endothermic reaction at 369 °C, representing cellulose. These thermal events showed the interactions among the primary components of this lignocellulosic residue [78,79].

The biomass underwent four distinct stages of weight loss, with the initial stage occurring below 140 °C, the second within the 250–300 °C range, the third at 380 °C, and the fourth at 570 °C, as evident in the DTG curves depicted in Figure 3. During the first stage, a minor weight loss of approximately 10% was observed from room temperature to 140 °C, primarily attributed to the evaporation of water and physically adsorbed moisture [75]. Furthermore, the observed mass loss between 200 and 300 °C may be associated with the decomposition of hemicelluloses and pectins, which are more thermally degradable [80,81]. The decomposition stages that occurred between 300 and 380 °C were attributed to the degradation of cellulose [76]. As pointed out by Yang et al. (2007) [71], in contrast to hemicellulose, cellulose is composed of an unbranched, long polymer of glucose with a highly organised and robust structure, which contributes to its enhanced thermal stability. As for lignin, its decomposition began between 380 and 570 °C, lignin is an aromatic heteropolymer compound composed of phenyl propane units. These units impart increased thermal stability to lignin, necessitating higher temperatures for its degradation [76,82].

2.7. X-ray Diffraction (XRD) Analysis

The crystallinity index (CrI) is often used to assess the crystalline fraction of cellulose materials and monitor the changes resulting from various physical–chemical and biological treatments. This parameter is based on distinguishing between a crystalline phase and an amorphous phase of the material. The XRD diffraction patterns of holocellulose and cellulose shown in Figure 4 exhibit three distinct diffraction peaks near the 2θ angles of 16°, 22°, and 34°, revealing that the crystalline structure of the samples was cellulose, as described in previous literature [83]. However, α-cellulose exhibits a broader 2θ peak ranging from 15.0° to 25.0°; these variations may be possibly due to the highly amorphous nature of the synthesised cellulose [73].

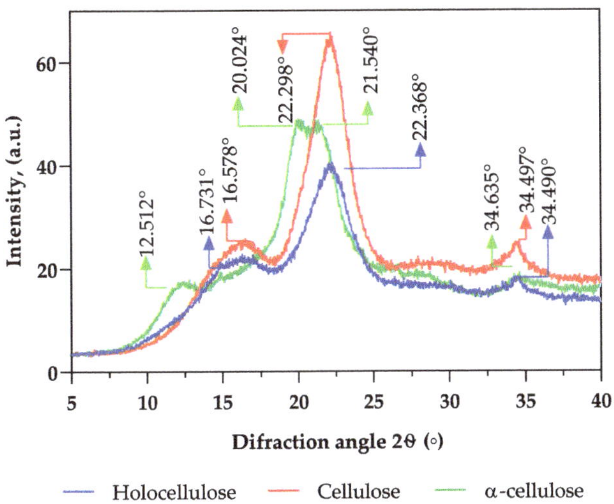

Figure 4. X-ray diffraction patterns of separated components from pine bark, specifically hemicellulose, cellulose, and α-cellulose.

Concerning the crystallinity index, both holocellulose and cellulose showed similar percentages, at approximately 25.5%. On the other hand, α-cellulose displayed the highest crystallinity index at 41%. The high crystallinity of the α-cellulose sample seemed to indicate that the depolymerisation of hemicellulose and delignification were successful during the extraction process [84].

3. Materials and Methods

3.1. Sampling

In the northern region of Portugal, specifically in Valença, the pine bark sourced from *P. pinaster* was gathered in May 2021. These trees were part of an experimental and certified plantation situated in the forest area of the Minho region. It is noteworthy that the collected pine bark originated from trees that were precisely 21 years old. Pine bark samples were collected by making a circular cut in an area of the main trunk of the freshly cut tree, about 1.30 cm above the ground.

3.2. Sample Preparation

The pine bark was washed with distilled water several times to remove dirt, lichens, and resin. The bark was dried at 40 °C for 48 h and subsequently milled and sieved at an amplitude of 0.2 for 1 min (Analysette 3 PRO, Fritsch, Idar-Oberstein, Germany) to select the particles from 200 to 850 μm diameter. Finally, the ground pine bark was kept in sealed bags and stored in a dry and dark place until further analysis.

3.3. Experimental Design

The experimental design followed to study the composition of pine bark is shown in Figure 5.

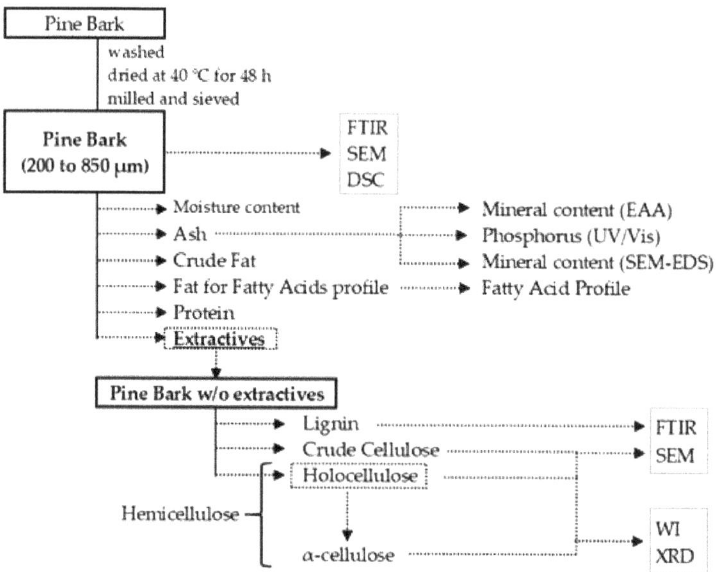

Figure 5. Experimental design of the study.

3.4. Chemical Composition of Pine Bark

The pine bark was analysed for moisture, ash, proteins, and fat contents using the AOAC procedures [85]. The moisture was determined by drying in an oven at 103 °C until constant weight (AOAC 930.04); the ash content was determined via incineration at 550 °C (AOAC 930.05); the crude protein content (N x 6.25) of the samples was estimated using the

Kjeldahl method (AOAC 978.04); and the crude fat was determined by extracting a known weight of ground sample with petroleum ether using a Soxhlet apparatus (AOAC 920.39).

The analysis of the chemical composition of the pine bark also included the determination of extractives using the solvents toluene–ethanol (T:E), ethanol (E), and deionised water (H), as well as crude cellulose, α-cellulose, hemicelluloses, and lignin.

Extractive content: The samples were extracted successively with toluene–ethanol 95% (2:1 by volume) for 6 h, ethanol 95% for 4 h, and distilled water for 2 h in a Soxhlet apparatus. The extractive content was determined from the extracts' dry matter at 105 °C and reported as a percentage of the sample dry matter according to the ASTM D 1105-96 method (2001) [36] and Ona et al. (1995) [86].

Lignin: The acid-soluble and acid-insoluble lignin contents were determined based on a protocol adapted from the method described by Sluiter et al. (2008) [87]. This procedure uses a two-step acid hydrolysis to fractionate the biomass into forms that are more easily quantified. In the first stage of acid hydrolysis, 3 mL of 72% sulphuric acid was added to 300 mg of the sample, and the mixture was homogenised for 1 min. The mixture was placed in a water bath set at 30 ± 3 °C and incubated for 60 ± 5 min, with stirring every 5 to 10 min. The tube was removed from the water bath, and the sample was quantitatively transferred to an autoclave flask. The acid concentration in the flask was diluted to 3.5% by adding distilled water until the mixture weighed 89.21 g. In the second step of acid hydrolysis, the sample was autoclaved for 1 h at 121 °C. The hydrolysate was allowed to cool slowly to near room temperature. Subsequently, the samples were vacuum-filtered through a glass fibre filter conditioned at 575 °C (GF/C, Whatman, Maidstone, England). The filter was washed with 3.5% sulphuric acid without exceeding the volume of 100 mL. The filtrate was transferred to a 100 mL volumetric flask, and the volume was completed with 3.5% sulphuric acid.

Acid-soluble lignin: The absorbance was measured at 240 nm against 3.5% sulphuric acid (Hitachi U 1100 Spectrophotometer, Tokyo, Japan). Absorbance should be in the range of 0.7–1.0.

Acid-insoluble lignin: The filter with the residue was washed with at least 50 mL of distilled water, dried at 105 ± 3 °C to a constant weight (>4 h), cooled, and weighed to the nearest 0.1 mg. The filter was placed in a muffle at 575 ± 25 °C for 24 ± 6 h, cooled, and weighed to the nearest 0.1 mg. Acid-insoluble material can also include ash and protein, which must be considered during gravimetric analysis.

Crude cellulose: Of the different cellulose determination methods proposed in the literature, the one involving a mixture of acetic acid and nitric acid has the advantage of being chlorine-free, reducing the environmental impact associated with that compound. Furthermore, this methodology has the advantage of being able to achieve a high degree of delignification and removal of noncellulose polysaccharides in a single-step procedure [88]. Based on the method described by Sun et al. (2004) [89], to 0.5 g of pine bark sample, 50 mL of an 80% acetic acid/70% nitric acid mixture (v/v, 10:1) was added in a reflux flask. The mixture was heated to the boiling point and refluxed for 25 min with periodic stirring. Subsequently, the hot mixture was filtered under vacuum, and the filter was washed with 5 mL of the hot acidic mixture. Sequentially, the filter was washed with hot water until it reached a neutral pH, several times with ethanol (95%), and with ethyl ether. Finally, the solid was dried at 105 ± 3 °C until it reached a constant weight.

Holocellulose: In a 250 mL Erlenmeyer flask, 180 mL of 0.2% sodium acetate (pH 3.5), 0.5 mL of acetic acid, and 1.2 g of sodium chlorite were added to 3 g of sample. The mixture was heated in a water bath at 70 °C. After 30 min, 0.5 mL of acetic acid and 1.2 g of sodium chlorite were added. After each successive hour, fresh portions of 0.5 mL of acetic acid and 1.2 g of sodium chlorite were added. The delignification process degrades some of the polysaccharides, and the application of excessive chlorination should be avoided. The addition of 0.5 mL of acetic acid and 1.2 g of sodium chlorite was repeated until the bark sample was completely separated from the lignin. Chlorination usually takes 6 to 8 h, and the sample can be left without adding acetic acid and sodium chlorite in the water bath overnight. At the end of 24 h of reaction, the sample was allowed to cool, the holocellulose was filtered through filter paper in a Buchner funnel, and washed with distilled water until the yellow colour (the holocellulose colour is white) and the chlorine dioxide odour had been removed. The filter was washed with acetone, dried at 105 °C for 24 h, cooled, and weighed. Holocellulose may contain some lignin, which should be considered during gravimetric analysis.

α-Cellulose: The α-cellulose was determined following holocellulose hydrolysis. In a 200 mL beaker, 25 mL of 17.5% sodium hydroxide was added to 1 g of the sample (holocellulose) and stirred. After 4 min, the sample was smashed for 5 min. After 16 min, 25 mL of distilled water was added, and the sample was stirred for 1 min. After 16 min, 25 mL of distilled water was added, and the sample was stirred for 1 min. After 5 min, it was filtered under vacuum through a glass filter conditioned at 105 °C. The filter was washed with distilled water until it reached a neutral pH. A 40 mL volume of 10% acetic acid was added and kept for 5 min. The filter was washed again with distilled water until it reached a neutral pH, and it was dried at 105 °C until it reached a constant weight [86].

Hemicellulose: Hemicellulose was calculated by subtracting α-cellulose from holocellulose. Data are expressed in weight percentage.

The determination of moisture, ash, crude protein, crude fat, and crude cellulose was carried out in the washed, milled, sieved, and dried pine bark sample (sample preparation), lignin and holocellulose in the sample after removing extractives, and α-cellulose in the holocellulose sample.

Degree of Whiteness: The degree of whiteness of holocellulose, cellulose, and α-cellulose was measured using a colourimeter (CR-400, Konica Minolta, Ramsey, NJ, USA). The CIELab colour system value (L*, a*, b*) was used to estimate the degree of whiteness, where L* is lightness, a* is redness, and b* is yellowness. The L* value ranges from 0 to 100 (black to white). The value of +a* (positive) describes a red shift, and −a* (negative) indicates a green shift; the value of +b* (positive) describes a yellow shift, and −b* (negative) indicates a blue shift.

3.5. Fatty Acid Composition

Fatty acids were quantified via gas–liquid chromatography with flame ionisation detection (GC–FID)/capillary column as previously described by Pereira et al. (2011) [90]. First, the transesterification of the fatty acids was carried out from the previously extracted fat. For this, 5 mL of methanol–sulphuric acid–toluene 2:1:1 (*v:v*) was added, and the reaction occurred for at least 12 h in a bath at 50 °C and 160 rpm; then, 3 mL of deionised water was added, fatty acid methyl esters (FAMEs) were recovered with 3 mL of diethyl ether by shaking in the vortex, and the upper phase was passed through a micro-column of sodium sulphate anhydrous, in order to remove any aqueous residue; the sample was recovered in a vial with Teflon, and before injection, the sample was filtered with 0.2 μm nylon filter. The FAMEs were analysed with a DANI model GC 1000 instrument equipped with a split/splitless injector, a flame ionisation detector (FID), and a Macherey–Nagel column (30 m × 0.32 mm ID × 0.25 μm df), according to Barros et al. (2010) [91]. FAME identification was carried out by comparing the relative retention times with standards. The results were recorded and processed using CSW 1.7 software (DataApex 1.7) and expressed in relative percentage of each fatty acid.

3.6. Mineral Analysis

The mineral analysis was carried out on the ash of the sample. To determine the concentration of cadmium (Cd), chromium (Cr), copper (Cu), lead (Pb), nickel (Ni), phosphorus (P), and zinc (Zn), the ashes were hot-dissolved with nitric acid and the resulting solution filtered. The phosphorus content in the sample was determined via molecular absorption spectrophotometry (MAS; Hitachi; U 1100 Spectrophotometer, Tokyo, Japan) using the ascorbic acid method. In this method, in an acid medium, the ammonium molybdate and antimony potassium tartrate react with orthophosphate to form phosphomolybdic acid, which is reduced to intensely coloured molybdenum blue by ascorbic acid (SMEWW 4500-P E) [92]. The concentration of the remaining minerals in that set was determined via atomic absorption spectrometry (AAE) with an air-acetylene flame (Varian AA300, Mulgrave, VIC, Australia) (SMEWW 3111 B) [92]. The remaining elemental chemical characterisation of the ash was performed directly on the ash via scanning electron microscopy (SEM-EDS) using a Hitachi S4100 microscope equipped with a detector for energy-dispersive spectroscopy, with the elemental distribution being partially quantified using semiquantitative methods for spectrum analysis (EDS-EDAX with Bruker AXS detector; software: Quantax 200, Esprit version 1.9).

3.7. Structural Characterisation of *Pinus pinaster* Cortex and Its Individual Fractions

Fourier transform infrared spectroscopy (FTIR): To analyse the chemical groups and bonding configurations present in the cortex of Pinus pinaster, as well as in its isolated fractions (cellulose, α-cellulose, holocellulose, and lignin), FTIR was employed. The FTIR analysis was conducted using a Thermo Scientific infrared spectrometer (Nicolet iS5 FTIR, Waltham, MA, USA) equipped with a diamond composite attenuated total reflectance (ATR) cell. The FTIR spectra were generated at a resolution of 4 cm^{-1}, comprising a total of 32 scans and covering the frequency range from 4000 to 600 cm^{-1}.

Differential Scanning Calorimetry (DSC): The thermogravimetric analysis of the initial material (*Pinus pinaster* cortex) was conducted using a DSC (Perkin Elmer, Norwalk, CT, USA, DSC-7). The analysis was performed in the temperature range of 25 to 600 °C at a linear heating rate of 10 °C per minute.

Scanning Electron Microscopy: The micrographs of the initial material (*Pinus pinaster* cortex), along with its distinct fractions (cellulose, alpha-cellulose, holocellulose, and lignin), were captured utilising a field-emission scanning electron microscope (FEG-SEM) (Jeol JSM-7001F, Tokyo, Japan). The imagery was acquired employing a 5 kV voltage, at a 400 times magnification. To prevent electric charge buildup during observation, the samples intended for qualitative examination were precoated with an Au-Pd alloy layer.

X-ray diffractometry: A conventional X-ray diffractometer (Bruker D8 Advance DaVinci model, located in Karlsruhe, Germany) was employed for XRD analysis, using Cu-Kα radiation filtered through Ni (with a wavelength of λ = 0.15418 nm), generated at 40 kV and 30 mA. A linear detector (Lynxeye 1-D) was used to record the datasets in the range of 5° to 60° (2θ) with 0.02° (2θ) increments. Each data point was recorded at intervals of 0.5 s, and the rotation rate was maintained at 15 rotations per minute. The Rietveld refinement was carried out using Diffrac.eva v5.2 software, developed by Bruker AXS in Karlsruhe, Germany. In this process, the fundamental parameter approach was applied for phase quantification. The crystallinity index (CrI) was determined by calculating the intensities in the crystalline and amorphous regions following Equation (1) [83], representing the intensities in the crystalline and amorphous regions at peak located at approximately 2θ = 21.54–22.368° and 12.512–16.731°, respectively:

$$CrI\ (\%) = \frac{I_{002} - I_{am}}{I_{002}} \times 100 \quad (1)$$

where I_{002} = maximum intensity of the lattice diffraction, and I_{am} represents the intensity of the amorphous diffraction.

4. Conclusions

The residues and by-products derived from pine are a significant reservoir of valuable biocompounds that hold substantial industrial value. Their extraction through the application of the biorefinery approach not only aids in their recuperation but also plays a pivotal role in advancing the principles of the circular economy. Total extractive materials with a higher value, approximately 10% w/w of the sample by mass, show potential use in a biorefinery. After lipophilic and hydrophilic extraction, the dry pine bark, approximately 90%, has high potential use as a raw material in industry and biorefineries. The findings indicate that the *Pinus pinaster* bark is primarily composed of lignin (51.55%) in terms of macromolecular compounds, followed by holocellulose (46.09%) and cellulose (23.21%). In the degradation stage, the *Pinus pinaster* bark exhibited a 10% reduction in mass when exposed to a temperature of 140 °C. Regarding the crystallinity index, both holocellulose and cellulose demonstrated comparable values, approximately around 25.5%. In contrast, α-cellulose displayed the highest crystallinity index, reaching 41%. This study shows that the low content of heavy metals in the pine bark may qualify this residue for various types of uses.

Author Contributions: Conceptualisation and investigation, D.B. and É.F.; writing—original draft preparation, D.B., É.F. and M.J.; writing—review and editing, L.B., J.I.A.-E. and P.P.; supervision, M.V.V. All authors have read and agreed to the published version of the manuscript.

Funding: The authors are grateful to the Foundation for Science and Technology (FCT, Portugal) for financial support through national funds FCT/MCTES to the CISAS (UIDB/05937/2020 and UIDP/05937/2020), including the postdoc grant of the M.J, as well as CIMO (UIDB/00690/2020 and UIDP/00690/2020) and SusTEC (LA/P/0007/2020); Diana Barros thanks FCT—SFRH/BD/146720/2019 PhD scholarship co-financed by Fundo Social Europeu (FSE) through the Programa Operacional Regional Norte 2020; L. Barros thanks the national funding by FCT, P.I., through the institutional scientific employment program contract. J. I. Alonso-Esteban thanks the Ministry of Universities (Spain) and the European Union-Next Generation EU Funds for the Margarita Salas grant (UCM/572725/2021). This work was financed by the European Regional Development Fund (ERDF) through the Regional Operation Program North 2020, within the scope of the R&D Enterprise Co-promotion project PICAR—Funcionalização de produtos cárneos curados com extratos de córtex de pinheiro-bravo do Minho (*Pinus pinaster* subsp. *atlantica*) (PICAR—Functionalisation of meat products with maritime pine cortex extracts (NORTE-01-0247-FEDER-069584).

Data Availability Statement: Data are contained within the article.

Acknowledgments: The authors thank CentroPinus for the donation of pine bark for this study.

Conflicts of Interest: The authors declare no conflict of interest.

References

1. Dönmez, I.; Hemming, J.; Willför, S. Bark Extractives and Suberin Monomers from *Arbutus andrachne* and *Platanus orientalis*. *BioResources* **2016**, *11*, 2809–2819. [CrossRef]
2. Magalhães, S.; Moreira, A.; Almeida, R.; Cruz, P.F.; Alves, L.; Costa, C.; Mendes, C.; Medronho, B.; Romano, A.; Carvalho, M.d.G.; et al. Acacia Wood Fractionation Using Deep Eutectic Solvents: Extraction, Recovery, and Characterization of the Different Fractions. *ACS Omega* **2022**, *7*, 26005–26014. [CrossRef] [PubMed]
3. Agarwal, C.; Hofmann, T.; Vršanská, M.; Schlosserová, N.; Visi-Rajczi, E.; Voběrková, S.; Pásztory, Z. In vitro antioxidant and antibacterial activities with polyphenolic profiling of wild cherry, the European larch and sweet chestnut tree bark. *Eur. Food Res. Technol.* **2021**, *247*, 2355–2370. [CrossRef]
4. Aspé, E.; Fernández, K. The effect of different extraction techniques on extraction yield, total phenolic, and anti-radical capacity of extracts from *Pinus radiata* Bark. *Ind. Crops Prod.* **2011**, *34*, 838–844. [CrossRef]
5. Ku, C.S.; Jang, J.P.; Mun, S.P. Exploitation of polyphenol-rich pine barks for potent antioxidant activity. *J. Wood Sci.* **2007**, *53*, 524–528. [CrossRef]
6. Sillero, L.; Prado, R.; Andrés, M.A.; Labidi, J. Characterisation of bark of six species from mixed Atlantic forest. *Ind. Crops Prod.* **2019**, *137*, 276–284. [CrossRef]
7. Palanisamy, S.; Kalimuthu, M.; Nagarajan, R.; Fernandes Marlet, J.M.; Santulli, C. Physical, Chemical, and Mechanical Characterization of Natural Bark Fibers (NBFs) Reinforced Polymer Composites: A Bibliographic Review. *Fibers* **2023**, *11*, 13. [CrossRef]
8. Wagner, K.; Roth, C.; Willför, S.; Musso, M.; Petutschnigg, A.; Oostingh, G.; Schnabel, T. Identification of Antimicrobial Compounds in Different Hydrophilic Larch Bark Extracts. *Bioresources* **2019**, *14*, 5807–5815. [CrossRef]
9. Kulikowska, D.; Bernat, K. Waste Willow-Bark from Salicylate Extraction Successfully Reused as an Amendment for Sewage Sludge Composting. *Sustainability* **2021**, *13*, 6771. [CrossRef]
10. Simões, R.; Pimentel, C.; Ferreira-Dias, S.; Miranda, I.; Pereira, H. Phytochemical characterization of phloem in maritime pine and stone pine in three sites in Portugal. *Heliyon* **2021**, *7*, e06718. [CrossRef]
11. Pásztory, Z.; Mohácsiné, I.; Gorbacheva, G.; Börcsök, Z. The utilization of tree bark. *BioResources* **2016**, *11*, 7859–7888. [CrossRef]
12. Zhao, X.; Qi, F.; Liu, D. Hierarchy Nano- and Ultrastructure of Lignocellulose and Its Impact on the Bioconversion of Cellulose. In *Nanotechnology for Bioenergy and Biofuel Production*; Springer: Cham, Switzerland, 2017.

13. Davison, B.H.; Parks, J.; Davis, M.F.; Donohoe, B.S. Plant Cell Walls: Basics of Structure, Chemistry, Accessibility and the Influence on Conversion. In *Aqueous Pretreatment of Plant Biomass for Biological and Chemical Conversion to Fuels and Chemicals*; John Wiley & Sons, Ltd.: Hoboken, NJ, USA, 2013; pp. 23–38.
14. Zhang, J.; Cai, D.; Qin, Y.; Liu, D.; Zhao, X. High value-added monomer chemicals and functional bio-based materials derived from polymeric components of lignocellulose by organosolv fractionation. *Biofuels Bioprod. Biorefining* **2020**, *14*, 371–401. [CrossRef]
15. Aspe, E.; Fernández, K. Comparison of phenolic extracts obtained of Pinus radiata bark from pulp and paper industry and sawmill industry. *Maderas. Cienc. Y Tecnol.* **2010**, *13*, 243–252. [CrossRef]
16. Miranda, I.; Gominho, J.; Mirra, I.; Pereira, H. Chemical characterization of barks from Picea abies and Pinus sylvestris after fractioning into different particle sizes. *Ind. Crops Prod.* **2012**, *36*, 395–400. [CrossRef]
17. Fengel, D.; Wegener, G. Constituents of Bark. In *Wood: Chemistry, Ultrastructure, Reactions*; Wegener, D.F.G., Ed.; Walter de Gruyter: Berlin, Germany, 1984; pp. 240–267.
18. Pereira, H.; Graça, J.; Rodrigues, J. Wood Chemistry in Relation to Quality. *Cheminform* **2004**, *35*, 53–83. [CrossRef]
19. Chupin, L.; Maunu, S.L.; Reynaud, S.; Pizzi, A.; Charrier, B.; Charrier-El Bouhtoury, F. Microwave assisted extraction of maritime pine (*Pinus pinaster*) bark: Impact of particle size and characterization. *Ind. Crops Prod.* **2015**, *65*, 142–149. [CrossRef]
20. Venkatesan, T.; Choi, Y.-W.; Kim, Y.-K. Impact of Different Extraction Solvents on Phenolic Content and Antioxidant Potential of *Pinus densiflora* Bark Extract. *BioMed Res. Int.* **2019**, *2019*, 3520675. [CrossRef]
21. Mármol, I.; Quero, J.; Jiménez-Moreno, N.; Rodríguez-Yoldi, M.J.; Ancín-Azpilicueta, C. A systematic review of the potential uses of pine bark in food industry and health care. *Trends Food Sci. Technol.* **2019**, *88*, 558–566. [CrossRef]
22. Ferreira-Santos, P.; Genisheva, Z.; Pereira, R.N.; Teixeira, J.A.; Rocha, C.M.R. Moderate Electric Fields as a Potential Tool for Sustainable Recovery of Phenolic Compounds from Pinus pinaster Bark. *ACS Sustain. Chem. Eng.* **2019**, *7*, 8816–8826. [CrossRef]
23. Seabra, I.J.; Dias, A.M.A.; Braga, M.E.M.; de Sousa, H.C. High pressure solvent extraction of maritime pine bark: Study of fractionation, solvent flow rate and solvent composition. *J. Supercrit. Fluids* **2012**, *62*, 135–148. [CrossRef]
24. Deutz, P. Circular Economy. In *International Encyclopedia of Human Geography*, 2nd ed.; Kobayashi, A., Ed.; Elsevier: Oxford, UK, 2020; pp. 193–201.
25. Feng, S.; Cheng, S.; Yuan, Z.; Leitch, M.; Xu, C. Valorization of bark for chemicals and materials: A review. *Renew. Sustain. Energy Rev.* **2013**, *26*, 560–578. [CrossRef]
26. Vázquez, G.; Antorrena, G.; Parajó, J.C. Studies on the utilization of Pinus pinaster bark. *Wood Sci. Technol.* **1987**, *21*, 65–74. [CrossRef]
27. Fradinho, D.M.; Neto, C.P.; Evtuguin, D.; Jorge, F.C.; Irle, M.A.; Gil, M.H.; Pedrosa de Jesus, J. Chemical characterisation of bark and of alkaline bark extracts from maritime pine grown in Portugal. *Ind. Crops Prod.* **2002**, *16*, 23–32. [CrossRef]
28. Santos, J.; Pereira, J.; Ferreira, N.; Paiva, N.; Ferra, J.; Magalhães, F.D.; Martins, J.M.; Dulyanska, Y.; Carvalho, L.H. Valorisation of non-timber by-products from maritime pine (*Pinus pinaster*, Ait) for particleboard production. *Ind. Crops Prod.* **2021**, *168*, 113581. [CrossRef]
29. Elmas, G.; Yilgor, N. Chemical and thermal characterizations of Pinus sylvestris and Pinus pinaster. *BioResources* **2020**, *15*, 3604–3620. [CrossRef]
30. Alonso-Esteban, J.I.; Carocho, M.; Barros, D.; Velho, M.V.; Heleno, S.; Barros, L. Chemical composition and industrial applications of Maritime pine (Pinus pinaster Ait.) bark and other non-wood parts. *Rev. Environ. Sci. Bio/Technol.* **2022**, *21*, 583–633. [CrossRef]
31. Nunes, E.; Quilhó, T.; Pereira, H. Anatomy and Chemical Composition of Pinus Pinaster Bark. *IAWA J.* **1996**, *17*, 141–150. [CrossRef]
32. Ferreira-Santos, P.; Genisheva, Z.; Botelho, C.; Santos, J.; Ramos, C.; Teixeira, J.A.; Rocha, C.M.R. Unravelling the Biological Potential of Pinus pinaster Bark Extracts. *Antioxidants* **2020**, *9*, 334. [CrossRef]
33. Vieito, C.; Pires, P.; Fernandes, É.; Vaz-Velho, M. Chemical characterization of pine bark (Pinus pinaster Aiton subsp. atlantica), antioxidant properties and phenolic profile of its extracts. *Milleninum* **2019**, *2*, 79–87. [CrossRef]
34. Maekawa, E.; Ichizawa, T.; Koshijima, T. An Evaluation of the Acid-Soluble Lignin Determination in Analyses of Lignin by the Sulfuric Acid Method. *J. Wood Chem. Technol.* **1989**, *9*, 549–567. [CrossRef]
35. Schutyser, W.; Renders, T.; Van Den Bosch, S.; Koelewijn, S.F.; Beckham, G.T.; Sels, B.F. Chemicals from lignin: An interplay of lignocellulose fractionation, depolymerisation, and upgrading. *Chem. Soc. Rev.* **2018**, *47*, 852–908. [CrossRef] [PubMed]
36. *D1105-96*; Standard Test Method for Preparation of Extractive-Free Wood. ASTM: West Conshohocken, PA, USA, 2001.
37. Fernandes, C.S. Following Green Chemistry Principles for the Valorization of Lignocellulose Waste. On the Extraction and Characterization of Lignin from Maritime Pine Sawdust with Novel Deep Eutectic Solvents. Master's Thesis, FCTUC, Coimbra, Portugal, 2020.
38. Vangeel, T.; Neiva, D.M.; Quilhó, T.; Costa, R.A.; Sousa, V.; Sels, B.F.; Pereira, H. Tree bark characterization envisioning an integrated use in a biorefinery. *Biomass Convers. Biorefinery* **2023**, *13*, 2029–2043. [CrossRef]
39. Willför, S.M.; Smeds, A.I.; Holmbom, B.R. Chromatographic analysis of lignans. *J. Chromatogr. A* **2006**, *1112*, 64–77. [CrossRef] [PubMed]
40. Sunardi; Istikowati, W.T.; Sari, D.I. Extraction of α-cellulose from Eleocharis dulcis Holocellulose using NaOH and KOH. *J. Phys. Conf. Ser.* **2019**, *1397*, 012031. [CrossRef]
41. Arnata, I.; Suprihatin, S.; Fahma, F.; Richana, N.; Sunarti, T. Cellulose Production from Sago Frond with Alkaline Delignification and Bleaching on Various Types of Bleach Agents. *Orient. J. Chem.* **2019**, *35*, 8–19. [CrossRef]

42. Sousa, J.; Ramos, P.; Freire, C.; Silva, A.; Silvestre, A. Chemical Composition of Lipophilic Bark Extracts from Pinus pinaster and Pinus pinea Cultivated in Portugal. *Appl. Sci.* **2018**, *8*, 2575. [CrossRef]
43. Kurtça, M.; Tumen, I. Investigation into Seasonal Variations of Chemical Compounds of Maritime Pine (Pinus Pinaster Ait.) Grown in Turkey. *Fresenius Environ. Bull.* **2020**, *29*, 8156–8167.
44. Bożym, M.; Gendek, A.; Siemiątkowski, G.; Aniszewska, M.; Malaťák, J. Assessment of the Composition of Forest Waste in Terms of Its Further Use. *Materials* **2021**, *14*, 973. [CrossRef]
45. Kabata-Pendias, A. *Trace Elements in Soils and Plants*, 4th ed.; CRC Press: Boca Raton, FL, USA, 2010; pp. 1–520.
46. Narayan, O.P.; Kumar, P.; Yadav, B.; Dua, M.; Johri, A.K. Sulfur nutrition and its role in plant growth and development. *Plant Signal Behav.* **2022**, *2022*, 2030082. [CrossRef]
47. Zhao, F.-j.; Tausz, M.; De Kok, L.J. Role of Sulfur for Plant Production in Agricultural and Natural Ecosystems. In *Sulfur Metabolism in Phototrophic Organisms*; Hell, R., Dahl, C., Knaff, D., Leustek, T., Eds.; Springer: Dordrecht, The Netherlands, 2008; pp. 417–435.
48. Chorianopoulou, S.N.; Bouranis, D.L. The Role of Sulfur in Agronomic Biofortification with Essential Micronutrients. *Plants* **2022**, *11*, 1979. [CrossRef]
49. Cetin, M.; Sevik, H.; Cobanoglu, O. Ca, Cu, and Li in washed and unwashed specimens of needles, bark, and branches of the blue spruce (Picea pungens) in the city of Ankara. *Environ. Sci. Pollut. Res.* **2020**, *27*, 21816–21825. [CrossRef] [PubMed]
50. Poikolainen, J. Sulphur and heavy metal concentrations in Scots pine bark in northern Finland and the Kola peninsula. *Water Air Soil Pollut.* **1997**, *93*, 395–408. [CrossRef]
51. Kirchner, P.; Biondi, F.; Edwards, R.; McConnell, J.R. Variability of trace metal concentrations in Jeffrey pine (*Pinus jeffreyi*) tree rings from the Tahoe Basin, California, USA. *J. For. Res.* **2008**, *13*, 347–356. [CrossRef]
52. Liu, F.; Wienke, C.; Fiencke, C.; Guo, J.; Dong, R.; Pfeiffer, E.M. Biofilter with mixture of pine bark and expanded clay as packing material for methane treatment in lab-scale experiment and field-scale implementation. *Environ. Sci. Pollut. Res. Int.* **2018**, *25*, 31297–31306. [CrossRef] [PubMed]
53. Sukhdolgor, J.; Badamtsetseg, S.; Adyakhuu, D. Chemical Composition and Amount of Macro and Microelements of Pine (*Pinus silvestris* L) and Larch (*Larix sibirica* Ldb) Trees in Mongolia. *Mong. J. Biol. Sci.* **2003**, *1*, 81–83. [CrossRef]
54. Aniszewska, M.; Gendek, A.; Drożdżek, M.; Bożym, M.; Wojdalski, J. Physicochemical properties of seed extraction residues and their potential uses in energy production. *Rocz. Ochr. Srod.* **2017**, *19*, 302–334.
55. Coşkun, M. Toxic Metals in the Austrian Pine (*Pinus Nigra*) Bark in the Thrace Region, Turkey. *Environ. Monit. Assess.* **2006**, *121*, 173–179. [CrossRef]
56. Harju, L.; Saarela, K.E.; Rajander, J.; Lill, J.O.; Lindroos, A.; Heselius, S.J. Environmental monitoring of trace elements in bark of Scots pine by thick-target PIXE. *Nucl. Instrum. Methods Phys. Res. Sect. B Beam Interact. Mater. At.* **2002**, *189*, 163–167. [CrossRef]
57. Lippo, H.; Poikolainen, J.; Kubin, E. The use of moss, lichen and pine bark in the nationwide monitoring of atmospheric heavy metal deposition in Finland. *Water Air Soil Pollut.* **1995**, *85*, 2241–2246. [CrossRef]
58. Ots, K.; Mandre, M. Monitoring of heavy metals uptake and allocation in Pinus sylvestris organs in alkalised soil. *Environ. Monit. Assess.* **2012**, *184*, 4105–4117. [CrossRef]
59. Wang, X.; Chen, H.; Feng, X.; Zhang, Q.; Labbé, N.; Kim, K.; Huang, J.; Ragauskas, A.J.; Wang, S.; Zhang, Y. Isolation and characterization of lignocellulosic nanofibers from four kinds of organosolv-fractionated lignocellulosic materials. *Wood Sci. Technol.* **2020**, *54*, 503–517. [CrossRef]
60. Shangdiar, S.; Lin, Y.-C.; Ponnusamy, V.K.; Wu, T.-Y. Pretreatment of lignocellulosic biomass from sugar bagasse under microwave assisted dilute acid hydrolysis for biobutanol production. *Bioresour. Technol.* **2022**, *361*, 127724. [CrossRef] [PubMed]
61. Lin, Y.-C.; Shangdiar, S.; Chen, S.-C.; Cho, C.-A.; Datov, A. A study on different parameters affecting the saccharification rate of Typha orientalis pretreated with Ionic Liquids (ILs) and microwave irradiation for bioethanol production by using response surface methodology. *Catal. Commun.* **2021**, *150*, 106265. [CrossRef]
62. Ramos-Andrés, M.; Aguilera-Torre, B.; García-Serna, J. Production of purified hemicellulose-pectin fractions of different molecular weight from discarded carrots by hydrothermal treatment followed by multistep ultrafiltration/diafiltration. *J. Clean. Prod.* **2021**, *321*, 128923. [CrossRef]
63. Sharma, S.; Nargotra, P.; Sharma, V.; Bangotra, R.; Kaur, M.; Kapoor, N.; Paul, S.; Bajaj, B.K. Nanobiocatalysts for efficacious bioconversion of ionic liquid pretreated sugarcane tops biomass to biofuel. *Bioresour. Technol.* **2021**, *333*, 125191. [CrossRef]
64. Zhuang, J.; Li, M.; Pu, Y.; Ragauskas, A.J.; Yoo, C.G. Observation of Potential Contaminants in Processed Biomass Using Fourier Transform Infrared Spectroscopy. *Appl. Sci.* **2020**, *10*, 4345. [CrossRef]
65. Hurtubise, F.G.; Krassig, H. Classification of Fine Structural Characteristics in Cellulose by Infared Spectroscopy. Use of Potassium Bromide Pellet Technique. *Anal. Chem.* **1960**, *32*, 177–181. [CrossRef]
66. Boeriu, C.G.; Bravo, D.; Gosselink, R.J.A.; van Dam, J.E.G. Characterisation of structure-dependent functional properties of lignin with infrared spectroscopy. *Ind. Crops Prod.* **2004**, *20*, 205–218. [CrossRef]
67. González-Muñoz, M.J.; Santos, V.; Parajó, J.C. Purification of oligosaccharides obtained from Pinus pinaster hemicelluloses by diafiltration. *Desalination Water Treat.* **2011**, *27*, 48–53. [CrossRef]
68. Tao, J.; Hosseinaei, O.; Delbeck, L.; Kim, P.; Harper, D.; Bozell, J.; Rials, T.; Labbé, N. Effects of organosolv fractionation time on thermal and chemical properties of lignins. *RSC Adv.* **2016**, *6*, 79228–79235. [CrossRef]
69. Rivas, S.; Raspolli-Galletti, A.M.; Antonetti, C.; Santos, V.; Parajó, J.C. Sustainable conversion of Pinus pinaster wood into biofuel precursors: A biorefinery approach. *Fuel* **2016**, *164*, 51–58. [CrossRef]

70. Margellou, A.G.; Pappa, C.P.; Psochia, E.A.; Petala, M.D.; Triantafyllidis, K.S. Mild isolation and characterization of surface lignin from hydrothermally pretreated lignocellulosic forestry and agro-industrial waste biomass. *Sustain. Chem. Pharm.* **2023**, *33*, 101056. [CrossRef]
71. Yang, X.; Berglund, L.A. Structural and Ecofriendly Holocellulose Materials from Wood: Microscale Fibers and Nanoscale Fibrils. *Adv. Mater.* **2021**, *33*, 2001118. [CrossRef] [PubMed]
72. Pacheco Pinilla, C.; Avila, C.; Reyes, G.; Aguayo, M.; Rojas, O.; Aguayo, M.G. Characterization of Residues from Chilean Blueberry Bushes: A Potential Source of Cellulose. *BioResources* **2018**, *13*, 7345–7359. [CrossRef]
73. Adeleye, O.A.; Bamiro, O.A.; Albalawi, D.A.; Alotaibi, A.S.; Iqbal, H.; Sanyaolu, S.; Femi-Oyewo, M.N.; Sodeinde, K.O.; Yahaya, Z.S.; Thiripuranathar, G.; et al. Characterizations of Alpha-Cellulose and Microcrystalline Cellulose Isolated from Cocoa Pod Husk as a Potential Pharmaceutical Excipient. *Materials* **2022**, *15*, 5992. [CrossRef]
74. Bortoluz, J.; Cemin, A.; Bonetto, L.R.; Ferrarini, F.; Esteves, V.I.; Giovanela, M. Isolation, characterization and valorization of lignin from Pinus elliottii sawdust as a low-cost biosorbent for zinc removal. *Cellulose* **2019**, *26*, 4895–4908. [CrossRef]
75. Volli, V.; Gollakota, A.R.K.; Shu, C.-M. Comparative studies on thermochemical behavior and kinetics of lignocellulosic biomass residues using TG-FTIR and Py-GC/MS. *Sci. Total Environ.* **2021**, *792*, 148392. [CrossRef]
76. Sulaiman, M.; Rabbani, F.A.; Iqbal, T.; Kazmi, M.A.; Yasin, S.; Mujtaba, M.A.; Kalam, M.A.; Almomani, F. Impact of eco-friendly chemical pretreatment on physicochemical and surface mechanical properties of sustainable lignocellulosic agricultural waste. *Algal Res.* **2023**, *71*, 103051. [CrossRef]
77. Balotin, G.; De Almeida, J.; da Silva, R.M.; Carvalho, W.A.; Carvalho, C.T.; Rodrigues, R. Upgrading catalytic efficiency of activated carbons by tailoring lignocellulosic biomass waste for sustainable conversion of glycerol to solketal. *Mol. Catal.* **2023**, *538*, 112976. [CrossRef]
78. Ufodike, C.O.; Eze, V.O.; Ahmed, M.F.; Oluwalowo, A.; Park, J.G.; Okoli, O.I.; Wang, H. Evaluation of the inter-particle interference of cellulose and lignin in lignocellulosic materials. *Int. J. Biol. Macromol.* **2020**, *147*, 762–767. [CrossRef]
79. Yang, H.; Yan, R.; Chen, H.; Lee, D.H.; Zheng, C. Characteristics of hemicellulose, cellulose and lignin pyrolysis. *Fuel* **2007**, *86*, 1781–1788. [CrossRef]
80. Gupta, G.K.; Ram, M.; Bala, R.; Kapur, M.; Mondal, M.K. Pyrolysis of chemically treated corncob for biochar production and its application in Cr(VI) removal. *Environ. Prog. Sustain. Energy* **2018**, *37*, 1606–1617. [CrossRef]
81. Boonsuk, P.; Sukolrat, A.; Bourkaew, S.; Kaewtatip, K.; Chantarak, S.; Kelarakis, A.; Chaibundit, C. Structure-properties relationships in alkaline treated rice husk reinforced thermoplastic cassava starch biocomposites. *Int. J. Biol. Macromol.* **2021**, *167*, 130–140. [CrossRef]
82. Chin, D.W.K.; Lim, S.; Pang, Y.L.; Lim, C.H.; Lee, K.M. Two-staged acid hydrolysis on ethylene glycol pretreated degraded oil palm empty fruit bunch for sugar based substrate recovery. *Bioresour. Technol.* **2019**, *292*, 121967. [CrossRef]
83. Wang, X.; Wang, P.; Su, Y.; Wang, Q.; Ling, Z.; Yong, Q. Supramolecular Deconstruction of Bamboo Holocellulose via Hydrothermal Treatment for Highly Efficient Enzymatic Conversion at Low Enzyme Dosage. *Int. J. Mol. Sci.* **2022**, *23*, 11829. [CrossRef]
84. Chiang, A.K.M.; Ng, L.Y.; Ng, C.Y.; Lim, Y.P.; Mahmoudi, E.; Tan, L.S.; Mah, S.K. Conversion of palm oil empty fruit bunches to highly stable and fluorescent graphene oxide quantum dots: An eco-friendly approach. *Mater. Chem. Phys.* **2023**, *309*, 128433. [CrossRef]
85. AOAC. *Official Methods of Analysis*, 15th ed.; Association of Official Analytical Chemists: Rockville, MD, USA, 1995.
86. Ona, T.; Sonoda, T.; Shibata, M.; Fukazawa, K. Small-Scale Method to Determine the Content of Wood Components from Multiple Eucalyptus Samples. *Tappi J.* **1995**, *78*, 121–126.
87. Sluiter, A.; Hames, B.; Ruiz, R.; Scarlata, C.; Sluiter, J.; Templeton, D.; Crocker, D. Determination of structural carbohydrates and lignin in biomass. In *Laboratory Analytical Procedure (LAP)*; National Renewable Energy Laboratory: Golden, CO, USA, 2008.
88. Alcaraz, L.; Saquinga, D.N.; López, F.; Lima, L.D.; Alguacil, F.J.; Escudero, E.; López, F.A. Application of a Low-Cost Cellulose-Based Bioadsorbent for the Effective Recovery of Terbium Ions from Aqueous Solutions. *Metals* **2020**, *10*, 1641. [CrossRef]
89. Sun, J.X.; Sun, X.F.; Zhao, H.; Sun, R.C. Isolation and characterization of cellulose from sugarcane bagasse. *Polym. Degrad. Stab.* **2004**, *84*, 331–339. [CrossRef]
90. Pereira, C.; Barros, L.; Carvalho, A.M.; Ferreira, I.C.F.R. Nutritional composition and bioactive properties of commonly consumed wild greens: Potential sources for new trends in modern diets. *Food Res. Int.* **2011**, *44*, 2634–2640. [CrossRef]
91. Barros, L.; Carvalho, A.M.; Sá Morais, J.; Ferreira, I.C.F.R. Strawberry-tree, blackthorn and rose fruits: Detailed characterisation in nutrients and phytochemicals with antioxidant properties. *Food Chem.* **2010**, *120*, 247–254. [CrossRef]
92. APHA/AWWA/WEF. *Standard Methods for the Examination of Water and Wastewater*; American Water Works Association: Denver, CO, USA, 2017.

plants

MDPI

Article

Effect of Elicitor Treatments on Quality Attributes in Blueberry: Implications of Cultivar and Environmental Conditions

Gabriele Cola [1], Beatrice Cavenago [1], Claudio Sebastiano Gardana [2] and Anna Spinardi [1,*]

[1] Department of Agricultural and Environmental Sciences (DISAA), Università Degli Studi Di Milano, 20133 Milan, Italy; gabriele.cola@unimi.it (G.C.); beatrice.cavenago@unimi.it (B.C.)

[2] Department of Food, Environmental and Nutritional Sciences (DeFENS), Università Degli Studi Di Milano, 20133 Milan, Italy; claudio.gardana@unimi.it

* Correspondence: anna.spinardi@unimi.it

Abstract: Elicitors of plant defence responses can trigger defence mechanisms that are able to protect plant tissues from biotic or abiotic stresses. Since one defence response involves the activation of secondary metabolites' biosynthesis, the purpose of this study was to evaluate the effect of chitosan and melatonin pre-harvest treatments on the quality and the nutritional parameters of the fruits of blueberry (*Vaccinium corymbosum* L.). Across the two years of experiment, three different cultivars (cv.s. 'Cosmopolitan', 'Hortblue Poppins' and 'Legacy') were treated with 1% chitosan or 100 μM melatonin every two weeks during the ripening season and ripe fruits were progressively harvested and analysed. The treatment with both elicitors had only slight effects on dry matter, soluble solids content, titratable acidity and pH, with a cultivar-dependent response. On the other hand, elicitors significantly affected the levels of phenylpropanoid and antioxidant compounds in all cvs. in both years, with a higher accumulation of total anthocyanins and phenolics and the enhancement of the antioxidant capacity, with positive effects on the nutraceutical quality of fruits. The anthocyanin profile in terms of both absolute concentrations and the relative proportion of single anthocyanins was affected by both harvest year and cv., highlighting the role of the genetic background in the plant response to environmental conditions (with particular reference to summer heat stress) and to elicitor treatments.

Keywords: chitosan; melatonin; southern highbush; northern highbush; phenolic compounds

1. Introduction

Plant disease resistance is dependent on both pre-existing physical and chemical barriers associated with plant-innate immunity and plant-induced immunity, where signalling molecules activate the defence response to protect plant tissues from damage caused by biotic or abiotic stresses [1,2]. In the absence of any pathogen attack, these defence mechanisms may be induced by physical or chemical elicitation [3,4], resulting in enhanced resistance to pathogens and diseases and tolerance to abiotic stress and adversity. In fact, defence priming supports a faster, stronger and more sustained response to stresses [5–7].

From a sustainability point of view, induced resistance also has a positive impact on the environment due to the absence of toxic residues on the plant, on the fruits and in the environment. Furthermore, the induced resistance reduces the application of pesticides.

Regarding the time of application, when compared to postharvest approaches, preharvest elicitor treatments are expected to favour a better-coordinated allocation of resources for both plant metabolism and defence, as in-field fruits are connected to the plant, and secondary metabolites' biosynthesis is not entirely based on the energetic resources present in detached plant tissues as it is at postharvest.

Although elicitors were initially used to increase plant tolerance to pathogens, the mechanism of induced resistance has been found to be involved in the activation of pathways of the plant's secondary metabolism [8–10].

As secondary metabolites play an important role in the adaptation of plants to stress conditions, elicitors that simulate different biotic and abiotic stress conditions may trigger the plant biochemical system towards the increasing accumulation of secondary metabolites, including phenolic compounds [11–13]. For this reason, elicitors can be considered an interesting tool to obtain plants with a higher polyphenol content with positive consequences on the nutraceutical content of the fruit and vegetables [14].

Highbush blueberry (*Vaccinium corymbosum* L.) has gained popularity owing to its unique taste and its high content of bioactive compounds, such as phenolic compounds, with possible beneficial health properties. These health benefits, mainly due to the antioxidant activity, have been extensively described [15,16]. As a consequence, more interest has been placed on developing cultural management practices to enhance fruit polyphenolic levels and the exploitation of processing by-products as potential polyphenolic sources for the nutraceutical market. The possibility of further increasing the content of bioactive substances and, in parallel, improving the resistance of plants and fruits to adversity would further increase their nutritional and market value.

Among GRAS (generally recognized as safe) substances used as elicitors of plant defence response, chitosan is one of the most promising compounds studied. Chitosan is an inexpensive natural polysaccharide with high antimicrobial activity against a wide range of pathogenic and spoilage microorganisms, including fungi and bacteria. It shows a vast potential in many different areas, ranging from agriculture, as a possible elicitor of plant defence responses and as seed coating agent, to food systems as an additive, the cosmetic industry as a skin care and hair care ingredient and medicine as a food supplement with a hypocholesterolemic effect [17–20].

Many publications have reported the antifungal and elicitor effects of chitosan, ranging from in vitro studies [21–23] to applications in postharvest storage life and as a natural antifungal edible coating [3,24–26].

Numerous works have been conducted using treatments with chitosan in postharvest, but on the contrary, there are few reports in which this substance is used in preharvest. The effects of preharvest treatments with chitosan have been studied on perennial fruit trees like jujube [27], table grape [28], grapevine [29], apple [30], peach [31,32], sweet cherry [33], date palm [34] and apricot [35], as well as on horticultural crops such as tomato [36,37] and muskmelon [38].

Another elicitor that, in recent years, has attracted attention and received much interest is melatonin. Melatonin is an indolamine with amphiphilic characteristics acting as multifunctional signalling molecule. In plants, it is involved in the regulation of many different developmental processes regarding vegetative growth and reproductive behaviour [39,40]. This molecule promotes seed viability and germination, can stimulate root growth-effecting root apparatus architecture, can control floral transition and delay flowering and can improve fruit yield and crop quality by regulating fruit ripening. These effects are linked to its high antioxidant and free radical scavenging activity that protects tissues from oxidative stress and plays a pivotal role in defending plants against environmental stresses and in plant responses to harmful environmental factors [41,42] Melatonin also can enhance cellular antioxidant defence mechanisms by stimulating antioxidant enzymes and regenerating endogenous antioxidants [43]. It may also stimulate the biosynthesis of bioactive compounds and molecules that protect membrane structure against peroxidation [44–46].

To date, reports on preharvest melatonin treatments have investigated sweet cherry [47,48], date palm [49], apricot [50], pomegranate [51,52], japanese plum [53], Nanhong pear [46], eggplant [54] and tomato [55].

The aim of this study was to evaluate the effect of preharvest treatments with chitosan and melatonin on fruit quality traits and the nutritional parameters of three different blueberry cultivars. Particular interest was devoted to the accumulation of bioactive compounds and to the anthocyanin profile. To our knowledge, this is the first report on the effects of preharvest treatments with chitosan and with melatonin on blueberry plants and fruits.

2. Results

2.1. Environmental Factors

The analysis of thermal anomalies of 2021 and 2022 in comparison with the climatic normal from 1991 to 2020 highlights the peculiar features of the two seasons. Table 1 shows the monthly number of anomalies for daily minimum and maximum temperature. Based on the statistical distribution of historical data, each day is classified as characterized by no anomaly (NORM), mild positive anomaly (PAM), strong positive anomaly (PAS), mild negative anomaly (NAM) and strong negative anomaly (NAS). The 1991–2020 monthly average of each day is reported in brackets, showing the average distribution of anomaly along the twelve months compared to the norm. In this way, it is possible to understand when the month of a specific year is characterized by a peculiar behaviour differing from the average.

Table 1. Monthly count of daily thermal anomalies in 2021 and 2022. Based on the statistical distribution of historical data, each day is classified as characterized by no anomaly (NORM), mild positive anomaly (PAM), strong positive anomaly (PAS), mild negative anomaly (NAM) or strong negative anomaly (NAS). The 1991–2020 monthly average of anomaly is reported in brackets.

	2021									
Month	Minimum Daily Temperature					Maximum Daily Temperature				
	Nas	Nam	Norm	Pam	Pas	Nas	Nam	Norm	Pam	Pas
January	0 (0.7)	2 (4.0)	22 (21.7)	7 (3.9)	0 (0.7)	0 (0.2)	2 (5)	25 (21.1)	4 (3.9)	0 (0.8)
February	0 (1.0)	2 (3.1)	8 (19.5)	13 (4.0)	5 (0.4)	0 (0.7)	1 (3.9)	22 (19.1)	2 (3.8)	3 (0.5)
March	1 (0.6)	6 (4.3)	19 (20.9)	5 (4.9)	0 (0.3)	0 (0.9)	4 (4.1)	22 (21.1)	5 (4.5)	0 (0.4)
April	2 (1.2)	6 (3.6)	19 (20.6)	3 (4.3)	0 (0.3)	0 (0.4)	10 (4.9)	17 (19.6)	1 (4.8)	2 (0.3)
May	0 (0.8)	6 (4.1)	25 (21.2)	0 (4.8)	0 (0.1)	1 (0.9)	3 (4.2)	27 (21.3)	0 (4.3)	0 (0.3)
June	0 (0.7)	0 (3.8)	21 (20.9)	8 (4.2)	1 (0.4)	0 (0.9)	0 (3.8)	28 (20.4)	2 (4.7)	0 (0.2)
July	0 (0.9)	2 (4.4)	26 (20.9)	3 (4.5)	0 (0.3)	1 (1.1)	4 (3.7)	26 (21.8)	0 (4.1)	0 (0.3)
August	0 (0.8)	6 (4.3)	22 (20.7)	1 (5.0)	2 (0.2)	2 (0.9)	4 (3.6)	21 (22.3)	3 (3.7)	1 (0.5)
September	0 (0.8)	0 (4.3)	21 (20.0)	8 (4.8)	1 (0.1)	0 (0.8)	1 (3.9)	26 (20.4)	3 (4.7)	0 (0.3)
October	0 (0.7)	10 (4.1)	16 (21.4)	3 (4.6)	2 (0.2)	0 (0.7)	3 (4.1)	27 (21.3)	1 (4.3)	0 (0.6)
November	0 (0.4)	5 (4.8)	16 (19.4)	9 (5.0)	0 (0.4)	0 (0.7)	4 (3.8)	23 (21)	3 (3.9)	0 (0.6)
December	0 (0.4)	3 (4.4)	23 (21.2)	4 (4.1)	1 (0.9)	0 (0.7)	7 (4.1)	23 (21.2)	1 (4.4)	0 (0.6)
	2022									
Month	Minimum daily temperature					Maximum daily temperature				
	Nas	Nam	Norm	Pam	Pas	Nas	Nam	Norm	Pam	Pas
January	0 (0.7)	4 (4.0)	24 (21.7)	1 (3.9)	2 (0.7)	0 (0.2)	4 (5)	24 (21.1)	1 (3.9)	2 (0.8)
February	0 (1.0)	2 (3.1)	18 (19.5)	5 (4.0)	3 (0.4)	0 (0.7)	2 (3.9)	18 (19.1)	5 (3.8)	3 (0.5)
March	2 (0.6)	9 (4.3)	16 (20.9)	2 (4.9)	2 (0.3)	2 (0.9)	9 (4.1)	16 (21.1)	2 (4.5)	2 (0.4)
April	1 (1.2)	6 (3.6)	22 (20.6)	1 (4.3)	0 (0.3)	1 (0.4)	6 (4.9)	22 (19.6)	1 (4.8)	0 (0.3)
May	0 (0.8)	0 (4.1)	12 (21.2)	13 (4.8)	6 (0.1)	0 (0.9)	0 (4.2)	12 (21.3)	13 (4.3)	6 (0.3)
June	0 (0.7)	1 (3.8)	11 (20.9)	17 (4.2)	1 (0.4)	0 (0.9)	1 (3.8)	11 (20.4)	17 (4.7)	1 (0.2)
July	0 (0.9)	1 (4.4)	16 (20.9)	12 (4.5)	2 (0.3)	0 (1.1)	1 (3.7)	16 (21.8)	12 (4.1)	2 (0.3)
August	0 (0.8)	0 (4.3)	19 (20.7)	12 (5.0)	0 (0.2)	0 (0.9)	0 (3.6)	19 (22.3)	12 (3.7)	0 (0.5)
September	1 (0.8)	5 (4.3)	13 (20.0)	8 (4.8)	3 (0.1)	1 (0.8)	5 (3.9)	13 (20.4)	8 (4.7)	3 (0.3)
October	0 (0.7)	0 (4.1)	16 (21.4)	9 (4.6)	6 (0.2)	0 (0.7)	0 (4.1)	11 (21.3)	11 (4.3)	9 (0.6)
November	0 (0.4)	0 (4.8)	23 (19.4)	5 (5.0)	2 (0.4)	0 (0.7)	1 (3.8)	19 (21)	9 (3.9)	1 (0.6)
December	0 (0.4)	0 (4.4)	15 (21.2)	8 (4.1)	8 (0.9)	1 (0.7)	3 (4.1)	20 (21.2)	7 (4.4)	0 (0.6)

Overall, 2021 was an average year. January was warm, with particular reference to minimum temperature, while February and March showed average thermal conditions. The spring was mildly cold, as testified by the anomalies of April (on both minimum and maximum temperature) and May (with six mild negative anomalies of minimum temperature and the absence of positive anomalies of maximum temperature). The month of June was a particularly average month with 28 normal days of maximum temperature, while July had no cases of positive anomaly of maximum temperature. August and September were normal months, except for the minimum temperatures of September, characterized by a relevant number of mild positive anomalies. The final part of the year was characterized by very normal maximum temperatures, while in the case of minimum temperature October had a large number of mildly negative anomalies and November of mild positive ones.

The year 2022 was strongly characterized by warm conditions. More in detail, January was characterized by very normal conditions, February exhibited warm conditions and March was very variable. After mild cold conditions in April, warming characterized the following part of the year, with a decrease in the number of normal days and an increase in the number of mild and strong positive anomalies, especially during summer.

The characterization of the environmental resources for blueberry was performed by considering global solar radiation (GSR), the accumulation of growing degree days with a base temperature of 7 °C (GDD7), the accumulation of hourly thermal resources (NHH) and hourly thermal excess (HHH, the latter two based on the normal heat hour approach [56] (Figure 1), focusing on the January–July period of 2021 and 2022. With reference to the availability of solar radiation (GSR), the period January–July 2022 was characterized by +6% of GSR when compared to 2021, with lower levels of radiation only in March (−7%) and June (−3%). The accumulation of GDD7 was higher in 2022 (+15%). The warming of the April–July period (+17%) drove this change. As explained in the Methods section, GDD7 does not consider the negative effects of high temperature. The normal heat hours accumulation shows a decrease in thermal resources that are useful for blueberry in 2022 (−7%), with a strong effect in June and July (−7% and −21%, respectively). This decrease is caused by thermal excess, as shown by the accumulation of HHH that characterizes the period of May–July (+90% in 2022).

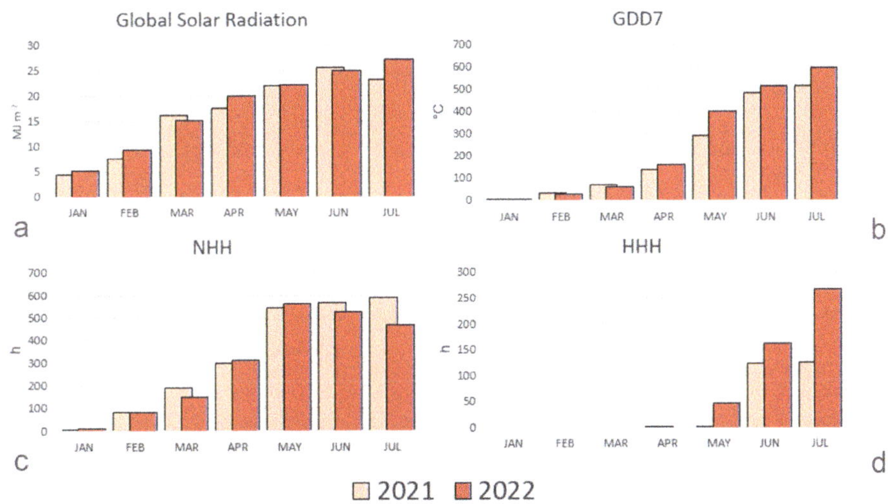

Figure 1. Monthly course of agrometeorological indices for 2021 and 2022. GSR = Global Solar Radiation [MJ m^{-2}], GDD7 = accumulation of growing degree days with base temperature of 7 °C [°C], NHH = accumulation of hourly thermal resources [h], HHH = accumulation of hourly thermal excess [h].

2.2. Fruit Quality Parameters

2.2.1. Dry Matter

Dry matter ranged from 14.6 to 16.2% in 'Cosmopolitan' (Figure 2a,b), from 17.8 to 19.9% in 'Hortblue Poppins' ('H. Poppins') (Figure 2c,d), and from 17.3 to 17.8% in 'Legacy' (Figure 2e,f). In 2022, the untreated fruit of 'Cosmopolitan' had significantly lower dry matter percentage than 'Legacy' and 'H. Poppins'. Differences between subsequent years were recorded in 'H. Poppins' which increased dry matter by 18% and 15% in control and chitosan-treated berries, respectively, and in melatonin-treated fruit of 'Legacy' (+15%). The percentage was affected in 2021 in 'H. Poppins' by chitosan (+15%) and in 2022 in 'Legacy' by melatonin (+15%).

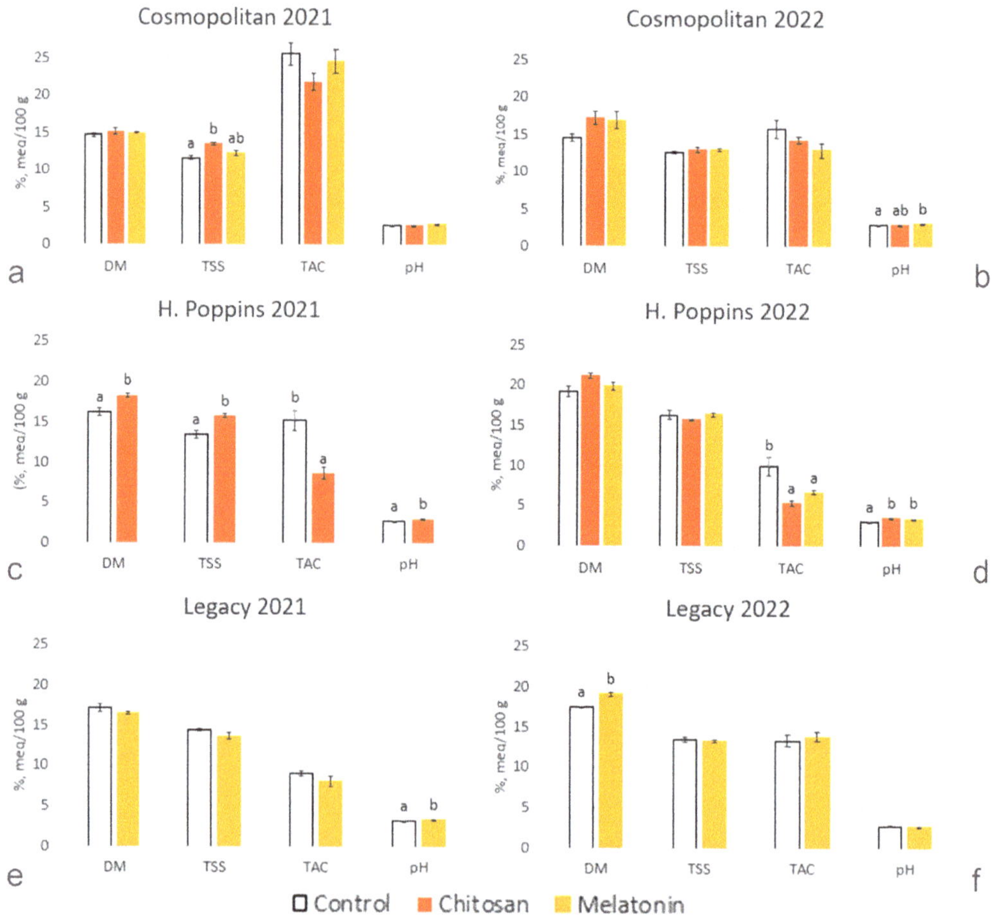

Figure 2. Effect of elicitor treatments on dry matter (DM, %), total soluble solids (TSS, %), total titratable acidity (TAC, meq/100 g FW, milliequivalents per 100 grams of fresh weight) and pH in cvs. 'Cosmopolitan', 'H. Poppins' and 'Legacy' in two subsequent harvest years (2021 and 2022). Values represent means ± SE (n = 4). Different letters indicate statistical differences ($p \leq 0.05$) between treatments for each cultivar in each year.

2.2.2. Total Soluble Solids

As for dry matter, the content of total soluble solids was lowest in control berries of 'Cosmopolitan' in 2021 (Figure 2a) and significantly higher in those of 'H. Poppins'

(Figure 2c) and of 'Legacy' (Figure 2e). An increase was recorded only in the untreated fruit of 'H. Poppins' in 2022 compared to 2021. An effect of treatments was only evident in 2021 with chitosan increasing the total soluble solids content in 'Cosmopolitan' (+16%) and 'H. Poppins' (+18%) (Figure 2a,c).

2.2.3. Total Titratable Acidity

Considering control berries, 'Cosmopolitan' was the cv. which displayed the highest titratable acidity with an average of 20.7 meq 100 g FW^{-1} (Figure 2a,b). The lowest and significantly different level of acidity was showed by 'Legacy' in 2021 (Figure 2e) and by 'H. Poppins' in 2022 (Figure 2d) with 9.1 meq 100 g FW^{-1} and 9.9 meq 100 g FW^{-1}, respectively. Still in untreated berries, there were significantly lower levels in titratable acidity across years in 'Cosmopolitan' and 'H. Poppins' (−38% and −35%, respectively) and an enhancement in 'Legacy' (+46%). A similar trend was also observed after treatments: in the second year of harvest, in 'Cosmopolitan', levels were reduced by 34% and by 47% in the berries treated with chitosan and melatonin, respectively, and levels were reduced in 'H. Poppins' by 37% in the berries treated with chitosan. On the other hand, melatonin-treated fruit of 'Legacy' showed a 72% increase in titratable acidity in 2022 compared to 2021. Elicitor treatments had no effect on 'Cosmopolitan' and 'Legacy', while 'H. Poppins' responded with a decrease in acidity in 2021 by 43% after chitosan treatment (Figure 2c) and in 2022 by 45% and 32% after chitosan and melatonin treatments, respectively (Figure 2d).

2.2.4. pH

The pH trend is linked to that of acidity, with few differences (Figure 2). Considering untreated berries, the lowest pH was shown by 'Cosmopolitan' in 2021 (Figure 2a) and by 'Legacy' in 2022 (Figure 2f), although the latter was not significantly different from 'Cosmopolitan'. Over the years, the trend is the same as shown by titratable acidity, with higher pH in 2022 in berries of 'Cosmopolitan' and 'H. Poppins' and lower pH in those of 'Legacy'. In response to elicitor treatments, there was an increase in pH in all cvs.; in 2021, in treated an increase was observed in berries of 'H. Poppins' and 'Legacy' (Figure 2c,e), and in 2022, an increase was observed in 'Cosmopolitan' after melatonin treatment and in 'H. Poppins' after melatonin and chitosan treatments (Figure 2b,d).

2.2.5. Total Phenolics

Control berries of the three cvs. were significantly different in phenols in 2021, i.e., content was the lowest in 'Cosmopolitan' (170 mg/100 g FW) (Figure 3a) and the highest in 'Legacy' (257 mg/100 g FW) (Figure 3e). In 2022, on the other hand, there were no differences among cvs. (Figure 3b,d,f). Consequently, untreated 'Cosmopolitan' and 'H. Poppins' accumulated more phenolic compounds in 2022 than in 2021 (+55 and +24%, respectively), while phenolic levels remained unchanged in 'Legacy'. The same trends were present in the case of treated berries. Across years, in 'Cosmopolitan', phenolic levels increased by 49 and 64% in the fruits treated with chitosan and melatonin, respectively, and phenolic levels increased in 'H. Poppins' by 21% in the berries treated with chitosan. Generally, elicitor treatments increased the total phenolic content. Chitosan stimulated phenolic accumulation in both years in 'Cosmopolitan' (+19% in 2021 and +18% in 2022) and in 'H. Poppins' (+12% in 2021 and +10% in 2022), whereas in 2022, melatonin induced higher levels in 'H. Poppins' and in 'Legacy' (+9% and +13%, respectively). In 'Cosmopolitan' fruits treated with melatonin, in both years, the levels of total phenolics were intermediate between those of the control and the chitosan treatment and not significantly different. Considering total phenolics across years, all fruits of 'Cosmopolitan' and 'H. Poppins', untreated and treated, markedly increased in their phenolic contents, whereas no changes were observed in 'Legacy'.

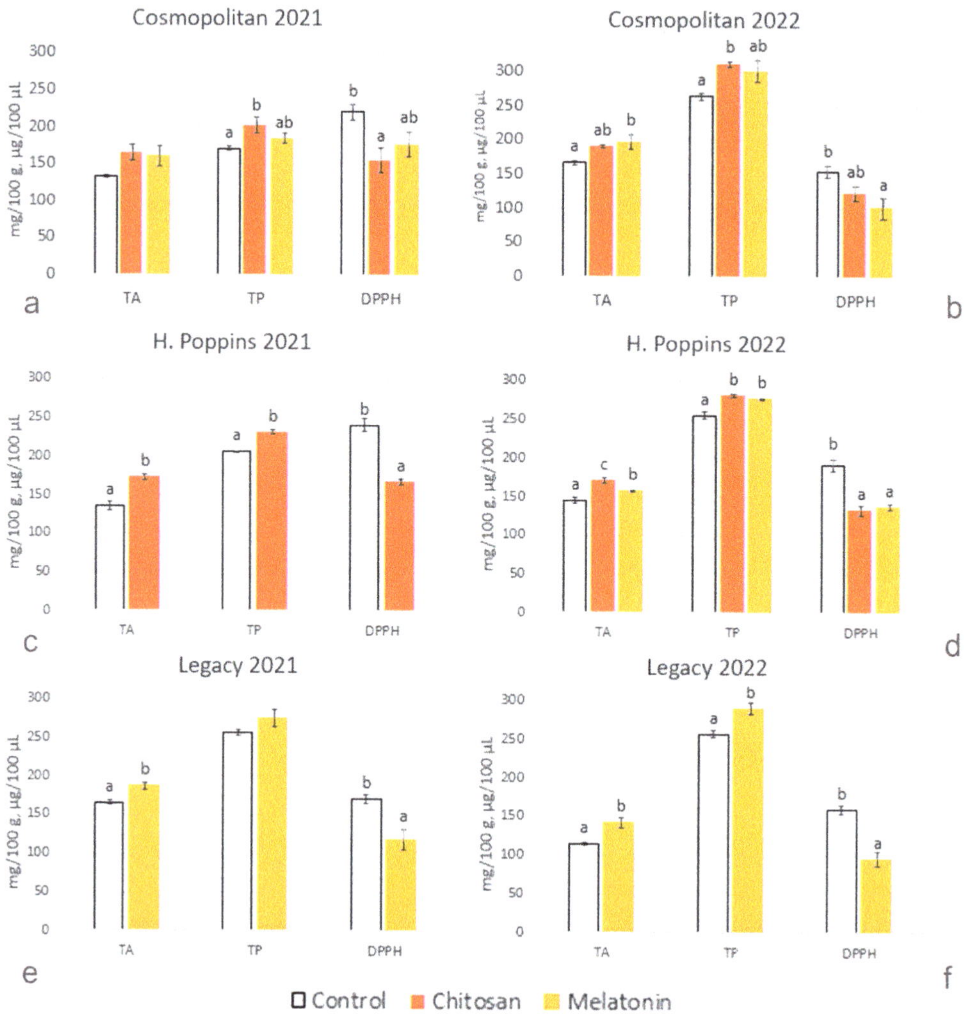

Figure 3. Effect of elicitor treatments on total anthocyanin content (AT, mg/100 g FW), total polyphenol content (TP, mg/100 g FW) and antioxidant activity (DPPH, IC 50%, µg/100 µL) in cvs. 'Cosmopolitan', 'H. Poppins' and 'Legacy' in two subsequent harvest years (2021 and 2022). Values represent means ± SE (n = 4). Different letters indicate statistical differences ($p \leq 0.05$) between treatments for each cultivar in each year.

2.2.6. Total Anthocyanins

Comparing total anthocyanins of the untreated cvs., in 2021, the fruits of 'Legacy' showed the highest content (166 mg/100 g FW) (Figure 3e), significantly greater than those of 'Cosmopolitan' and 'H. Poppins' (132 mg/100 g FW and 135 mg/100 g FW, respectively) (Figure 3a,c). Conversely, in 2022, 'Cosmopolitan' berries were the richest (166 mg/100 g FW) (Figure 3b) and 'Legacy' berries the poorest (114 mg/100 g FW) (Figure 3f). Consequently, comparing control fruits in the two years, an increase in anthocyanin content was found in 'Cosmopolitan', and a decrease was found in 'Legacy', while the levels remained stable in 'H. Poppins'. The positive effects of elicitor treatments were evident in all the cvs. in both years except for 'Cosmopolitan' in the first sampling year. Chitosan positively affected the anthocyanin content in 'H. Poppins' by 28% in 2021 and by 18% in 2022, while in 2022, the 'Cosmopolitan' treated berries showed intermediate levels not significantly different from the control nor from the melatonin-treated fruits. Melatonin stimulated anthocyanin accumulation in 'Legacy' in both years, i.e., by 13% in 2021 and by 25% in 2022. In 2022, melatonin treatment resulted in anthocyanin levels significantly increasing by 18% in 'Cosmopolitan' and 9% in 'H. Poppins'. Considering the differences in total anthocyanin levels between years, untreated 'Cosmopolitan' berries showed an increase in 2022 compared to 2021, while the treated ones did not. Anthocyanin levels of treated berries of 'Legacy' and 'H. Poppins', in contrast, mirrored those of control fruits, with a decrease and no change, respectively, between the two years.

2.2.7. Antioxidant Activity

The antioxidant capacity evaluated by DPPH is expressed as IC50. Lower IC50 values in the DPPH radical scavenging activity determination are related to reductions in sample concentrations required to scavenge the DPPH radical by 50%, indicating an improvement in the antioxidant capacity of the fruit phenolic extract.

Considering the antioxidant activity in untreated berries, in 2021, fruit of 'Legacy' showed significantly higher activity (171 μg/100 μL) (Figure 3e) than those of 'Cosmopolitan' (220 μg/100 μL) (Figure 3a) and 'H. Poppins' (241 μg/100 μL) (Figure 3c). On the other hand, in 2022, both 'Cosmopolitan' and 'Legacy' showed stronger antioxidant capacity (155 μg/100 μL and 160 μg/100 μL, respectively) (Figure 3b,f) than 'H. Poppins' (192 μg/100 μL) (Figure 3d).

Comparing the values across years, in 2022, the antioxidant capacities of untreated 'Cosmopolitan' and 'H. Poppins' were significantly higher (+29% and +20%) than in 2021, while that of untreated 'Legacy' did not change.

Elicitor treatments markedly enhanced the antioxidant activity in 'H. Poppins' (+30% in 2021 after chitosan treatment and +31% and +28% after chitosan and melatonin treatments, respectively, in 2022) and 'Legacy' (+31% in 2021 and +40% in 2022), while in 'Cosmopolitan', significant differences were reached by chitosan in 2021 (+30%) and by melatonin in 2022 (+35%).

2.2.8. Anthocyanin Profile

The total amounts of anthocyanins, calculated as the sum of the individual anthocyanins measured by the chromatographic profile, are consistent with the results obtained by the spectrophotometrical analysis. The identified anthocyanins were monogalactosides (gal), monoarabinosides (ara) and monoglucosides (glu) of delphinidin (Dp), cyanidin (Cy), petunidin (Pt), peonidin (Pn) and malvidin (Mv). In 'Cosmopolitan', the acetylated forms of Dp, Cy, Pt, Pn and Mv were detected, but these were lacking in 'H. Poppins' and 'Legacy' (Supplementary Tables S1 and S2). To better understand the variations in the anthocyanin composition, the identified anthocyanin compounds were grouped on the basis of the anthocyanidin present in the molecular structure. In all cvs., the predominant anthocyanidins were the tri-hydroxylated (in the B ring) Mv followed by Dp and Pt. The di-hydroxylated Cy and Pn were present in minor absolute and relative amounts.

Considering the anthocyanins of untreated berries, the highest amounts of Mv and Dp were detected in 2021 in 'Legacy' (693 μg/g FW) (Figure 4e) and 'Cosmopolitan' (420 μg/g FW) (Figure 4a), respectively, and in 2022, the highest amounts were detected in 'Cosmopolitan' (600 μg/g FW) (Figure 4b) and in 'H. Poppins' (341 μg/g FW) (Figure 4d). In 2021, Pt reached the highest levels in 'Legacy' (277 μg/g FW) (Figure 4e), and in 2022, Pt reached the highest levels in 'H. Poppins' (233 μg/g FW) (Figure 4d). In both years, the cv. richest in di-hydroxylated Cy and Pn was 'H. Poppins', with 55 and 17 μg/g FW, respectively, in 2021 (Figure 4c) and 60 and 22 μg/g FW in 2022 (Figure 4d). After elicitor treatments, the plant responses were different depending on the cv. and the harvest year. In 'Cosmopolitan', in 2021, only the content of Pt in chitosan-treated berries increased, while in 2022, the levels of Dp, Pt and Cy were enhanced by chitosan and melatonin and those of Pn only by melatonin. 'H. Poppins' responded to elicitors by increasing the content of all different classes of anthocyanidins in both years, except Cy in 2021. In 'Legacy', melatonin stimulated the accumulation of Mv, Cy and Pn in 2021, whereas in 2022, all types of anthocyanidins increased except Pn (Figure 4e,f).

In addition to the absolute contents, it is useful to express anthocyanin composition in relative terms. Considering the relative proportion of the single anthocyanins, the effects of the treatments with elicitors depended on the cv. and on the harvest year, similar to the case for the absolute concentrations but in different terms. In 'Cosmopolitan', in 2021, chitosan increased the percentage of Pt (Figure 5a), and in 2022, chitosan decreased that of the acetylated forms and of Pn (Figure 5b). In chitosan-treated 'H. Poppins' berries, all relative proportions of anthocyanidins were affected in 2021, with increases in tri-hydroxylated forms (sum of Mv, Dp, Pt), in methoxylated forms (sum of Mv, Pt, Pn) and in Mv and decreases in the di-hydroxylated forms (sum of Cy, Pn), Dp, Cy and Pt (Figure 5c). In 2022, however, chitosan only affected Pt with a decrease and Pn with an increase (Figure 5d). In 2021, the effect of melatonin on 'Legacy' consisted of enhanced proportions of the tri-hydroxylated and methoxylated forms, Pn and Mv and decreased proportions of di-hydroxylated anthocyanidins, Dp, Cy and Pt (Figure 5e). In contrast, in 2022, no variation was recorded for the treated berries (Figure 5f).

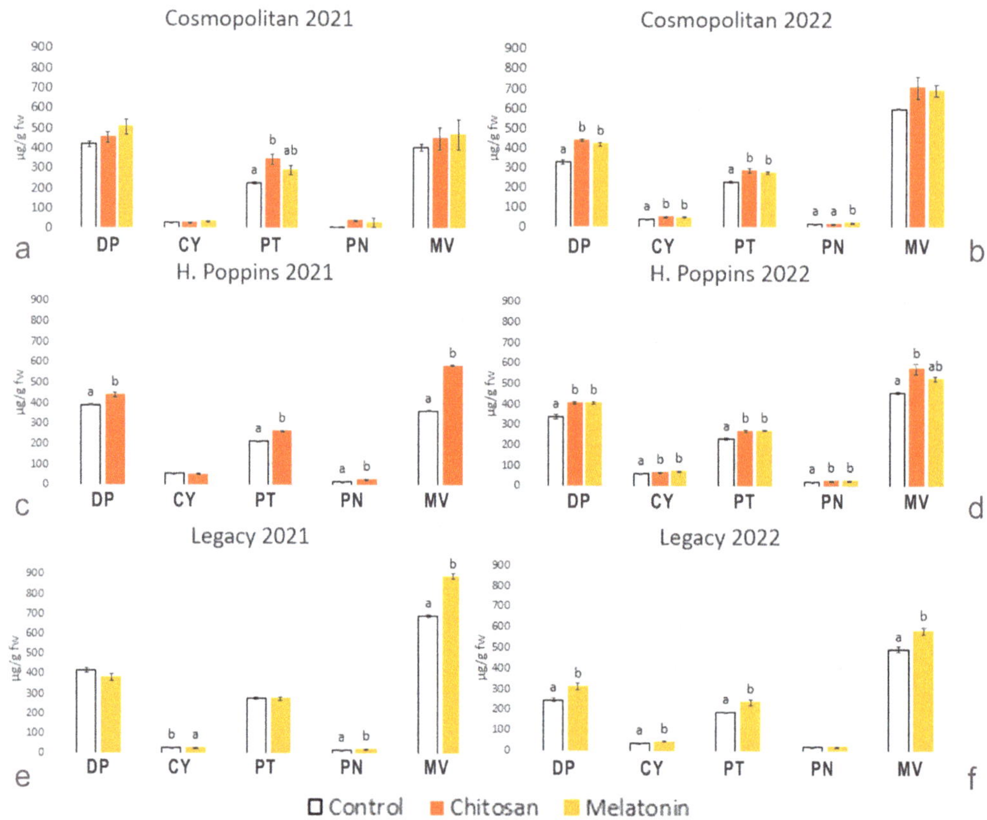

Figure 4. Effect of elicitor treatments on total amounts of individual anthocyanins (clustered into the representative anthocyanidin classes) in cvs. 'Cosmopolitan', 'H. Poppins' and 'Legacy' in two subsequent harvest years (2021 and 2022). Values represent means ± SE (n = 4). Different letters indicate statistical differences ($p \leq 0.05$) between treatments for each cultivar in each year.

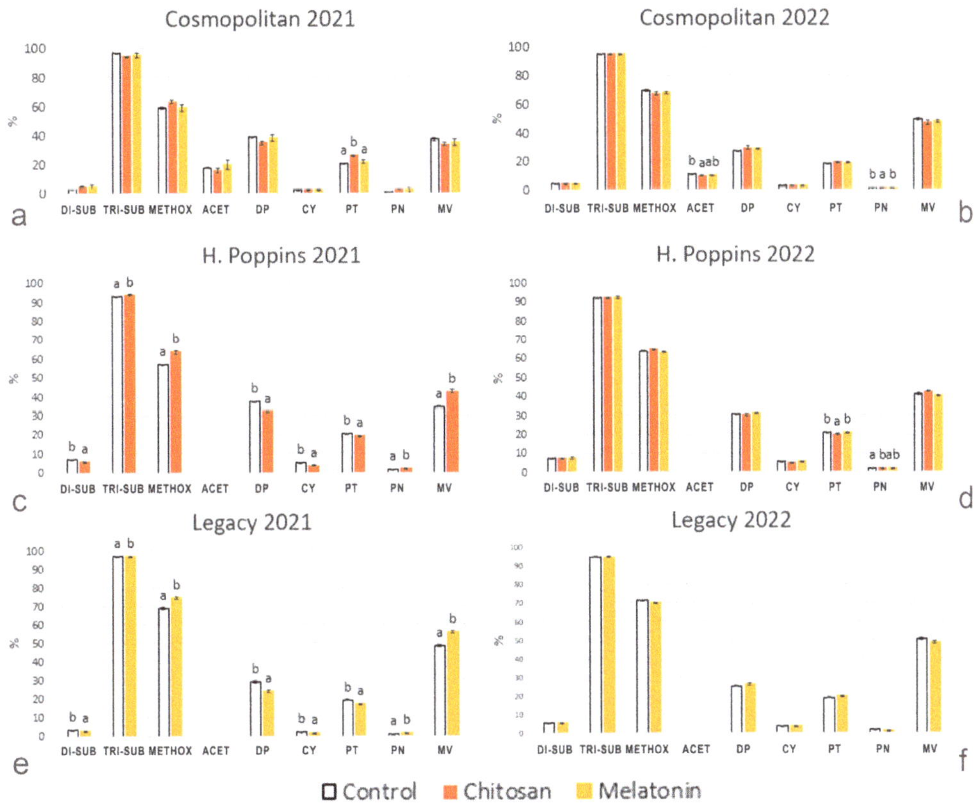

Figure 5. Effect of elicitor treatments on relative proportion of individual anthocyanins (clustered into the representative anthocyanidin classes) and of di-hydroxylated (DI-SUB), tri-hydroxylated (TRI-SUB), methoxylated (METHOX) and acetylated (ACET) forms in cv. 'Cosmopolitan', 'H. Poppins' and 'Legacy' in two subsequent harvest years (2021 and 2022). Values represent means ± SE (n = 4). Different letters indicate statistical differences ($p \leq 0.05$) between treatments for each cultivar in each year.

3. Discussion

Cropping systems are evolving towards agronomic strategies that can reduce the environmental impact and improve the sustainability of their production. Elicitors can act as priming stimuli to which a plant responds by acquiring a memory that boosts induced defences and resistance against a broad range of pathogens. This acquired protection could reduce the need for synthetic fungicides. Chemical elicitors such as chitosan and melatonin are non-toxic natural compounds and represent an attractive option for sustainable and organic agriculture, minimizing the impact on human health and the environment and contributing to the prevention of the development of pathogen resistance to certain fungicides, while covering the demand for residue-free plant products. Moreover, these substances can generally induce low-cost metabolic changes in the plant, including the synthesis of different classes of secondary metabolites such as phenolic compounds that contribute to the enhancement of antioxidant activity and nutritional quality, which leads to a longer storage potential and shelf life [14]. In relation to diet and human health, polyphenols are antioxidant bioactive compounds that are important in the prevention of chronic diseases such as cardiovascular disorders, cancers, diabetes and neurodegenerative diseases [57]. Therefore, elicitor treatments may represent a sustainable and inexpensive strategy to increase the nutraceutical quality of fruits. In our study, chitosan and melatonin proved to be effective in stimulating different blueberry plant responses. Regarding the effect of elicitor treatments on the quality parameters related to the ripening stage, 'Cosmopolitan' generally did not change these parameters and only total soluble solids were higher in 2021, while pH slightly decreased in 2022. On the other hand, 'H. Poppins' reacted more extensively to chitosan treatment. In 2021, all parameters were affected, i.e., dry matter and total soluble solids increased, while titratable acids diminished, altering pH, and in 2022, chitosan similarly affected titratable acidity by lowering it and pH by raising it. These data agree with previous studies that reported slight effects of chitosan total soluble solids and titratable acidity in blackberry [58]. The preharvest treatments of this study were conducted with a water-soluble chitosan. On the contrary, in many postharvest research studies, fruits were dipped in an acidic solution of a practical-grade chitosan, which forms a semi-permeable film on the surface of the fruit, decreasing gaseous exchange with a positive effect on titratable acidity and giving a glossy appearance to the fruit. In blueberries, a postharvest chitosan coating delayed ripening as indicated by lower respiration rate, higher total soluble solids content and titratable acidity [59]. Also, when used in preharvest, this formulation showed viscous properties that affect the internal gas composition of fruit by producing a film coating on the surface and maintaining a higher titratable acidity, as reported by Gayed et al. on peaches [32]. This property of chitosan as a film-forming solution and edible coating is absent in treatments with water-soluble chitosan, which, however, is more practical when sprayed and does not alter the appearance of the fruit. As for chitosan, melatonin did not exert in 'Cosmopolitan' any effect on ripening parameters in both years. In the same way, in 2021 and 2022, melatonin-treated 'Legacy' berries increased only the pH and the dry matter, respectively. On the other hand, in 2022, the response of 'H. Poppins' to melatonin was the same as shown after chitosan treatment, linked to a decrease in titratable acidity accompanied by a pH raise.

Differently from quality/ripening parameters, elicitor treatments markedly affected the levels of phenylpropanoid and antioxidant compounds. In particular, total anthocyanins and total phenolics accumulation were stimulated in all cvs. in both years. Considering chitosan treatment, in 'Cosmopolitan', chitosan increased total phenolics and influenced total anthocyanin content and antioxidant capacity positively, although not significantly. In 'H. Poppins', the effect of chitosan was more pronounced, and all the parameters linked to nutraceutical content, i.e., total phenolic and anthocyanin contents and antioxidant capacity, were enhanced. Melatonin caused similar responses to chitosan treatment in 'Cosmopolitan' in both years and in 'H. Poppins' in 2022, enhancing phenolic and anthocyanin accumulation and increasing antioxidant potential. The melatonin-treated fruit of 'Legacy' also showed a significant improvement in nutraceutical and antioxidant properties.

Comparing the two harvesting years, the behaviour of 'Cosmopolitan' and 'H. Poppins' was similar. In 2022, the berries of both cvs. showed, compared to 2021, a significant increase in total phenolics, accompanied by an enhancement in antioxidant capacity and a considerable reduction in titratable acidity linked to a raise in pH. On the contrary, in 'Legacy' in 2022, the levels of phenolic compounds and antioxidant power remained unchanged compared to 2021, but the amounts of total anthocyanins declined. Likewise, Feliziani et al. reported differences in the effectiveness of preharvest treatments with inducers in controlling postharvest decay between two strawberry varieties due to possible activation of cultivar-specific pathways [33]. Different responses to a specific elicitor treatment were also observed among six clones of Monastrell grape, suggesting the need to conduct preliminary exploratory field research to evaluate the possible variation in responses between clones from the same variety and between seasons, since clone–environment interactions may also exist [60]. Moreover, genetic background affected the responses to other elicitors such as benzothiadiazole in mango [61] and jasmonic acid seed treatment in tomato [62].

Regarding the impact of elicitor treatments on the anthocyanin profile, a positive effect is shown for total anthocyanin accumulation. The effects of chitosan and melatonin were quite comparable and, generally, the amounts of all anthocyanin classes improved in both years, largely in 2022. In addition to absolute concentrations, it is useful to evaluate the composition of anthocyanin in relative terms. Considering the relative proportion of individual anthocyanins, responses to elicitor treatments were different among cvs. and even more markedly different between years. 'Cosmopolitan' generally did not change in anthocyanin proportions after treatments in both years, while in 2021, an enhancement in the proportion of tri-substituted and methoxylated molecular structures was observed in 'H. Poppins' and 'Legacy', mainly caused by a steep increase in Mv that partially occurred at the expenses of Dp and Pt. Moreover, Cy decreased, while Pn slightly increased. On the contrary, in 2022, both 'H. Poppins' and 'Legacy' did not change the anthocyanin profile in response to treatments. In particular, treatments did not cause the marked increase in Mv seen in 2021 and, consequently, no significant decreases in the proportions of the other tri-substituted Dp and Pt were observed.

These contradictory data obtained in the two growing years are mainly attributable to the different climatic conditions of the two harvest years. In 2022, the higher temperatures experienced by the plants during the last ripening period determined a heat stress. Metabolic processes such as respiration and photosynthesis are very sensitive to temperature increases. Heat stress, which is often accompanied by drought, may significantly decrease stomatal conductance and cause higher respiration rates at the expense of organic acids, the main substrates of respiration in blueberry fruit [63]. In fact, the higher temperatures recorded in 2022 compared to 2021 resulted in a significant reduction in the acidity of the fruit of 'Cosmopolitan' and 'H. Poppins', consequently increasing the pH value. Similar effects are reported in several studies on grape [64,65]. Conversely, these cvs. showed no changes in anthocyanin content and higher phenolic levels in 2022, unlike grapes, which generally respond to high temperatures (over 30 °C) by lowering anthocyanin content [63]. 'Legacy' showed the opposite behaviour with respect to 'Cosmopolitan' and 'H. Poppins' and showed higher acidity and lower anthocyanin content in 2022 compared to 2021. In 'Legacy', heat stress may not affect respiration but rather anthocyanin metabolism.

Blueberries are typically classified into different groups depending on their chilling requirements for the onset of flowering. Highbush blueberry varieties with high chilling requirements, typically grown in cold winter regions, are named northern highbush. Instead, cvs. with low chilling requirements (less than about 550 chilling hours, <7 °C) are named southern highbush and are interspecific hybrids derived from crosses between the northern highbush and *Vaccinium* species both native and adapted to the southeastern United States with lower chilling requirements, such as *V. darrowii* Camp. 'Cosmopolitan' and 'H. Poppins' are northern highbush blueberries, whereas 'Legacy' belongs to the southern group [66].

These different genetic backgrounds affected both heat stress responses and elicitor treatment effects. A marked difference concerns the acidity of the fruit in the two different years. Decreases in organic acid content due to high temperatures were likely associated in the two northern highbush blueberries with higher respiration rates, in contrast to the southern highbush 'Legacy'. Modifications in organic acid levels may be associated with different respiration rates but also with different plant resource allocations. In fact, in 2022, characterized by higher summer temperatures and heat stress compared to 2021, 'Cosmopolitan' and 'H. Poppins' had a higher total phenolic accumulation and total anthocyanin content remained unchanged, while 'Legacy' showed no changes in levels of phenols and allocated fewer resources to the accumulation of anthocyanins. However, a different response to elicitor treatment was shown by 'H. Poppins' and 'Legacy', which could imply a different allocation of resources. The response to treatment in 'H. Poppins' is a marked increase in the content of anthocyanins and total phenolics accompanied by a decrease in acidity, while in 'Legacy', the increase in the phenolic compounds is not associated with a variation in acidity.

In berries, apart from total anthocyanin concentrations, changes in the contents and relative proportions of individual anthocyanins were related to high temperature stress.

In 'Cosmopolitan' and in 'H. Poppins', compositional changes related to heat stress were associated with the increased proportions of Mv and reduced proportions of Dp, as reported also by many authors in grapevine [64,65,67]. Moreover, in accordance with these changes, Timmers et al. [68] observed that across highbush blueberry genotypes, the percentages of Dp and Mv glycosides mirrored each other in the relative proportion during the harvest season, with an increasing trend in Mv percentage during almost the entire harvest period and a parallel decrease in Dp. These two forms of anthocyanins are related because they are found within the same section of the anthocyanin biosynthetic pathway. In fact, Dp glycosides undergo two-step methylation to form Mv glycosides. The intermediate in this two-step methylation is Pt, which decreased in relative content in 2022 compared to 2021 in 'Cosmopolitan' but not in 'H. Poppins'.

In the warmer year of 2022, 'Legacy' showed an increase in the relative levels of the methoxylated forms and also, but not reaching significance, in Mv and a decrease in Dp proportions in untreated berries. On the other hand, it showed a decrease in relative levels of the methoxylated forms and Mv and a parallel increase in the proportions of Dp in treated fruit. 'Cosmopolitan' and 'H. Poppins' showed also increased proportions of methoxylated pigments in 2022 regardless of treatment. Berries of 'Cosmopolitan' accumulated acetylated anthocyanins in higher proportions under heat stress conditions. A similar response to high summer temperatures was reported in grape. Three grape varieties under a two-crop-a-year cultivation system accumulated significantly higher levels of methoxylated anthocyanins, and one variety also accumulated acetylated anthocyanins in summer compared to winter [69]. A progressive increase in the proportion of the more stable tri-substituted and methoxylated pigments occurs also during ripening [70].

Considering the anthocyanin content in absolute terms, in 'Cosmopolitan' and 'H. Poppins', the changes across years in Mv and Dp reflected those shown in relative terms and are in line with that published by Venios et al. [64]. Moreover, in all the fruits of the two cvs., the accumulation of Cy was higher under conditions of greater thermal stress, i.e., in 2022, as reported in Kyoho grape by Cheng et al. [71].

The antioxidant capacity mirrored the contents of anthocyanins and phenolic compounds and was always higher in treated berries, regardless of cv. and harvest year, and it is linked to enhanced fruit quality. The positive effect of elicitor treatments on the accumulation of nutraceutical antioxidant compounds did not lead to significantly different IC50 values only in the case of 'Cosmopolitan' in 2021, although the trend is similar to the other responses to treatment. Moreover, similar beneficial effects on phenolic content and antioxidant activity were obtained with melatonin preharvest application on Merlot grape [45], sweet cherry [72] and Elvira blackberry [73]. The stimulation of phenol and anthocyanin accumulation and an increase in antioxidant activity after preharvest treatment with chitosan was also reported by He et al. on strawberry [74], by Lo Piccolo (2023) on raspberries [75] and by Griñán et al. on pomegranate [76]. Antioxidant phytochemicals play an important role in the prevention and treatment of chronic diseases, and the consumption of fruits with high antioxidant capacity is, therefore, beneficial [77].

4. Materials and Methods

4.1. Chemicals

Standards of cyanidin (Cy)-, delphinidin (Dp)-, petunidin (Pt)-, peonidin (Pn)-, malvidin (Mv)- and their 3-O-glucoside (glc), Cy-, Pn-, Pt-, Mv-3-O-galactoside (gal) and Cy-arabinoside (Cy-ara) were purchased from Polyphenols Laboratory (Sandnes, Norway). Methanol, acetonitrile, phosphoric, trifluoroacetic acid (TFA) and melatonin were from Merck (Darmstadt, Germany). Tween was from Sigma-Aldrich (Saint Louis, MO, USA), and chitosan was provided by SKL Fertilizzanti (Chieti, Italy). Soil for acidophilic plants was from Vegetal Radic Acid (Tercomposti S.P.A., Calvisano, Italy). Water was from Milli-Q apparatus (Millipore, Milford, MA, USA).

4.2. Plant Material and Treatments

Blueberry plants (*Vaccinium corymbosum* L.) were grown in Arcagna (45.33992° N; 9.45209° E) in Northern Lombardy (Italy) at 83 m above sea level. Seven-year-old plants of three different cultivars, 'Cosmopolitan', 'Hortblue Poppins' ('H. Poppins') and 'Legacy' were grown in 70 L pots in a soil for acidophilic plants, optimal for blueberry growth. Conventional farming practices and micro-irrigation were carried out in the field. Berry samples were collected 24 h after elicitor spray treatments with 1% chitosan or 100 μM melatonin + Tween 0.01%. Treatments were performed every two weeks starting from the appearance of fruit pigmentation throughout the growing season (in a total of 3 applications, with each application performed until dripping). Fruits from plants treated with water were also collected as a control. 'Cosmopolitan', an early ripening variety, is harvested in Northern Italy in mid-to-late June, whereas 'H. Poppins' and 'Legacy', mid-season varieties, ripen between the end of June and the first week of July. In 2021, fully pigmented ripe berries of 'Cosmopolitan' were collected on June 11th, of 'H. Poppins' on 21 June and of 'Legacy' on 2 July. In 2022, 'Cosmopolitan' and 'H. Poppins' were sampled on 27 June and 'Legacy' on 30 June. After elicitor application, four treated plants and four control plants were randomly selected for sampling. Plants were uniform in vegetative development and size. Consequently, the pool of berries harvested from the same plant represented one biological replicate for each cultivar, i.e., each sample subjected to chemical analysis was taken from the fruit pool of a single plant. Of each cv., the four treated and the four untreated plants provided the four control and the four treatment samples. In 2021, chitosan was applied to 'Cosmopolitan' and 'H. Poppins' and melatonin to 'Legacy'. In 2022, the treatments were replicated as in 2021 and a treatment with melatonin to 'H. Poppins' was added to the experimental plan.

The total number of samples was 60 (in 2021, 4 individual plants treated or not with chitosan for 'H. Poppins', 4 individual plants treated or not with melatonin for 'Legacy' and 4 individual plants treated or not with chitosan or melatonin for 'Cosmopolitan'; in 2022, 4 individual plants treated or not with chitosan or melatonin for 'Cosmopolitan' and 'H. Poppins' and 4 individual plants treated or not with melatonin for 'Legacy'). Harvested fruits were then weighed, placed in plastic bags and stored at −80 °C until laboratory analysis.

4.3. Thermal and Radiative Analysis

The characterization of the environmental conditions across the two experimental seasons was based on the analysis of meteorological data collected by the weather station of Sant'Angelo Lodigiano, belonging to the ARPA Lombardia meteorological network [78]. The station is 8 km from the experimental site and provides a long time series of data, allowing a robust analysis of seasonal anomalies. The agreement between Sant'Angelo Lodigiano and the experimental farm data was previously verified with specific monitoring campaigns and subsequent comparison activities. Seasons 2021 and 2022 were characterized based on the climatic normal of 1991–2020. The characterization focuses on radiative and thermal analysis, since the drip irrigation system provided the optimal water supply during the two experimental seasons.

The thermal course of the two seasons was described by means of an anomaly classification for maximum and minimum daily temperature. The temperature (T) of each day of the year was classified based on the climatic normal statistics as follows:

1. NAS—Strong negative anomaly: $T < AVG - 2SD$
2. NAM—Mild negative anomaly: $AVG - 2SD \leq T < AVG - 1SD$
3. NORM—No anomaly: $AVG - 2SD \leq T < AVG - 1SD$
4. PAM—Mild positive anomaly: $AVG - 1SD \leq T < AVG + 1SD$
5. PAS—Strong positive anomaly: $T \geq AVG + 2SD$

where AVG and SD are the average and the standard deviation of the dataset of temperature for each day of the year in the 1991–2020 period.

The evaluation of the environmental resources and limitations for blueberry across the two seasons was based on the calculation of the following indices:

- GSR—Summation of daily global solar radiation [MJ m^{-2}]
- GDD7—Accumulation of growing degree days with base temperature 7 °C [79] [GDD]. GDD represents the thermal resources that are useful for the development of blueberry. However, the absence of upper thresholds leads this index to the overestimation of the positive effect of high temperatures.
- NHH—Accumulation of normal heat hours [56,80], with a cardinal minimum of 7 °C [79], an optimal range between 20 °C and 26 °C [81] and a cardinal maximum of 32 °C [82] [NHH]. This approach weighs the hourly temperature by considering under-optimal, optimal and over-optimal temperatures.
- HHH—Accumulation of thermal excess [56,80] based on the same parameterization of NHH [HHH]. This is related to the previous index and accounts for the conditions of thermal excess.

4.4. *Fruit Quality Parameters*

4.4.1. Dry Matter Content

To measure the dry matter content of the samples, six intact berries for each sample were freeze dried and weighed before and after the procedure until no more decrease in weight was recorded.

4.4.2. Total Soluble Solids

Total soluble solids, expressed as °Brix, were determined by a hand refractometer (Atago mod., N1, Tokyo, Japan) on the juice obtained from squeezing 5 g of berries.

4.4.3. Total Titratable Acidity and pH

Titratable acidity, expressed as meq 100 g FW^{-1}, was measured by titrating 100 mL samples of 10% extracted juice with 0.1 M NaOH to pH 8.2 with an automatic sample titrator (TitroMatic 2S-3B, Crison Instruments, Barcelona, Spain).

4.4.4. Total Phenolics and Total Anthocyanins Analysis

Approximately 10 g of frozen berries of equal size were ground in 20 mL of cold extraction solution containing EtOH/HCOOH/H_2O (25/2/73). The homogenate was placed in an ultrasonic bath for 15 min. The suspension was centrifuged at 10,000× *g* for 20 min at 4 °C, and the supernatant was recovered. The pellet was resuspended with 10 mL of extraction solution and treated as described above. Then, the supernatants were combined, and the volume was adjusted to 40 mL by extraction solution. All extracts were stored at −80 °C before spectrophotometric analysis.

Total phenolic contents were determined following the Folin–Ciocalteau method. Briefly, 0.5 mL of Folin–Ciocalteau reagent, 4.45 mL of distilled water and 2 mL of 10% Na_2CO_3 were added to 0.05 mL of extract. The solution was immediately diluted to a final volume of 10 mL with distilled water. The optical density, after 90 min, was measured at 700 nm on a 10S UV-vis spectrophotometer (Thermo-Scientific, Waltham, MA, USA). Results were expressed as milligrams of gallic acid per 100 g FW. For all evaluations, duplicate reactions per replicate were carried out. Total anthocyanins were estimated on a 10S UV-vis spectrophotometer (Thermo-Scientific, Waltham, MA, USA) by measuring the absorption peak of the anthocyanin pigments at 530 nm and expressed as cyanidin 3-glucoside using a molar extinction coefficient of 26.900 L mol^{-1} cm^{-1} and reported in milligrams per 100 g FW.

4.4.5. Antioxidant Activity

The free radical scavenging capacity of blueberry extracts was evaluated by the DPPH method. Four different dilutions of each extract obtained for total phenolic and anthocyanin analysis were tested, adding 430 µL of a DPPH solution (296 µM) to each sample diluted in ethanol [83]. The absorbances of the solutions were measured at 515 nm after incubation in the dark at 25 °C for 60 min. The scavenging rate was calculated as IC 50% for each of the extracts with the following formula:

$$IC\% = \frac{Asample - Ablank}{Acontrol - Ablank} \times 100$$

Asample: absorbance of the extract;
Acontrol: absorbance of the control reaction (reagents except the extract);
Ablank: ethanol absorbance.

4.4.6. Anthocyanin Determination by UHPLC-DAD-HR-MS

The extract obtained for the analysis of total anthocyanins was also used, after dilution, for the qualitative and quantitative determination of individual anthocyanins. The chromatographic analysis was performed by a UHPLC Vanquish Flex (Thermo) coupled to a Vanquish HL PDA (Thermo) and an HR-MS Orbitrap mod. Q Exactive model Focus (Thermo) equipped with an HESI-II probe for ESI. The separation was carried out using a 2.6 µm Kinetex C_{18} column (150 mm × 4.6 mm, Phenomenex, Torrence, CA, USA) protected with a guard column at 1.7 mL/min and a flow rate split 5:1 before exposure to an electrospray ionization (ESI) source. The column and sample were maintained at 45 and 20 °C, respectively. The eluents were (A) 0.2% TFA in water and (B) acetonitrile 0.2% TFA in water (35:65, *v*/*v*). The linear gradient was 0–15 min 14% B; 15–25 min from 14% to 20% B; 25–35 min from 20 to 32% B; 35–45 min from 32 to 50% B; 45–48 min 50 to 90% B; and 90% for 3 min. The HR-MS operative conditions were as follows: spray voltage +4.0 kV, sheath gas flow rate 60 (arbitrary units), auxiliary gas flow rate 20 (arbitrary units), capillary temperature 350 °C, capillary voltage +30 V, S-lens +80 V, and heater temperature 130 °C. The MS data were processed using Xcalibur 4.1 Software (Thermo Scientific, Rodano, MI, Italy). Mass spectrometry was performed through the infusion of a 5 µg/mL cyanidin-3-O-glucoside solution. The calibration curve in the range of 0.2–20 µg/mL was used for ACN quantification. Peaks were identified by evaluating accurate mass, fragments obtained in the collision cell, and the online UV spectra (220–650 nm).

4.5. Statistical Analysis

All data were subjected to ANOVA performed in IBM SPSS Statistics software, version 25 (SPSS Inc., Chicago, IL, USA), using general linear model univariate analysis with treatments, harvesting year, or cultivar as fixed factors. Significant differences among means were calculated by Tukey's post hoc test. Differences at $p \leq 0.05$ were considered as significant. Additional information is reported in the figure legends.

5. Conclusions

The results of the present study indicate that the preharvest spraying of melatonin and chitosan had a beneficial impact on improving the quality of blueberry in terms of nutraceutical content related to the content of anthocyanin and phenols and in terms of the antioxidant activity of fruit. The effect of elicitor treatments on other quality attributes, i.e., titratable acidity and anthocyanin composition, also depended on climatic factors and on genetic background, as diverse cvs. responded variously in harvest years characterized by different climatic conditions. The consumption of berries with increased polyphenol content and antioxidant activity has a beneficial effect on human health, with a positive impact on the prevention of different chronic diseases. To our knowledge, this is the first report showing the impact of preharvest treatments with melatonin and chitosan on blueberry fruit quality at harvest. The objective in using these elicitors is to minimize the impact on human health and the environment of using conventional agrochemicals to prevent fruit spoilage by fungal pathogens, obtaining residue-free fruits and, as shown in this research, improving quality attributes. This sustainable strategy can be successfully adopted, keeping in mind that some differences in the response to treatments are linked to the different genetic backgrounds of the blueberry varieties in relation also to the different climatic parameters to which they are subjected.

Supplementary Materials: The following supporting information can be downloaded at: https://www.mdpi.com/article/10.3390/plants13081105/s1, Table S1: Statistics of Anthocyanins (Anthocyanin profile); Table S2: Chromatogram integrated at 515 nm of blueberry fruits (*Vaccinium corymbosum* L.) cv 'Cosmopolitan'.

Author Contributions: Conceptualization, G.C. and A.S.; Methodology, G.C., C.S.G. and A.S.; Formal analysis, B.C. and C.S.G.; Data curation, C.G, G.C. and A.S.; Writing—original draft, A.S.; Writing—review & editing, G.C. and A.S. All authors have read and agreed to the published version of the manuscript.

Funding: This research received no external funding. The authors acknowledge the support of the APC central fund of the University of Milan.

Data Availability Statement: Data are contained within the article and Supplementary Materials.

Conflicts of Interest: The authors declare no conflict of interest.

References

1. Abdul Malik, N.A.; Kumar, I.S.; Nadarajah, K. Elicitor and receptor molecules: Orchestrators of plant defense and immunity. *Int. J. Mol. Sci.* **2020**, *21*, 963. [CrossRef]
2. Reglinski, T.; Havis, N.; Rees, H.J.; de Jong, H. The practical role of induced resistance for crop protection. *Phytopathology®* **2023**, *113*, 719–731. [CrossRef]
3. Palou, L.; Ali, A.; Fallik, E.; Romanazzi, G. GRAS, plant-and animal-derived compounds as alternatives to conventional fungicides for the control of postharvest diseases of fresh horticultural produce. *Postharvest. Biol. Technol.* **2016**, *122*, 41–52. [CrossRef]
4. Desmedt, W.; Kudjordjie, E.N.; Chavan, S.N.; Desmet, S.; Nicolaisen, M.; Vanholme, B.; Vestergård, M.; Kyndt, T. Distinct chemical resistance-inducing stimuli result in common transcriptional, metabolic, and nematode community signatures in rice root and rhizosphere. *J. Exp. Bot.* **2022**, *73*, 7564–7581. [CrossRef] [PubMed]
5. Desmedt, W.; Vanholme, B.; Kyndt, T. Plant defense priming in the field: A review. In *Recent Highlights in the Discovery and Optimization of Crop Protection Products*; Academic Press: Cambridge, MA, USA, 2021; pp. 87–124.
6. Martinez-Medina, A.; Flors, V.; Heil, M.; Mauch-Mani, B.; Pieterse, C.M.; Pozo, M.J.; Ton, J.; van Dam, N.M.; Conrath, U. Recognizing plant defense priming. *Trends Plant Sci.* **2016**, *21*, 818–822. [CrossRef] [PubMed]
7. Tripathi, D.K.; Singh, V.P.; Brestič, M.; Deshmukh, R.; Vaculik, M. Priming-mediated abiotic stress management in plants: Recent avenues and future directions. *Plant Stress* **2022**, *5*, 100097. [CrossRef]
8. Wisniewski, M.; Droby, S.; Norelli, J.; Liu, J.; Schena, L. Alternative management technologies for postharvest disease control: The journey from simplicity to complexity. *Postharvest. Biol. Technol.* **2016**, *122*, 3–10. [CrossRef]
9. Zhang, W.; Jiang, W. UV treatment improved the quality of postharvest fruits and vegetables by inducing resistance. *Trends Food Sci. Technol.* **2019**, *92*, 71–80. [CrossRef]
10. Ahuja, I.; Kissen, R.; Bones, A.M. Phytoalexins in defense against pathogens. *Trends Plant Sci.* **2012**, *17*, 73–90. [CrossRef]
11. La Camera, S.; Gouzerh, G.; Dhondt, S.; Hoffmann, L.; Fritig, B.; Legrand, M.; Heitz, T. Metabolic reprogramming in plant innate immunity: The contributions of phenylpropanoid and oxylipin pathways. *Immunol. Rev.* **2004**, *198*, 267–284. [CrossRef]
12. Yadav, V.; Wang, Z.; Wei, C.; Amo, A.; Ahmed, B.; Yang, X.; Zhang, X. Phenylpropanoid pathway engineering: An emerging approach towards plant defense. *Pathogens* **2020**, *9*, 312. [CrossRef]
13. Desmedt, W.; Jonckheere, W.; Nguyen, V.H.; Ameye, M.; De Zutter, N.; De Kock, K.; Debode, J.; Van Leeuwen, T.; Audenaert, K.; Vanholme, B.; et al. The phenylpropanoid pathway inhibitor piperonylic acid induces broad-spectrum pest and disease resistance in plants. *Plant Cell Environ.* **2021**, *44*, 3122–3139. [CrossRef]
14. Ruiz-García, Y.; Gómez-Plaza, E. Elicitors: A tool for improving fruit phenolic content. *Agriculture* **2013**, *3*, 33–52. [CrossRef]
15. Kalt, W.; Cassidy, A.; Howard, L.R.; Krikorian, R.; Stull, A.J.; Tremblay, F.; Zamora-Ros, R. Recent research on the health benefits of blueberries and their anthocyanins. *Adv. Nutr.* **2020**, *11*, 224–236. [CrossRef]
16. Silva, S.; Costa, E.M.; Veiga, M.; Morais, R.M.; Calhau, C.; Pintado, M. Health promoting properties of blueberries: A review. *Crit. Rev. Food Sci.* **2020**, *60*, 181–200. [CrossRef]
17. Rabea, E.I.; Badawy, M.E.T.; Stevens, C.V.; Smagghe, G.; Steurbaut, W. Chitosan as antimicrobial agent: Applications and mode of action. *Biomacromolecules* **2003**, *4*, 1457–1465. [CrossRef]
18. Li, K.; Xing, R.; Liu, S.; Li, P. Chitin and chitosan fragments responsible for plant elicitor and growth stimulator. *J. Agric. Food Chem.* **2020**, *68*, 12203–12211. [CrossRef]
19. Aranaz, I.; Acosta, N.; Civera, C.; Elorza, B.; Mingo, J.; Castro, C.; De los Llanos Gandìa, M.; Heras Caballero, A. Cosmetics and cosmeceutical applications of chitin, chitosan and their derivatives. *Polymers* **2018**, *10*, 213. [CrossRef]
20. Swiatkiewicz, S.; Swiatkiewicz, M.; Arczewska-Wlosek, A.; Jozefiak, D. Chitosan and its oligosaccharide derivatives (chito-oligosaccharides) as feed supplements in poultry and swine nutrition. *J. Anim. Physiol. Anim. Nutr.* **2015**, *99*, 1–12. [CrossRef]

21. Ferri, M.; Tassoni, A. Chitosan as elicitor of health beneficial secondary metabolites in in vitro plant cell cultures. In *Handbook of Chitosan Research and Applications*; Mackay, R.G., Tait, J.M., Eds.; Nova Science Publishers: Hauppauge, NY, USA, 2012; pp. 389–414.
22. Coelho, N.; Romano, A. Impact of chitosan on plant tissue culture: Recent applications. *Plant Cell Tissue Organ Cult. (PCTOC)* **2022**, *148*, 1–13. [CrossRef]
23. Gutiérrez-Martínez, P.; Bautista-Baños, S.; Berúmen-Varela, G.; Ramos-Guerrero, A.; Hernández-Ibañez, A.M. In vitro response of Colletotrichum to chitosan. Effect on incidence and quality on tropical fruit. Enzymatic expression in mango. *Acta Agronómica* **2017**, *66*, 282–289. [CrossRef]
24. Bautista-Baños, S.; Hernandez-Lauzardo, A.N.; Velazquez-Del Valle, M.G.; Hernández-López, M.; Barka, E.A.; Bosquez-Molina, E.; Wilson, C.L. Chitosan as a potential natural compound to control pre and postharvest diseases of horticultural commodities. *Crop Prot.* **2006**, *25*, 108–118. [CrossRef]
25. Gutiérrez-Martínez, P.; Ramos-Guerrero, A.; Rodríguez-Pereida, C.; Coronado-Partida, L.; Angulo-Parra, J.; González-Estrada, R. Chitosan for postharvest disinfection of fruits and vegetables. In *Postharvest Disinfection of Fruits and Vegetables*; Academic Press: Cambridge, MA, USA, 2018; pp. 231–241.
26. Xoca-Orozco, L.Á.; Cuellar-Torres, E.A.; González-Morales, S.; Gutiérrez-Martínez, P.; López-García, U.; Herrera-Estrella, L.; Vega-Arreguìn, J.; Chacón-López, A. Transcriptomic analysis of avocado hass (*Persea americana* Mill) in the interaction system fruit-chitosan-Colletotrichum. *Front. Plant Sci.* **2017**, *8*, 956. [CrossRef]
27. Yan, J.Q.; Cao, J.K.; Jiang, W.B.; Zhao, Y.M. Effects of preharvest oligochitosan sprays on portharvest fungal disease, storage quality, and defense responses in jujube (*Zizyphus jujuba* Mill. cv. Dongzao) fruit. *Sci. Hortic.* **2012**, *142*, 196–204. [CrossRef]
28. Romanazzi, G.; Nigro, E.; Ippolito, A.; Di Venere, D. Effects of pre- and postharvest chitosan treatments to control storage grey mold of table grapes. *J. Food Sci.* **2002**, *67*, 1862–1867. [CrossRef]
29. Dos Santos, N.S.T.; Athayde Aguiar, A.J.A.; de Oliveira, C.E.V.; Verissimo de Sales, C.; de Melo E Silva, S.; Sousa da Silva, R.; Stamford, T.C.M.; de Souza, E.L. Efficacy of the application of a coating composed of chitosan and *Origanum vulgare* L. essential oil to control *Rhizopus stolonifer* and *Aspergillus niger* in grapes (*Vitis labrusca* L.). *Food Microbiol.* **2012**, *32*, 345–353. [CrossRef]
30. Felipini, R.B.; Boneti, J.I.; Katsurayama, Y.; Neto, A.C.R.; Veleirinho, B.; Maraschin, M.; Di Piero, R.M. Apple scab control and activation of plant defence responses using potassium phosphite and chitosan. *Eur. J. Plant Pathol.* **2016**, *145*, 929–939. [CrossRef]
31. Giacalone, G.; Chiabrando, V. Effect of preharvest and postharvest application of chitosan coating on storage quality of nectarines. *Acta Hortic.* **2015**, *1084*, 675–680. [CrossRef]
32. Gayed, A.A.N.A.; Shaarawi, S.A.M.A.; Elkhishen, M.A.; Elsherbini, N.R.M. Pre-harvest application of calcium chloride and chitosan on fruit quality and storability of 'Early Swelling'peach during cold storage. *Cienc. Agrotec.* **2017**, *41*, 220–231. [CrossRef]
33. Feliziani, E.; Santini, M.; Landi, L.; Romanazzi, G. Pre-and postharvest treatment with alternatives to synthetic fungicides to control postharvest decay of sweet cherry. *Postharvest. Biol. Technol.* **2013**, *78*, 133–138. [CrossRef]
34. Ahmed, Z.F.; Alblooshi, S.S.; Kaur, N.; Maqsood, S.; Schmeda-Hirschmann, G. Synergistic effect of preharvest spray application of natural elicitors on storage life and bioactive compounds of date palm (*Phoenix dactylifera* L.; cv. Khesab). *Horticulturae* **2021**, *7*, 145. [CrossRef]
35. Cui, K.; Shu, C.; Zhao, H.; Fan, X.; Cao, J.; Jiang, W. Preharvest chitosan oligochitosan and salicylic acid treatments enhance phenol metabolism and maintain the postharvest quality of apricots (*Prunus armeniaca* L.). *Sci. Hortic.* **2020**, *267*, 109334. [CrossRef]
36. Migliori, C.A.; Salvati, L.; Di Cesare, L.F.; Scalzo, R.L.; Parisi, M. Effects of preharvest applications of natural antimicrobial products on tomato fruit decay and quality during long-term storage. *Sci. Hortic.* **2017**, *222*, 193–202. [CrossRef]
37. Zheng, J.; Chen, H.; Wang, T.; Mustafa, G.; Liu, L.; Wang, Q.; Shao, Z. Quality improvement of tomato fruits by preharvest application of chitosan oligosaccharide. *Horticulturae* **2023**, *9*, 300. [CrossRef]
38. Li, Z.; Xue, S.; Xu, X.; Wang, B.; Zheng, X.; Li, B.; Xie, P.; Bi, Y.; Prusky, D. Preharvest multiple sprays with chitosan accelerate the deposition of suberin poly phenolic at wound sites of harvested muskmelons. *Postharvest. Biol. Technol.* **2021**, *179*, 111565. [CrossRef]
39. Tan, D.X.; Hardeland, R.; Manchester, L.C.; Korkmaz, A.; Ma, S.; Rosales-Corral, S.; Reiter, R.J. Functional roles of melatonin in plants, and perspectives in nutritional and agricultural science. *J. Exp. Bot.* **2021**, *63*, 577–597. [CrossRef] [PubMed]
40. Wang, K.; Xing, Q.; Ahammed, G.J.; Zhou, J. Functions and prospects of melatonin in plant growth, yield, and quality. *J. Exp. Bot.* **2022**, *73*, 5928–5946. [CrossRef]
41. Godoy, F.; Olivos-Hernández, K.; Stange, C.; Handford, M. Abiotic stress in crop species: Improving tolerance by applying plant metabolites. *Plants* **2021**, *10*, 186. [CrossRef]
42. Zhang, N.; Sun, Q.; Zhang, H.; Cao, Y.; Weeda, S.; Ren, S.; Guo, Y.D. Roles of melatonin in abiotic stress resistance in plants. *J. Exp. Bot.* **2015**, *66*, 647–656. [CrossRef]
43. Khan, A.; Numan, M.; Khan, A.L.; Lee, I.J.; Imran, M.; Asaf, S.; Al-Harrasi, A. Melatonin: Awakening the defense mechanisms during plant oxidative stress. *Plants* **2020**, *9*, 407. [CrossRef]
44. Sun, H.; Cao, X.; Wang, X.; Zhang, W.; Li, W.; Wang, X.; Liu, S.; Lyu, D. RBOH-dependent hydrogen peroxide signaling mediates melatonin-induced anthocyanin biosynthesis in red pear fruit. *Plant Sci.* **2021**, *313*, 111093. [CrossRef] [PubMed]
45. Meng, J.F.; Yu, Y.; Shi, T.C.; Fu, Y.S.; Zhao, T.; Zhang, Z.W. Melatonin treatment of pre-veraison grape berries modifies phenolic components and antioxidant activity of grapes and wine. *Food Sci. Technol.* **2018**, *39*, 35–42. [CrossRef]

46. Sun, H.L.; Wang, X.Y.; Shang, Y.; Wang, X.Q.; Du, G.D.; Lü, D.G. Preharvest application of melatonin induces anthocyanin accumulation and related gene upregulation in red pear (*Pyrus ussuriensis*). *J. Integr. Agric.* **2021**, *20*, 2126–2137. [CrossRef]
47. Michailidis, M.; Tanou, G.; Sarrou, E.; Karagiannis, E.; Ganopoulos, I.; Martens, S.; Molassiotis, A. Pre- and post-harvest melatonin application boosted phenolic compounds accumulation and altered respiratory characters in sweet cherry fruit. *Front. Nutr.* **2021**, *8*, 695061. [CrossRef] [PubMed]
48. Carrión-Antolí, A.; Lorente-Mento, J.M.; Valverde, J.M.; Castillo, S.; Valero, D.; Serrano, M. Effects of melatonin treatment on sweet cherry tree yield and fruit quality. *Agronomy* **2021**, *12*, 3. [CrossRef]
49. Fekry, W.M.; Rashad, Y.M.; Alaraidh, I.A.; Mehany, T. Exogenous application of melatonin and methyl jasmonate as a pre-harvest treatment enhances growth of barhi date palm trees, prolongs storability, and maintains quality of their fruits under storage conditions. *Plants* **2021**, *11*, 96. [CrossRef] [PubMed]
50. Medina-Santamarina, J.; Zapata, P.J.; Valverde, J.M.; Valero, D.; Serrano, M.; Guillén, F. Melatonin treatment of apricot trees leads to maintenance of fruit quality attributes during storage at chilling and non-chilling temperatures. *Agronomy* **2021**, *11*, 917. [CrossRef]
51. Medina-Santamarina, J.; Serrano, M.; Lorente-Mento, J.M.; García-Pastor, M.E.; Zapata, P.J.; Valero, D.; Guillén, F. Melatonin treatment of pomegranate trees increases crop yield and quality parameters at harvest and during storage. *Agronomy* **2021**, *11*, 861. [CrossRef]
52. Lorente-Mento, J.M.; Guillén, F.; Castillo, S.; Martínez-Romero, D.; Valverde, J.M.; Valero, D.; Serrano, M. Melatonin treatment to pomegranate trees enhances fruit bioactive compounds and quality traits at harvest and during postharvest storage. *Antioxidants* **2021**, *10*, 820. [CrossRef]
53. Cortés-Montaña, D.; Bernalte-García, M.J.; Serradilla, M.J.; Velardo-Micharet, B. Optimal Preharvest Melatonin Applications to Enhance Endogenous Melatonin Content, Harvest and Postharvest Quality of Japanese Plum. *Agriculture* **2023**, *13*, 1318. [CrossRef]
54. Song, L.; Tan, Z.; Zhang, W.; Li, Q.; Jiang, Z.; Shen, S.; Luo, S.; Chen, X. Exogenous melatonin improves the chilling tolerance and preharvest fruit shelf life in eggplant by affecting ROS-and senescence-related processes. *Hortic. Plant J.* **2023**, *9*, 523–540. [CrossRef]
55. Li, S.; Cheng, Y.; Yan, R.; Liu, Y.; Huan, C.; Zheng, X. Preharvest spray with melatonin improves postharvest disease resistance in cherry tomato fruit. *Postharvest. Biol Technol.* **2022**, *193*, 112055. [CrossRef]
56. Mariani, L.; Parisi, S.G.; Cola, G.; Failla, O. Climate change in Europe and effects on thermal resources for crops. *Int. J. Biometeorol.* **2012**, *56*, 1123–1134. [CrossRef] [PubMed]
57. Scalbert, A.; Manach, C.; Morand, C.; Rémésy, C.; Jiménez, L. Dietary polyphenols and the prevention of diseases. *Crit. Rev. Food Sci.* **2005**, *45*, 287–306. [CrossRef] [PubMed]
58. Martínez-Camacho, J.E.; Guevara-González, R.G.; Rico-García, E.; Tovar-Pérez, E.G.; Torres-Pacheco, I. Delayed senescence and marketability index preservation of blackberry fruit by preharvest application of chitosan and salicylic acid. *Front. Plant Sci.* **2022**, *13*, 796393. [CrossRef] [PubMed]
59. Chiabrando, V.; Giacalone, G. Quality evaluation of blueberries coated with chitosan and sodium alginate during postharvest storage. *Int. Food Res. J.* **2017**, *24*, 1553–1561.
60. Ruiz-García, Y.; Romero-Cascales, I.; Bautista-Ortín, A.B.; Gil-Muñoz, R.; Martínez-Cutillas, A.; Gómez-Plaza, E. Increasing bioactive phenolic compounds in grapes: Response of six monastrell grape clones to benzothiadiazole and methyl jasmonate treatments. *Am. J. Enol. Viticult.* **2013**, *64*, 459–465. [CrossRef]
61. Pan, Y.G.; Liu, X.H. Effect of benzo-thiadiazole-7-carbothioic acid S-methyl ester (BTH) treatment on the resistant substance in postharvest mango fruits of different varieties. *Afr. J. Biotechnol.* **2011**, *10*, 15521. [CrossRef]
62. Smart, L.E.; Martin, J.L.; Limpalaër, M.; Bruce, T.J.; Pickett, J.A. Responses of herbivore and predatory mites to tomato plants exposed to jasmonic acid seed treatment. *J. Chem. Ecol.* **2013**, *39*, 1297–1300. [CrossRef]
63. Jiang, B.; Fang, X.; Fu, D.; Wu, W.; Han, Y.; Chen, H.; Liu, R.; Gao, H. Exogenous salicylic acid regulates organic acids metabolism in postharvest blueberry fruit. *Front. Plant Sci.* **2022**, *13*, 1024909. [CrossRef]
64. Venios, X.; Korkas, E.; Nisiotou, A.; Banilas, G. Grapevine responses to heat stress and global warming. *Plants* **2020**, *9*, 1754. [CrossRef] [PubMed]
65. Bernardo, S.; Dinis, L.T.; Machado, N.; Moutinho-Pereira, J. Grapevine abiotic stress assessment and search for sustainable adaptation strategies in Mediterranean-like climates. A review. *Agron. Sustain.* **2018**, *38*, 66. [CrossRef]
66. Campa, A.; Ferreira, J.J. Genetic diversity assessed by genotyping by sequencing (GBS) and for phenological traits in blueberry cultivars. *PLoS ONE* **2018**, *13*, e0206361. [CrossRef] [PubMed]
67. Tarara, J.M.; Lee, J.; Spayd, S.E.; Scagel, C.F. Berry temperature and solar radiation alter acylation, proportion, and concentration of anthocyanin in Merlot grapes. *Am. J. Enol. Viticult.* **2008**, *59*, 235–247. [CrossRef]
68. Timmers, M.A.; Grace, M.H.; Yousef, G.G.; Lila, M.A. Inter-and intra-seasonal changes in anthocyanin accumulation and global metabolite profiling of six blueberry genotypes. *J. Food Compos. Anal.* **2017**, *59*, 105–110. [CrossRef]
69. Zhu, L.; Huang, Y.; Zhang, Y.; Xu, C.; Lu, J.; Wang, Y. The growing season impacts the accumulation and composition of flavonoids in grape skins in two-crop-a-year viticulture. *J. Food Sci. Technol.* **2017**, *54*, 2861–2870. [CrossRef]
70. Spinardi, A.; Cola, G.; Gardana, C.S.; Mignani, I. Variation of anthocyanin content and profile throughout fruit development and ripening of highbush blueberry cultivars grown at two different altitudes. *Front. Plant Sci.* **2019**, *10*, 1045. [CrossRef]

71. Cheng, G.; Zhou, S.; Liu, J.; Feng, Q.; Wei, R.; Yu, H.; Wang, B.; Zhang, Y.; Bai, X. Widely targeted metabolomics provides new insights into the flavonoid metabolism in 'Kyoho'grapes under a two-crop-a-year cultivation system. *Sci. Hortic.* **2023**, *9*, 154.
72. Carrión-Antolí, A.; Martínez-Romero, D.; Guillén, F.; Zapata, P.J.; Serrano, M.; Valero, D. Melatonin pre-harvest treatments leads to maintenance of sweet cherry quality during storage by increasing antioxidant systems. *Front. Plant Sci.* **2022**, *13*, 863467. [CrossRef]
73. Shah, H.M.S.; Singh, Z.; Hasan, M.U.; Afrifa-Yamoah, E.; Woodward, A. Preharvest melatonin application alleviates red drupelet reversion, improves antioxidant potential and maintains postharvest quality of 'Elvira'blackberry. *Postharvest. Biol. Technol.* **2023**, *203*, 112418. [CrossRef]
74. He, Y.; Bose, S.K.; Wang, W.; Jia, X.; Lu, H.; Yin, H. Pre-harvest treatment of chitosan oligosaccharides improved strawberry fruit quality. *Int. J. Mol. Sci.* **2018**, *19*, 2194. [CrossRef] [PubMed]
75. Lo Piccolo, E.; Quattrocelli, P.; Becagli, M.; Cardelli, R.; El Horri, H.; Guidi, L.; Landi, M.; Pecchia, S. Can Chitosan Applications in Pre-and Post-Harvest Affect the Quality and Antioxidant Contents of Red Raspberries? *Sci. Hortic.* **2023**, *9*, 1135. [CrossRef]
76. Griñán, I.; Morales, D.; Collado-González, J.; Falcón-Rodríguez, A.B.; Torrecillas, A.; Martín-Palomo, M.J.; Galindo, A. Reducing incidence of peel physiopathies and increasing antioxidant activity in pomegranate fruit under different irrigation conditions by preharvest application of chitosan. *Sci. Hortic.* **2019**, *247*, 247–253. [CrossRef]
77. Zhang, Y.J.; Gan, R.Y.; Li, S.; Zhou, Y.; Li, A.N.; Xu, D.P.; Li, H.B. Antioxidant phytochemicals for the prevention and treatment of chronic diseases. *Molecules* **2015**, *20*, 21138–21156. [CrossRef]
78. ARPA Lombardia. 2023. Available online: https://www.arpalombardia.it/temi-ambientali/meteo-e-clima/form-richiesta-dati/ (accessed on 1 September 2023).
79. Nesmith, D.S. Fruit Development Period of Several Southern Highbush Blueberry Cultivars. *Int. J. Fruit Sci.* **2012**, *12*, 249–255. [CrossRef]
80. Cola, G.; Mariani, L.; Maghradze, D.; Failla, O. Changes in thermal resources and limitations for Georgian viticulture. *Aust. J. Grape Wine Res.* **2020**, *26*, 29–40. [CrossRef]
81. Gough, R.E. Climatic Requirements. In *The Highbush Blueberry and Its Management*; CRC Press: New York, NY, USA, 1993; pp. 67–74.
82. Yang, F.H.; Bryla, D.R.; Strik, B.C. Critical temperatures and heating times for fruit damage in northern highbush blueberry. *Hortic. Sci.* **2019**, *54*, 2231–2239. [CrossRef]
83. Schmidt, B.M.; Erdman, J.W., Jr.; Lila, M.A. Effects of food processing on blueberry antiproliferation and antioxidant activity. *J. Food Sci.* **2005**, *70*, s389–s394. [CrossRef]

MDPI AG
Grosspeteranlage 5
4052 Basel
Switzerland
Tel.: +41 61 683 77 34

Plants Editorial Office
E-mail: plants@mdpi.com
www.mdpi.com/journal/plants

www.ingramcontent.com/pod-product-compliance
Lightning Source LLC
Chambersburg PA
CBHW061548210926
51850CB00009B/2042
* 9 7 8 3 7 2 5 8 7 1 2 8 5 *